Editing for Writers

Lois Johnson Rew
San Jose State University

Prentice Hall
Upper Saddle River, New Jersey 07458

Library of Congress Cataloging-in-Publication Data

Rew, Lois Johnson.
    Editing for writers / by Lois Johnson Rew. — 1st ed.
       p.   cm.
    Includes bibliographical references and index.
    ISBN 0-13-749086-0 (alk. paper)
    1. Editing—Handbooks, manuals, etc. I. Title.
PN 162.R48 1999
808'.027—dc21   97-48712
                                       CIP

Editor-in-chief: Charlyce Jones Owen
Acquisitions editor: Maggie Barbieri
Editorial assistant: Joan Polk
Production liaison: Fran Russello
Editorial/production supervision: Bruce Hobart (Pine Tree Composition)
Cover director: Jane Conte
Cover designer: Bruce Kenselaar
Prepress and manufacturing buyer: Mary Ann Gloriande
Marketing manager: Rob Mejia

For permission to use copyrighted material, grateful acknowledgment is made to the copyright holders on page xxvi, which is hereby made part of this copyright page.

This book was set in 10/12 New Century Schoolbook by Pine Tree Composition, Inc., and was printed and bound by R.R. Donnelley Company & Sons. The cover was printed by Phoenix Color Corp.

© 1999 by Prentice-Hall, Inc.
Simon & Schuster/A Viacom Company
Upper Saddle River, New Jersey 07458

All rights reserved. No part of this book may be
reproduced, in any form or by any means,
without permission in writing from the publisher.

Printed in the United States of America
10 9 8 7 6 5 4 3 2 1

ISBN: 0-13-749086-0

Prentice-Hall International (UK) Limited, *London*
Prentice-Hall of Australia Pty. Limited, *Sydney*
Prentice-Hall Canada Inc., *Toronto*
Prentice-Hall Hispanoamericana, S.A., *Mexico*
Prentice-Hall of India Private Limited, *New Delhi*
Prentice-Hall of Japan, Inc., *Tokyo*
Simon & Schuster Asia Pte. Ltd., *Singapore*
Editora Prentice-Hall do Brasil, Ltda., *Rio de Janeiro*

For Ramona Rew Johnson
and
Karl Thomas Rew
with affection and admiration

Brief Contents

PART ONE   THE EDITING PROCESS AND THE EDITING PROFESSION

Chapter 1   Introduction to the Editing Process   1

Chapter 2   Levels of Editing   11

Chapter 3   Editor-Author Relationships   27

Chapter 4   Working as an Editor   37

PART TWO   SENTENCE-LEVEL EDITING

Chapter 5   Style, Sentence-Level Editing, and Editing Marks   52

Chapter 6   Final Editing or Proofreading   61

Chapter 7   Editing Copy and Constructing Style Sheets   78

Chapter 8   Capitalization, Spelling, Numbers, and Document Style   87

Chapter 9   A Review of Basic Grammar   99

Chapter 10 Punctuating Sentence Types   110

Chapter 11 Punctuating Within Sentences   138

Chapter 12 Choosing the Right Words   172

Chapter 13 Building Effective Sentences   196

PART THREE   DOCUMENT EDITING

Chapter 14 Developmental and Organizational Editing   242

Chapter 15 Graphics Editing   292

Chapter 16 Document Design   329

Chapter 17 Access Aids in Print and Online Documentation   365

Chapter 18 Editing Online Documentation   390

Chapter 19 International and Intercultural Issues for Editors   412

Contents

Preface   xix

PART ONE   THE EDITING PROCESS AND THE EDITING PROFESSION

Chapter 1   An Introduction to the Editing Process   1

   1.1   Why Do I Need Editing Skills?   3
   1.2   What Is Editing?   4
   1.3   What Is the Editing Process?   6
   1.4   How Can I Learn the Editing Process?   8
         EXERCISES   8
         References   10

Chapter 2   Levels of Editing   11

   2.1   Understanding Levels of Editing   12
   2.2   The Jet Propulsion Laboratory (JPL) Levels of Edit   13
   2.3   Accuracy and Readability   14
         Accuracy of Content   14
         Readability and Correctness of Text   16
   2.4   Planning an Editing Task   23
         Checklist for Efficient Editing   23
         EXERCISES   25
         References   26

Chapter 3   Editor-Author Relationships   27

   3.1   Authors and Their Responsibilities   28
   3.2   Editors and Their Responsibilities   31

3.3 Making Editor-Author Relationships Work Well 34
    EXERCISES 35
    References 36

Chapter 4 Working as an Editor 37

4.1 Understanding the Abilities and Skills of an Editor 38
4.2 Clarifying the Duties of an Editor 39
    Editing for Applicability 39
    Editing for Readability 40
    Editing for Usability 41
    Editing for Ethical and Legal Exposures 42
    Managing Projects 42
    Maintaining a Collegial and Collaborative Environment 43
4.3 Enhancing the Editing Job 43
    Learn the Style and Format Guidelines for Your Company or Field 43
    Collect Reference Materials 44
        Dictionaries / 44   Grammar / 45   Language, Usage, and Rhetoric / 45   Graphics and Document Design / 45
    Become an Expert in Writing and Editing Indexes 45
    Save and File Generic Portions of Written Documents 46
    Compile Abbreviation and Acronym Lists 46
    Establish a Glossary of Technical Terms 46
4.4 Flourishing Professionally 47
    Improve Your Communication Skills 47
    Increase Your Knowledge of the Industry 47
    Expand Your Professional Skills 48
    Learn Management Skills 48
    Keep Your Résumé and Portfolio Current 48
    EXERCISES 49
    References 51

PART TWO SENTENCE-LEVEL EDITING

Chapter 5 Style, Sentence-Level Editing, and Editing Marks 52

5.1 Style and Sentence-Level Editing 53
5.2 Definitions of Copyediting and Proofreading 54
5.3 Editing Marks 56

Copyediting Sample   58
Proofreading Sample   59
References   60

Chapter 6   Final Editing or Proofreading   61

6.1   Final Editing: The Art of Proofreading   62
6.2   The Importance of Final Editing   62
6.3   The Process of Final Editing   64
    EXERCISES   66
    References   77

Chapter 7   Editing Copy and Constructing Style Sheets   78

7.1   Developing Language Awareness   79
7.2   Learning a Process for Editing Copy   80
7.3   Constructing a Style Sheet   82
    How To Create a Style Sheet   82
    Example of a Style Sheet   83
7.4   Working Toward a Style Guide   83
    EXERCISES   85
    References   86

Chapter 8   Capitalization, Spelling, Numbers, and Document Style   87

8.1   Capitalization   88
    General Rules   88
    Specific Rules   89
8.2   Spelling   90
    EXERCISE   91
8.3   Numbers   92
    Numerals   92
    Words   93
    Number Use for International Readers   94
    EXERCISE   95
8.4   Document Style   96
    EXERCISES   97

Chapter 9   A Review of Basic Grammar   99

Diagnostic Quiz on Grammar

9.1   Definitions of Major Sentence Parts   101

9.2   Recognizing Sentence Parts   103
     EXERCISES   103

9.3   Definitions of Other Important Sentence Parts   105
     EXERCISE   107

9.4   Recognizing Subjects and Verbs in Complicated Sentences   108
     EXERCISE   109
     Reference   109

Chapter 10   Punctuating Sentence Types   110

10.1   Sentences, Fragments, and Run-ons   111

10.2   Compound Sentences   112
     Definitions   *113*
     Punctuation Rule for Compound Sentences   *114*
     Compound Sentences with Colons   *115*
     EXERCISES   115

10.3   Complex Sentences with Subordinators   121
     Definitions   *122*
     Punctuation Rule   *123*
     EXERCISES   124

10.4   Complex Sentences with Relative Pronouns   128
     Dependent Clauses Used as Adjectives   *128*
     Dependent Clauses Used as Nouns   *129*
     Punctuation Guidelines   *129*
     EXERCISES   130

Chapter 11   Punctuating Within Sentences   138

11.1   Abbreviations and Acronyms   139
     Common Abbreviations That Usually Require a Period   *139*
     Abbreviations That Do Not Require a Period   *140*
     EXERCISE   141

11.2   Apostrophes   143
     To Show Possession   *143*
     To Show Omission   *144*
     To Prevent Misreading   *144*

EXERCISE   144

11.3 Colons   146

To Introduce a Formal List or Series, Especially One Including the Word following   146
To Introduce a Word Group or Clause That Summarizes or Explains the First Word Group   147
To Introduce a Long or Formal Quotation   147
To Show Typographical Distinctions or Divisions   147
EXERCISE   147

11.4 Commas with Introductory and Interruptor Words and Phrases   149

Introductory Words   149
Introductory Phrases   149
Interrupter Words and Phrases   150
EXERCISE   150

11.5 Commas and Semicolons in a Series   151

Commas   151
Semicolons   152
EXERCISE   152

11.6 Dashes, Commas, Parentheses, and Brackets   153

To Indicate Parenthetical Information   154
To Show a Break in Thought, Emphatic Idea, or Summary   154
For Clarity When Commas Are Already Used in the Sentence   155
To Set Off Material Not Integral to the Sentence or Not Part of a Quotation   155
EXERCISE   155

11.7 Hyphens   156

To Indicate Syllable Breaks   156
In Compound Words   157
In Compound Words Used as Adjectives   157
In Compound Words Used as Nouns   158
Do Not Use Hyphens in the Following Cases   158
EXERCISES   159

11.8 Quotation Marks and Ellipses   163

Setting Off Someone's Exact Words   163
Identifying New Technical Terms or Words Discussed as Words   164
Identifying Parts of a Larger Published Work   164
Using Single Quotation Marks or No Quotation Marks   164
Using Ellipses and Quotation Marks   165
EXERCISES   165
Reference   171

Chapter 12 Choosing the Right Words   172

- 12.1 Ensuring Accurate Terminology   174
  - EXERCISES   175
- 12.2 Choosing Concrete and Denotative Language   177
  - *Concrete and Abstract Words*   *177*
  - EXERCISES   179
  - *Denotative and Connotative Meanings*   *180*
  - EXERCISE   181
- 12.3 Avoiding Euphemistic and Sexist Terms   182
  - *Euphemisms*   *182*
  - EXERCISE   182
  - *Sexist Language*   *183*
  - EXERCISES   185
- 12.4 Replacing Pompous Words   188
  - EXERCISES   188
- 12.5 Choosing Understandable and Usable Terms   190
  - EXERCISES   192
  - References   194

Chapter 13 Building Effective Sentences   196

- 13.1 Syntax, Style, and Reader Expectations   197
  - *Syntax*   *197*
  - *Style*   *197*
  - *Reader Expectations*   *198*
  - EXERCISE   199
- 13.2 Finding and Fixing Syntax Problems   200
  - *Verbs*   *200*
  - EXERCISE   201
  - *Pronouns*   *202*
    - Definitions of Terms / 202   Pronoun-Antecedent Agreement / 203
  - EXERCISE   204
    - Pronoun Reference / 205
  - EXERCISE   207
    - The Who-Whom-Whose Dilemma / 207
  - EXERCISE   209
  - *Modifiers*   *209*

Definitions of Terms / 209   How to Repair Sentences with Wrongly Placed Modifiers / 210   How to Repair Specific Modifier Problems / 210

EXERCISES   211

Wordiness   214

Redundancy / 215

EXERCISE   218

Long Modifier Strings / 219   Delay of Subject or Verb / 220

EXERCISE   221

Inappropriate Expletives / 223

EXERCISE   225

13.3   Improving Sentence Structure and Style   226

Choosing Active or Passive Voice   226

Definitions of Terms / 226   Reader Expectations and Active vs. Passive Voice / 227

EXERCISES   228

Locating the Action in the Verb   229

How to Spot the Hidden Action / 229   How to Verbalize the Action / 231

EXERCISE   231

Preferring Positive over Negative Constructions   232
Emphasizing by Position   233

EXERCISES   236

References   241

PART THREE   DOCUMENT EDITING

Chapter 14   Developmental and Organizational Editing   242

14.1   Developmental Editing   243

Understanding Long-term Memory   244
Using Short-term or Working Memory   246
Analyzing Readers   246

Reader Type / 246   Reader Tasks / 247   Reading Style / 248

Packaging Information to Meet Readers' Needs   248

Proposals / 249   Manuals, Procedures, and Instructions / 249   Reports / 250   Online Documentation / 251

Working with the Writer in a Developmental Edit   252

14.2   Organizational Editing   253

What Are the Methods of Organization?   253

Organization by Sequence / 254   Organization by Hierarchy / 254   Organizational Combinations / 255

What Does an Editor Need to Know about Organization? 256

How to Determine the Organization of a Written Piece / 257

EXERCISE 257

How to Evaluate an Outline / 258   How to Reorganize an Outline / 259

EXERCISE 259

How to Justify the Changes You Make / 260

EXERCISES 261

How Can an Editor Clarify the Chosen Organization of a Document? 264

Verbal Devices / 264

EXERCISES 268

Visual Devices / 271

EXERCISES 273

References 291

Chapter 15 Graphics Editing 292

15.1 Defining Key Graphics Terms 294
15.2 Understanding the Factors Involved in Choosing Graphics 295

The Purpose of the Document 296

GRAPHICS EXAMPLES 298

Photographs 298
  Editing Questions 298
Drawings 299
  Editing Questions 300
Pie Charts and Segmented Bar Graphs 301
  Editing Questions 302
Diagrams, Maps, and Charts 303
  Editing Questions 303
Flowcharts and Task-Breakdown Charts 305
  Editing Questions 306
Tables 308
  Editing Questions 309
Line Graphs, Bar Graphs, and Pictographs 310
  Editing Questions 310
Examples 312
  Editing Questions 313
Block, Schematic, and Wiring Diagrams 314
  Editing Questions 314

The Type and Experience of the Reader or Viewer   315
The Language and Cultural Background of the Reader   316
Content   317
The Medium   317
Time and Resources   318
Degrees of Visualization   319

15.3   Developmental Editing of Graphics: Major Questions   320

15.4   Copyediting Graphics: A Checklist   321

15.5   Proofreading Graphics: A Checklist   323

EXERCISES   324

References   328

Chapter 16   Document Design   329

16.1   Defining Document Design Terms   331

16.2   Understanding the Factors that Influence Document Design   331

Purpose   331
Needs of Readers   331
The Way Readers Will Read the Document   332
Document Types and the Conventions That Influence Them   333

16.3   Evaluating the Document Design and the Page Design   333

How Effective is the Design of the Whole Document?   334

Packaging / 334   Page-Level Considerations / 334   Principles of Page Design / 335

How Easy Is the Document To Read?   337

Line Length and Grid Divisions / 337   Alignment / 338   Type Size and Style / 338   Boxes, Rules, and Screens / 342   Color / 343

Can the Reader Find Information Easily?   343
Can the Reader Understand the Text Easily?   344

Chunking / 345   Queuing / 346   Filtering / 346   Visually Prominent Headings / 346   Use of Lists / 346   Use of Color / 348   Figures and Tables / 348

Will the Reader Be Able to Retain and Retrieve Information?   348

Chunking / 349   Presenting Information Both Verbally and Visually / 349   Using Repeated Patterns of Page Layout / 349   Presenting Questions at the End of an Information Unit / 349

Can The Reader Use This Document to Perform a Task?   350

16.4   Developmental Editing of Document Design: Major Questions   350

16.5   Copyediting Document Design: A Checklist   351

16.6   Proofreading Document Design: A Checklist   351

EXERCISES 352
References 364

Chapter 17 Access Aids in Print and Online Documentation 365

17.1 Tables of Contents and Menus 366
17.2 Indexes 369
  The Process of Indexing 372
  Indexing Online Documentation 373
17.3 Glossaries 375
17.4 Navigational Aids 377
  Headers, Footers, and Screen Orienters 377
  Maps and Flowcharts 379
  The Stack of Pages and Navigation Buttons 379
  The Table of Contents, the Home Location, and Bookmarks 380
  Editing Navigational Aids 380
17.5 Advance Organizers 380
17.6 Citation of Sources 382
  Common Citation Systems 383
    MLA / 383   The Chicago Manual of Style and APA / 383   CBE / 384
17.7 Developmental Editing of Access Aids: Major Questions 384
17.8 Copyediting Access Aids: A Checklist 385
17.9 Proofreading Access Aids: A Checklist 386
  EXERCISES 386
  References 389

Chapter 18 Editing Online Documentation 390

18.1 Electronic Approaches to Developing and Presenting Information 391
  Defining Key Terms 392
  Editing Information Online and Editing Online Information 392
  Understanding the Similarities and Differences between Print and Online Documentation 393
    Similarities / 393   Differences / 394
18.2 Readers and Users 395
  Who Are the Reader-Users? 395

Will the Information Be Printed Out or Read on Screen?   *396*
How Do Reader-Users Use Online Documentation?   *397*

18.3 Elements of Online Documentation   398
Diction, Syntax, and Punctuation   *398*
Diction / 399   Syntax / 399   Punctuation / 400
Topics   *400*
Size / 400   Contents / 400   Structure / 400
Links   *401*
Organization   *402*
Sequence / 402   Grid / 402   Hierarchy / 403   Partial Web / 404
Navigational and Access Aids   *405*
Design Elements of Screens and Windows   *405*

18.4 Developmental Editing of Online Documentation: Major Questions   406

18.5 Copyediting Online Documentation: A Checklist   407

18.6 Proofreading Online Documentation: A Checklist   408

EXERCISES   408

References   410

Chapter 19   International and Intercultural Issues for Editors   412

19.1 Recognizing the Importance of English   413
The Global Marketplace   *414*
International Englishes   *414*
Cultural and Language Diversity in the United States   *414*

19.2 Defining Key Terms   415

19.3 Clarifying Editors' Roles and Responsibilities   416
International and Intercultural Writers   *417*
International and Intercultural Readers   *418*

19.4 Understanding the Influence of Culture   419
Examples of Cultural Variables   *420*
Designing an International-User Analysis   *421*
Learning about Other Nations and Cultures   *422*
Written Information / 422   Oral Information / 423   Personal Experience / 423

19.5 Working Toward Global Communication: Guidelines for Editors   424
Diction Guidelines   *424*

Syntax and Punctuation Guidelines   *425*
Guidelines for Graphics, Technical Terms, and Design Issues   *427*

19.6 Cultural Editing: Major Categories   428
19.7 Developmental Editing of International Documents: Major Questions   429
19.8 Copyediting International Documents: A Checklist   429

Diction   *429*
Syntax and Punctuation   *430*
Graphics, Technical Terms, and Design Issues   *430*

19.9 Proofreading International Documents: A Checklist   431

EXERCISES   431

References   433

Glossary   **435**

Appendix   **447**

Index   **463**

Preface

To Individual Readers

Who Should Read This Book?

If you write on the job, and you want to improve your writing, this book is for you. It will teach you how to solve grammar and punctuation problems, and it will help you choose the right words and construct your sentences in the most effective way. It will help you organize documents, choose effective graphics, design documents on paper and online, and alert you to international and intercultural issues.

If you are a member of any of the following groups, this book will help you.

- **Workplace writers.** Whether you are in science, government, engineering, business, technology, manufacturing, or education, you probably have to write on the job. You also need editing skills.
- **College students.** Whether you are preparing for a profession in law, business, medicine, engineering, science or the like, you need top-notch communication skills. You can use this book for self-tutoring.
- **Continuing education students.** Whether you are taking formal classes toward a certificate or simply working on your own, you can improve your editing skills.
- **Workplace editors.** You may already be working as an editor or writer, or you may want to learn more about the profession. This book will introduce you to proofreading, copyediting, and document editing, and will give you many opportunities to apply what you have learned. The book includes 136 exercises.

Why Is This Book Called Editing for Writers?

Editing is one part of the total writing process; editing means reviewing and rereading what you have written, and polishing the sentences and paragraphs so that they communicate to the reader exactly what you—the writer—intend to say. You should always edit your own work because that work represents you to the reader, and you want to be regarded as articulate, accurate, and organized. In addition, you may be asked to "look at" or edit the documents of your coworkers. You can learn how to approach those documents, what marks you can use in editing, and what terminology you can use to talk about the problems you see. The Appendix includes answers to selected exercises, so you can check your work. With this book, you can also learn how to be a professional editor.

To Instructors

Editing for Writers is designed to provide current theory and practice in workplace editing; comments and explanations from working writers and editors; clear explanations of grammar, punctuation, diction, syntax, and organization; and many opportunities for students to practice the skills they are learning. This text can be used in any of the following classes:

- **Technical, business, or science writing courses** that prepare students for writing on the job. You can use *Editing for Writers* as an adjunct to a writing textbook, assigning exercises as you see a need for the particular skills either in the whole class or on an individual basis.
- **Editing courses.** This can be the primary text for a 10-week or 15-week term; there is more than enough material to engage the class for a lengthy period. The instructor's manual provides additional assignments that are especially appropriate for students seeking to become professional editors. But the text starts with the basics and allows the students to master sentence-level skills before tackling whole documents.
- **Continuing education courses** in technical or workplace communication. Whether in individual courses or as part of a certificate program, you can use *Editing for Writers* to provide continuing education students with skills they can apply immediately to work they are doing.
- **General composition classes** for college freshmen or sophomores. If your students need extensive practice in editing and rewriting, or if your composition course is geared toward workplace writing, *Editing for*

Writers can provide many opportunities for your students to improve their writing at the sentence and document level.

Purpose

The guiding premises of the text are:

1. All workplace writers—whether they are professional writers or professionals who write—need excellent editing skills. The ratio of writers to editors in most companies is about 15 to 1, and many companies have few or no editors. In fact, a 1995 Trends Survey by the Society for Technical Communication revealed that only 6 percent of technical communicators are editors. That low percentage of editors means that many writers are largely responsible for editing their own work; in addition, they must often edit for their coworkers.

2. While many students attracted to the technical communication profession are excellent writers with a good grasp of the basics of punctuation, grammar, diction, and syntax, few of them can articulate why they do what they do. To be good editors, they need the ability to explain or justify changes they make. Engineers, technicians, and scientists who must write and edit as part of their jobs are frequently uncomfortable with their level of skill with the language. English may be their second language, or they may simply need review of basic sentence-level skills in order to gain self-confidence.

3. Working writers and editors are unanimous in their belief that for successful editing, students need two things: absolute mastery of the basics of the language and good people skills. As one manager explained, "You have to have both the basics and the people skills. One without the other will not do it."

Content

For these three reasons, *Editing for Writers* is grounded in the basics and gives students hundreds of opportunities to master both sentence-level skills of grammar, punctuation, diction, and syntax, and document-level skills of organization, coherence, and providing access to information. It asks students to explain what they are doing and why they make the choices they do. The text is also incremental, beginning with easier applications in proofreading while teaching standard editing marks, moving to copyediting while reviewing grammatical problems, and finally addressing document editing, including editing for organization, graphics, document design, and access aids. The text includes 136 different kinds of exercises. Thus, this text provides both a solid review of language basics and an introduction to editing processes, levels of edit, and the dynamics of editor-writer relationships, with particular attention to those people skills required for good editing.

Organization

Editing for Writers has a three-part organization, providing maximum flexibility for instructors and students.

Part One: The Editing Process and the Editing Profession

Chapter 1 provides an overview of the process that experienced editors use, and Chapter 2 introduces students to the variety of ways editors decide "how much" editing to do on a specific document. Chapter 3 details the important people skills required of anyone who edits, with examples and recommendations from experienced managers and editors. Finally, Chapter 4 discusses editing as a profession and provides tips for professional editors. Part One thus provides context for editing, whether readers want to be professional editors or are simply writers who need to improve their skills.

Part Two: Sentence-Level Editing

The chapters in Part Two provide detailed explanations and exercises to give students ample opportunity to master sentence-level editing skills. Chapter 5 introduces and explains standard editing marks; subsequent chapters provide many opportunities to use those marks. Chapters in this section are organized from the simplest tasks (final editing or proofreading in Chapter 6) to those which are most complex (evaluating and revising syntax in Chapter 13). Most chapters end with comprehensive or cumulative exercises that allow students to apply all the sentence-level skills they have learned.

Part Three: Document Editing

With the growing movement in industry toward product teams that involve writers early in product development, students need to learn how to help writers develop documents and how to analyze and revise organization. Chapter 14, Developmental and Organizational Editing, provides explicit guidelines and nine short and long documents for practicing editing organization. Chapters 15 through 19 cover other document-level concerns: graphics, document design, access aids (table of contents, index, glossary, and so on), editing of online documentation, and international and intercultural concerns. The chapters in this section have extensive checklists that students and workplace writers can apply directly to documents.

This three-part organization allows instructors to concentrate on those areas that students need most. Depending on the level of the class, you can move through the text sequentially or can skip sections in which students already display competence. Students pursuing self-instruction can go directly to those areas in which they are weak. An Appendix provides selected answers, and the Instructor's Manual offers additional answers and suggested syllabi.

Research

This book is based on field research that began in fall 1994 and has continued to the present time. The research included on-site interviews with 10 professional editors and 3 managers of publication groups from 7 different companies. In addition, 50 workplace writers from 25 different companies answered a detailed questionnaire. Three freelance writers and editors also completed the questionnaire. These editors and writers worked in a variety of fields including computer hardware and software, telecommunications, networking, pharmaceuticals, instrumentation, environmental research, medical diagnostics, and transportation.

Throughout the text, I have quoted these writers and editors as they comment on important principles or give advice to readers. Each speaker is identified by name and area of work, but—because this job market is so volatile—not by company. However, the companies where I conducted research include the following:

Beta Text
Boole & Babbage, Inc.
Cadence Design Systems, Inc.

Cisco Systems
FORE Systems, Inc.
Hewlett-Packard

IBM	SK Writers
Lam Research	Style Guides by Design
Lotus Development Corp.	Sybase, Inc.
Mentor Graphics Corp.	Syva
Newbridge Networks Inc.	Tandem Computers
Novell	Triad Systems Corp.
Pacific Bell	Varian Oncology Systems
Rolm	Vintage Publications
SAIC	Write Method
Siemens Pyramid Information Systems, Inc.	

In addition to my field research with writers, managers, and editors, this book is based on extensive reading in the field and what I have learned about student needs during 17 years of teaching technical writing and 12 years of teaching technical editing at both the undergraduate and graduate levels. This material has been developed for and tested in the college classroom; it is further supported by my experience teaching a variety of seminars in industry, where clients have included Hewlett-Packard, Novell, Legato Systems, Wyse Technologies, and Becton Dickinson.

Editing is not a narrow professional niche inhabited by a few practitioners who have natural language skills. Editing is a basic component of the writing process, and all professionals who write (whether they are professional writers or not) need both exposure to the guidelines and practice in the skills that will enable them to edit their own writing or the writing of others.

Acknowledgments

In writing this book, I have been supported, challenged, and inspired by my students and colleagues at San Jose State University, by friends and clients in industry, by fellow teachers across the country, and by the staff at Prentice Hall. My thanks to all of you.

Research for this book began in 1994 during a sabbatical granted by the English Department and the College of Humanities and Arts at San Jose State University. At that time, the following writing professionals completed an extensive questionnaire about editing practices and needs. I appreciate the thoughtful and candid responses from Susie Albrecht, Mike Belef, Paula Bell, Maggie Black, Karen Boyett, Jane Bratun, Paul J. Briscoe, Marian Cochran, Tony Cyphers, Kevin DeYager, Andrea Dutra, Art Elser, Bruce England,

Jean Forsberg, Chrystal Groves Francois, John Gillis, Dan Greenwell, Lisa Gullicksen, Frank Guss, Linda Hoskins, Karen J. Kellerman, Connie Lamansky, Tricia Lyons, Susan Lytle, Amy L. McIntosh, Madhu Mitra, Sheila Moore, Diane Morabito, Michael O'Brien, Tina Ornduff, Ann C. Petersen, Kate Picher, Eric Radzinski, Patrick C. Ralph, Jenifer Renzel, Anne M. Rosenthal, Christopher Salander, Anastasia Saltabida, Phil San Diego, Sherri Sotnick, Della Stein, Beth Taylor, Joy Thomas, Liz Van Houten, Jeffrey Vargas, Teresa Marie Velasco, Durthy A. Washington, John Weiland, Nancy Woods, and Catherine Yaspo.

In addition, these editors, publications managers, and a writer/editor placement director agreed to long personal interviews, which helped shape the content of the book. Special thanks to Nancy Beckus, Judy Billingsley, Susan Borton, Ruth Chase, Martha Cover, Shirley Krestas, Yvonne Kucher, Jo Levy, Doug MacBeth, Kari Miller, Suzanne Moore, Cheri Porter, Anjali Puri, Sherri Sotnick, and Sylvia Thompson.

For their encouragement and high expectations, I am grateful to my San Jose State colleagues Bonnie Cox, Cindy Baer, Patricia Nichols, and Denise Murray. Bob McDermand of Clark Library and the librarians in Government Documents gave freely of their expertise. Mark Bussmann in the English office and my student assistants Linda Hoskins and Tammy Lewis helped solve numerous logistical problems.

My students in both writing and editing classes continue to enlighten and challenge me. Most of the exercises in this book have been tested in the classroom and refined by excellent student input. Extra applause for my graduate students, who critiqued chapters in Part Three.

Friends on campus and in industry helped me with insightful and thorough reviews of individual chapters. I am grateful to Cindy Aguilar, Rose Barry, Bonnie Cox, Shannon Daugherty, Jim Ferris, Marie Highby, Linda Hoskins, Janice Ishizaka, Brian Larzelere, Jo Levy, Doug MacBeth, Katherine McMurtrey, Madhu Mitra, Patricia Nichols, Anjali Puri, Regina Roman, and Sandra Sherrill.

Because I wanted to illustrate graphics principles with real graphics, I sought and received help and examples from my friends in industry and government. Thanks to Jane Bratun, Martha Cover, Tony Cyphers, Bruce England, Carol Gerich, Dan Greenwell, Jan Johnston-Tyler, Ellie Katsoudas, Elaine Keim, Brian Larzelere, Jo Levy, Doug MacBeth, Valerie Menager, Madhu Mitra, Thripura S. Naidu-P, Patricia Nichols, Jim Petersen, Allan Smirni, Sherri Sotnick, Sue Stull, Julie Thomas, and Wesley Wing.

Special thanks to Heidi King, who—from my rough sketches—designed the flow charts in Chapter 1 and most of the graphics in Chapter 16.

The writing and editing instructors who reviewed drafts of the text during development provided thoughtful critique and many valuable suggestions. My gratitude to Barbara Henry from West Virginia State College, Helen Hogan from Tarrant County Junior College, and Michael J. Kelly from

Slippery Rock University. In addition to his reviews, Thomas R. Williams from the University of Washington generously shared his expertise in cognitive psychology and helped me clarify explanations of theory. Many thanks.

At Prentice Hall, Maggie Barbieri has been a discerning and encouraging editor, and Joan Polk has helped with editorial details.

At home, my thanks to my husband, Bob, whose computer and engineering skills keep me going, and whose critical eye keeps me honest. Without him and his support, this book would not exist. Finally, this book is for our daughter, Ramona, and our son, Karl, who endured their mother's editing with patience and good cheer—and who survived.

Credits

Pages 17–21: Figure 2.1. Reprinted by permission of Jo Levy, San Jose, CA.

Pages 21–22: Figure 2.2. Reprinted by permission of Cadence Design Systems, Inc.

Pages 68–77: Exercises 3–6. Reprinted by permission of Wesley Wing.

Page 76: Table 1. Reprinted by permission of IEEE Transactions on Professional Communication.

Pages 165–169: Exercise 10. Reprinted by permission of Thripura S. Naidu-P.

Page 253: Figure 14.2. Reprinted by permission of Anjali Puri.

Page 274: Figure 14.4. Reprinted by permission of Bruce England.

Page 298: Figure 15.1. Reprinted by permission of 3 Com Corporation.

Pages 299 & 301: Figures 15.2 & 15.4. Reprinted by permission of Siemens Pyramid Information Systems, Inc.

Pages 307, 309, & 314: Figures 15.9 & 15.15 and Table 15.3. Reprinted by permission of FORE Systems, Inc.

Page 308: Figure 15.10. Reprinted by permission of Newbridge Networks.

Pages 312 & 313: Figures 15.13 & 15.14. Reprinted by permission of Cisco Systems.

Pages 315 & 379: Figures 15.16 & 17.12. Reprinted by permission of Cadence Design Systems, Inc.

Pages 359–363: Exercise 4. Reprinted by permission of Atmel Corporation.

Pages 367, 368, & 374: Figures 17.1, 17.2, 17.3, 17.7, & 17.8. © 1997 Lotus Development Corporation. Used with permission of Lotus Development Corporation. Lotus is a registered trademark and cc:Mail is a trademark of Lotus Development Corporation.

Page 421: Figure 19.1. Reprinted by permission of Blackwell Publishers, Inc.

1 An Introduction to the Editing Process

Contents of this chapter:

- Why Do I Need Editing Skills?
- What Is Editing?
- What Is the Editing Process?
- How Can I Learn the Editing Process?

If you have taken a business-writing or technical-writing course, or if you have written workplace documents, you know that the writing process is made up of three general activities: prewriting, writing, and revising. In scientific, business, and technical fields those three activities can be further subdivided as:

Prewriting
    planning
    gathering information
    considering legal and ethical responsibilities
    organizing
Writing
    writing the draft
    making information accessible
    designing the document
Revising
    reviewing and revising the document
    editing

In practice, many writers move back and forth among these activities—for example, gathering more information when they write or revising partial drafts before continuing to write. Under the pressure of deadlines, however, both students and workplace writers are sometimes forced to curtail the time they can spend reviewing, revising, and editing documents. In technical-writing classrooms, instructors often have little time to help students solve their problems with organization of text and with the unclear writing stemming from problems at the paragraph and sentence level. The result is that—in spite of increasing attention by colleges, companies, and government organizations—what we call business, scientific, and technical writing is still often considered "bad" writing. Instructions for programming a VCR, for example, are almost a national joke because they are poorly written and therefore incomprehensible to most adults.

    The premise of this book is that you can become a better writer by learning reviewing, revising, and editing skills. You can also become a successful editor by applying those skills to documents written by others. Whether you are in a college classroom or learning on your own, this book will help you. You may wish to learn editing skills to become a professional editor in publishing or in industry, but most often you will be working as a writer—either as a professional writer or as a professional who must write as part of your job responsibilities. Writing and editing are different activities, but they are both needed for successful communication. Writers edit, and editors write. Writers write, and editors edit. It is important to see the roles and the activities as both separate *and* connected.

Whether you are working mostly as a writer or as an editor, you need knowledge that will give you:

- confidence in the way you handle the English language
- the kind of language awareness that helps you spot potential problems in your own or others' documents
- competence in organizing documents to meet readers' needs
- understanding of how graphics and document design work with text
- the ability to explain or justify changes that you make

"Good editors need two basic skills," according to technical communications professor Charles Kemnitz. "They need to 'know what is right,' and they also need to know *how they know it's right*. The second skill . . . is much more important than the first. In addition to practical experience, theory gives editors the knowledge they need" (1994, 415).

1.1 Why Do I Need Editing Skills?

As a writer, do you really need revising and editing skills? In other words, can't you solve your editing problems by running software that checks spelling, readability, or style? Don't organizations have professional editors to review your documents for you?

The answer is yes and no. Yes, you need revising and editing skills because they will make you a better writer, and your ability to write well often influences your job performance. For example, Ted Kooser, a vice president and marketing executive at a life insurance company in Nebraska, says: "Having effective language skills can be quite powerful in a corporation" (Pandya 1994, 25A). Maggie Black, technical publications manager at a software company, says: "Editing skills are not just for editors. You can't be a good writer unless you have good editing skills." Yes, spelling and style checkers will solve some editing problems: Spelling checkers will catch most of your spelling errors, and style checkers will help eliminate such problems as overuse of the passive voice. Yes, some companies employ professional editors who are trained to review documents and work with you to solve editing problems.

But no, software programs will not guarantee a fault-free or well-written document. To understand what workplace writers need to know about revising and editing, I recently surveyed 50 people who work as writers in a variety of computer, networking, energy, telecommunications, and biotechnology companies. All those writers used a spelling checker on their documents. However, as Andrea Dutra, a senior writer at a company specializing in energy and transportation issues, noted: "A spelling checker is useful for most typos, but it does not replace editing and proofing tasks because it can't tell the right word from the wrong one." And although many writers had tried various style and grammar checkers, the unanimous verdict was that they were not useful. Bruce England, a software writer at a computer-systems manufacturer, said: "Grammar checkers are not sufficiently intuitive to be helpful." Joy Thomas, technical writer at a networking company, explained the problem: "Grammar checkers suggest not using so many long words, which is often unavoidable with highly technical documents."

Finally, professional editors will not free you from your own editing responsibility—partly because there are so few editors compared to the number of writers. Many companies, for example, have no editors, and others have a writer-to-editor ratio as high as 20 to 1. James B. Hansen, an information designer for 10 years, explains his experience this way: "Some years back I found myself the only technical writer at an instrument manufacturer in southern California. While I enjoyed the crazy challenges the job posed, working as a lone writer had some drawbacks. The biggest of these was the need to serve as my own editor. At professional gatherings, I learned that other writers often edited their own work—albeit reluctantly. So I began collecting editing tips" (1997, 14). Even good writers need an editor, and most often that editor must be you.

1.2 What Is Editing?

When your car is running in tiptop condition, you seldom think of its operation or its moving parts. As a driver, you simply use the car as a means to get to your destination. But when your car begins gasping, wheezing, or clattering, your attention shifts immediately to the car itself. "What's that noise?" you say to yourself, or "Now what's wrong?"

In the same way, organized, clear, and error-free writing is almost invisible. As a reader, you use such writing as a way to get at meaning: to learn how to do something or what something means. But if the writing is disorganized, inconsistent, or unclear, you must look at the writing itself, backtracking and trying to decipher the writer's meaning. Cars run smoothly when they are fine-tuned: when all the moving parts fit together exactly and are lubricated properly. Knowledge and experience are necessary to fine-tune a car's engine. Likewise, writing reads smoothly when it is revised and corrected so it is

- accurate in content
- appropriate to the purpose and reader
- organized in a meaningful way
- correctly worded
- clearly stated
- punctuated to help the reader understand

Editing involves fine-tuning the written word, and good editing is an important part of the total writing process. Some of that fine-tuning takes place at the sentence, paragraph, and individual screen level with text, but much of the fine-tuning involves looking at the whole document and assessing broader issues of purpose, reader, content, organization, format, and graphics. Editing, in other words, is the process of evaluating the readability, applicability, and usability of the document.

Readability is the level of comprehension that the reader can achieve. Readability is influenced by word choice, sentence length and construction, and the way that the words are presented on the page or screen. Chapters 4 through 16 provide much information on readability, both at the sentence level and the document level.

Applicability is the appropriateness or "fit" of the document to the purpose for which it is written and to the intended reader or readers. Chapters 14 and 19 deal with applicability.

Usability is the success of the document in enabling the readers to carry out a task. Usability is important because—except for memos and letters—instructions and manuals are the documents most frequently written. Usability is explained in Chapters 4, 15, and 18.

After the editing evaluation, an editor fixes errors, asks questions that promote revision, and suggests optional ways of organizing and presenting the material.

When you are editing, you need to become the reader's advocate; the primary job of an editor is to make the document understandable and usable for the reader. Thus, editing is far more than "fixing up the punctuation." Editing requires careful reading, good judgment, and a sense of the English language. Many times when you write, you must be your own editor—revising and correcting your drafts until they are as clear as you can make them. At other times—both in the classroom and on the job—you will be asked to "review" or "look at" writing done by your peers, and you will need good editing skills to be an effective reviewer. Sometimes you may even work as a professional editor, a person whose job is to work in collaboration with writers to perfect documents.

"Rather than expect any one person to bear the entire burden of effective documentation, the task is a joint responsibility," according to editor

Charles E. Beck. "For certain items, the manager must be the ultimate editor. For more routine matters, each writer may take a turn editing the work of another writer. For more complex projects (software documentation, procedures writing, and contracts), the organization may need to hire a formal editor for the project because of the time required" (1995, 339).

1.3 What Is the Editing Process?

There are many ways to describe the editing process, and the terms vary somewhat in publishing, journalism, and industry—all of which can deal with scientific and technical information. In general, experienced editors look at documents in this order:

1. organization and coherence
2. content, especially technical accuracy
3. language and style

This is a "top-down" approach to a document: starting with the most inclusive or general information and working to detail. This top-down approach also deals with editing in the broadest sense, a sense that includes what we often call review and revision. Ideally, a document will be reviewed by a manager or an editor for organization and coherence and will be revised by the writer. This is called *developmental editing*.

Following developmental editing, the document will be reviewed for accuracy of the content of text and graphics by subject-matter experts (SMEs) and will be revised again by the writer. This is called *technical review*. Sometimes the technical review and the developmental edit run concurrently.

The document will then be edited by a copyeditor, who will examine the language, word order, punctuation, capitalization, and spelling, and will fix errors or suggest changes. This is called *copyediting*. Again, the document will be revised by the writer.

Finally, after the document is in its final form, it will be read carefully to catch any remaining errors. This is called *proofreading*. See the flowchart in Figure 1.1 for a visual representation of this top-down process.

This multi-layered process—which for simplicity I call editing—ensures that the final document is accurate, readable, and usable. However, in the schedule-driven workplace, there is often not enough time for all the parts of this process to take place. Some of the experienced editors I interviewed in preparing to write this book had time for only one pass through a document, and during that pass they performed all the edits described above.

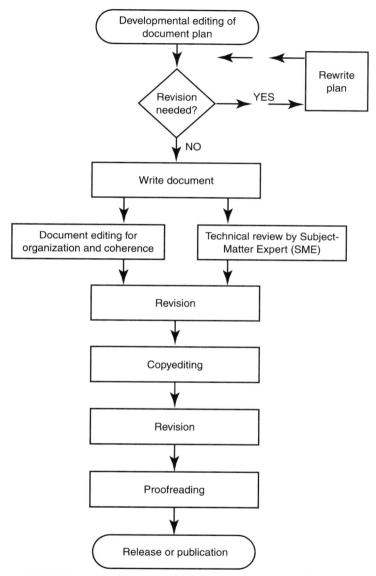

FIGURE 1.1 The "Top-Down" Editing Process in the Workplace

Sometimes, there was a review by subject-matter experts, and if there was time, the editor performed a quick proofread before the document's release.

Finally, in 13 of the 25 companies where I studied the editing process, there *were* no professional editors. (See the Preface for a list of companies.) Writers themselves were required to perform all editing functions, either by themselves or in collaboration with colleagues. In practical terms, this means that all writers need excellent editing skills.

1.4 How Can I Learn the Editing Process?

If the ideal editing process is "top-down"—that is, starting with the largest and most general topics and moving to details at the sentence and word level—by contrast, the easiest way to learn editing is "bottom-up." Errors in spelling, punctuation, and capitalization, for example, distract readers (even readers acting as editors) from concentrating on organization and effectiveness of presentation. Thus, it is more effective and efficient to learn the basics at the word and sentence level and then move on to larger matters of content, organization, and readability at the document level.

Therefore, after Part One of this book explains editing approaches and writer-editor relationships, Part Two begins with the details of capitalization, spelling, number usage, and punctuation. It gives you many chances to perfect your skills in editing at the sentence level, the processes called *proofreading* and *copyediting*.

In Part Three, the book takes up the larger matters of organization, coherence, and appropriateness for the reader, as well as guidelines for editing graphics, document design, online documentation, and international issues. All these matters are parts of *document editing*. See Figure 1.2 for a flowchart illustrating the "bottom-up" learning process.

If editing is new to you and you want to learn the basics, you might want to move directly to Part Two and work your way through the exercises at the sentence level. Then you can come back to Part One and read Chapters 2 through 4.

If you already have good language skills and are investigating editing as a potential career path, you might want to read Chapters 2, 3, and 4 next to understand editing within the context of the workplace. Use the chapters in Part Two to review and reinforce your sentence-level editing skills. Then move on to document editing in the chapters of Part Three.

EXERCISE 1

Assessing Your Current Editing Process

Assess and evaluate your current editing process on your own written documents. For the assessment, list the actions you take between generation of the first draft and completion of the final document. For example, do you always

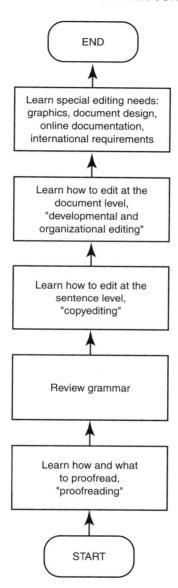

FIGURE 1.2 Learning the Editing Process from the "Bottom Up"

run a spelling checker? A style checker? Do you ask someone else to "look at" the document?

After your assessment, evaluate the success of each of the listed editing activities. How confident are you about your final documents? What kind of feedback do you get from teachers, managers, or peers?

EXERCISE 2

Evaluating Your Strengths and Weaknesses as a Writer

Evaluate—in writing—your current strengths and weaknesses as a writer. List the things you do well. List the areas in which you would like to improve.

EXERCISE 3

Learning the Terminology

Interview professional writers in your community to learn how they define the terms used in this chapter and carry out the activities. Choose any three of the following terms for your interview (or others about which you wish to learn more). In class compare the responses.

- readability
- applicability
- usability
- organization
- coherence
- developmental editing
- technical review
- copyediting
- proofreading

EXERCISE 4

Exploring the Field on the Internet

Browse the Internet to obtain an overview of editing issues and other concerns of workplace writers. You might begin with the Society for Technical Communication Web site at http://www.stc-va.org.

References

Beck, Charles E. 1995. Every writer needs an editor: A paradigm for writing in business, industry, and education. *Technical Communication* 42.2 (May).
Hansen, James B. 1997. Editing your own writing. *Intercom* 44.2 (February).
Kemnitz, Charles. 1994. Is all this theory necessary? *Technical Communication* 41.3 (August).
Pandya, Mukul. 1994. Business is no wasteland to modern working poets. *San Jose Mercury News.* Dec. 3.

2 Levels of Editing

Contents of this chapter:

- Understanding Levels of Editing
- The Jet Propulsion Laboratory (JPL) Levels of Edit
- Accuracy and Readability
- Planning an Editing Task

The writing of most workplace documents—whether proposals, reports, manuals, or online documentation—takes place under severe time and budget constraints. The proposal has a rigid due date; the report must go to upper management by Friday; the manual or online help must be ready when the product ships. In addition, the budget is tight, and only so much funding is available for review and editing.

These time and budget constraints play a big role in the editing process. Rarely do editors have enough time to perform the kind of thorough edit they would prefer. Thus, whether you are a professional editor, a colleague reviewing someone else's work, or a writer who must be your own editor, you will be faced with decisions about just how much editing you can "afford" to do. The need for this kind of decision-making has brought about the development of systems for levels of editing. There is no one optimum or standard system for levels of editing. Instead, writing departments or editors establish a system that works for them—one that provides a "safety net" for a specific writing situation.

2.1 Understanding Levels of Editing

Levels of editing (or, more commonly, *levels-of-edit*) systems define how thoroughly you edit a document or series of screens by referring to specific tasks you do or specific items you check.

Workplace writers and editors are keenly aware of their need to understand and apply levels-of-edit guidelines. Thirty-eight of the 50 writers I queried said that, in order to be effective editors, students should understand the levels of edit. A writer at a company specializing in design automation explained: "A critical skill is knowing the depth or level of editing needed." A senior medical writer commented: "I wish I had learned earlier what the different levels of editing are and when each is appropriate." And an editor for engineers who write about the design of high-level circuitry said: "The most critical skills for an editor to have are (1) the ability to assess a project and determine the appropriate level of edit and (2) the ability to gauge the receptiveness of a writer."

However, there is no one commonly accepted guide to the levels of edit, and—within the last five years—a shift of emphasis has even occurred within existing guidelines. To help you understand how levels of edit work, the next section summarizes the original levels-of-edit system.

2.2 The Jet Propulsion Laboratory (JPL) Levels of Edit

A comprehensive and detailed hierarchy of editorial tasks called *The Levels of Edit* was first written in 1976 for the Jet Propulsion Laboratory by Robert Van Buren and Mary Fran Buehler (2nd ed., 1980). Its purpose was to provide the documentation department with a systematic way to estimate the cost of the services it provided to authors. This system clarified for writers, editors, and managers the degree to which a specific document would be edited, based on time and budget constraints. It did not, however, include time for technical content reviews such as checking the accuracy of technical data, nor did it include dealing with translation or second-language problems.

First, the authors divided the editing task into nine types of edit. Then, they established five levels of edit to determine how many of the types of editing would be undertaken. A summary of the nine types of edit appears in Table 2.1; an explanation of the five levels of edit appears in Table 2.2.

Notice in Table 2.1 that the types of edit become more complex and time consuming as you move from type one to type nine. As Table 2.2 shows, at the highest level, the editor completes all the types of editing, whereas at the lowest level, the editor provides only a cursory examination of the document.

TABLE 2.1 Nine Types of Edit for the Jet Propulsion Laboratory

No.	Type	Explanation
1	coordination edit	planning, record keeping, scheduling
2	policy edit	adherence to management requirements; copyrights, trademarks, format, style
3	integrity edit	accuracy of references to figures, tables, and other sections
4	screening edit	minimum language edit: spelling, subject-verb agreement, complete sentences
5	copy clarification edit	instructions to printer or illustrator
6	format edit	physical display of headings, type fonts, column widths
7	mechanical style edit	consistency of format features: capitalization, abbreviations
8	language edit	complete language edit: grammar, punctuation, usage, parallelism, syntax
9	substantive edit	organization and completeness of content

TABLE 2.2 Five Levels of Edit for the Jet Propulsion Laboratory

Level	Types of Editing Completed
1	All nine types—highest level
2	First eight types—not substantive
3	First six types—not mechanical, language, or substantive
4	First four types only
5	First two types only—coordination and policy—lowest level

Ruth Chase, documentation manager and former editor at a software company, says, "When I set up our levels of edit, I used the JPL levels as a basis. The JPL levels are too rigid; I think most experienced editors would combine them." Sylvia Thompson, editor at an e-mail provider, notes, "Usually technical writing departments use some aspects of the JPL levels, and some levels go to production departments. This company does not have something as elaborate as the JPL levels."

In addition, the substantive editing included at the highest level of editing is very expensive and time consuming. As editor David Nadziejka notes (1996, 40), "When you are editing for substantive errors rather than typographical ones, the function is primarily intellectual rather than mechanical; substantive editing, in its requirements of time and knowledge, is closer to the author's process of originating the material than to the proofreader's process of checking for errors."

2.3 Accuracy and Readability

How do you decide on a level of edit given the reality that both time and budget are limited? Whether you are your own editor or someone else's, do you aim for accuracy of content or readability and correctness of the language? Can you have both? Under what terms? This section tackles these questions and offers some help in making your decisions.

Accuracy of Content

Few people in business, scientific, or technical fields would say that accuracy of information should be compromised because of time or budget constraints. Nevertheless, as freelance editor Bob Shaver notes (1995, 393): ". . .

everything has a price, including time." Shaver asks his customers what percent of accuracy they expect in their documentation, and "all customers respond with a 99 percent figure of accuracy or greater." Shaver then points out that an average page of text can have a density of 500 words. "Given that value," says Shaver, "permitted errors between 99 and 99.99 percent accuracies compute as follows per page."

Percent Accuracy	Permitted Errors
99	5 per page
99.5	2.5 per page
99.9	1 in 2 pages
99.99	1 in 20 pages

Shaver concludes: "Very quickly most customers agree that even the lowest permitted error rate is too high. What they expect for their documentation dollar is nothing less than perfection (393)."

Writers, editors, and technical reviewers would agree with those customers, so obviously technical review should be a major (perhaps *the* major) component in a levels-of-edit system. Nadziejka maintains that the "*technical quality*—not editorial quality—is the most important aspect of *technical* documents" (1995, 283).

There are three ways to help ensure technical accuracy, but all of them are costly in terms of time or money.

1. Include a technical editor as part of the team from the beginning. This person can serve as both a developmental editor (see Chapter 14 for an explanation of developmental editing) and either a usability tester for instructions and manuals or a fact checker for reports and journal articles. The key provision is involvement early in the project so that errors are caught early.

2. Send the document out for technical review either after an editorial review or concurrently with an editorial review. Remind the technical reviewers, who are subject-matter experts, that their job is to ensure technical accuracy, not to fix up the writing. With that reminder, reviewers will tap their real expertise, which is usually not in the niceties of phrasing or punctuation. In many companies, technical reviewers first analyze a document individually, noting questions, errors, and problem areas. Then the reviewers meet as a group (with the writer as a silent participant and notetaker) and solve disagreements of technical content or interpretation.

3. In a levels-of-edit system, make the editor's primary function the review of content. Nadziejka (1995) suggests three ways for the editor to edit content in reports:

1. Read and edit the abstract, introduction, and conclusion to ensure that they are not contradictory.
2. Read figures and tables and edit them for accuracy and for agreement with what the text says.
3. Study those parts of the text that summarize or state results and ensure that those results agree with the abstract and conclusion sections.

In manuals or instructions, however, editing of content must be approached differently. The best way to ensure accuracy of instructions is to user-test them. In other words, you should obtain the product, the software, or whatever the instructions are for, and work your way through the instructions step by step, whether you are installing, operating, troubleshooting, and so on. As you proceed, note omissions, ambiguities, errors, and problems in organization. See Chapters 4 and 14 for specifics.

Readability and Correctness of Text

If technical accuracy is paramount, how important are readability and the correctness of the language? Should you agree with software writer Frank Guss, who writes for expert users and who says, "Management here takes the position that customers don't notice minor errors in grammar and other mechanics and that emphasizing correct usage is not cost effective." Or should you echo the concern of freelance editor Liz Van Houten, who says, "What do you say to a client who has just found a typo in his million-dollar project (full color, glossy, 90 pages, 300,000 copies)? What if you marked the typo on the proof, but no longer have the marked proof? What if you did not see the blueline [camera-ready copy] and were therefore not able to catch the typo? What if you did see the blueline and missed it?" Poorly written or error-ridden documents reflect badly on a company because they reflect sloppy work; if the writing is bad, readers will often question the accuracy of content as well.

Two key considerations in determining how much time and money to spend on readability and correctness are the *purpose* of the document and the intended *audience*. For example, if the purpose is to make the company look good in an annual report or a proposal, and the audience consists of investors or a granting agency, then the readability and correctness of the writing are as important as the technical content. If, however, the purpose is to write installation instructions for a narrow audience of technicians who are already familiar with the product, then—while readability is important—grammar and punctuation may not be as critical. Nevertheless, you need to remember the results of the 99-percent rule discussed in section 2.3 and work toward the highest degree of language correctness you can achieve.

Companies solve levels-of-edit problems in many ways. Figures 2.1 and 2.2 show two different examples of levels-of-edit approaches. In both

CHAPTER 2 LEVELS OF EDITING **17**

examples, the levels of edit are in addition to a content review by technical personnel such as engineers or programmers.

In Figure 2.1, professional freelance editor Jo Levy explains to the writers with whom she works the levels of edit she has developed. She uses the pronoun "we" to emphasize that she and the writer are a team. The four levels—developmental, substantive, copyedit, and proof and production edit—ensure interaction between the writer and the editor at four different times while the document is being written. However, as Levy notes, sometimes two levels of edit must be combined to meet a deadline. If you must act as your own editor, you can also plan for editing at four different stages of your writing. One way to do this is to convert the statements under each level in Figure 2.1 to questions that you should ask.

As writers, your goal is to select information your reader needs to use the company's products and convey it in a package that puts the information at the reader's fingertips. The editorial process provides you with an informed and sympathetic first reader to help you achieve that goal.

Each level of edit focuses on different aspects of the work you perform to reach this goal. In the best of all possible worlds, software documentation goes through each level of edit in the order listed. Since we live in the real world, however, I will combine different levels of edit when the schedule requires it. For example, it is not unusual for me to combine a substantive edit with a copyedit. I may also take advantage of a developmental edit to point out a rule in the style guide that applies to your material.

1. The Developmental Edit—A View From 30,000 Feet

A developmental edit can assist you in deciding what information you must convey and how to package it for most effective delivery. It can help me decide how best to make use of the time available to edit your work as it progresses.

What I Need

A developmental edit requires the following material:

- Your documentation plan (and content plan, if one is available). A documentation plan is a plan for documenting the software. It usually describes all of the manuals and other units of documentation (information modules) you plan to deliver. A content plan describes the manual or other individual module you want me to edit. It should contain information about the target audience and list the tasks you intend to include.
- Some discussion, often both with you and the writing manager or another member of the project, about the project.

FIGURE 2.1 Levels of Edit

What I Do
Using this material, I

- Make sure the proposed information module contains the information needed to satisfy the intended audience.
- Make sure the proposed information module contains the information needed to satisfy the stated purpose.
- Examine the structure of the module for ease of use and maintainability.

What You Get

A developmental edit usually results in oral or written suggestions (sometimes both) about structure and organization.

I will probably spend some time talking with you about how to present the information required in the module. I will often ask you questions about the product. I may also ask questions about why you selected one organization method rather than another, or how you think your audience intends to use the module.

You will receive oral suggestions in a writer-editor conference, which will be arranged at a mutually convenient time. Writer-editor conferences are private and confidential—you and I will be the only people there.

You will receive written suggestions in the form of a memo. The memo will describe any organizational, structural, audience, or task definition issues I find. A copy of the memo is also given to your manager.

2. The Substantive Edit—Oooh, Come a Little Bit Closer

The substantive edit assumes that your overall structure (manuals in a library, chapters in a manual, topics in a help system, sections in a chapter) is well defined and will not substantially change. A substantive edit focuses on presentation and organization within that overall structure.

What I Need

A substantive edit requires a complete information module, such as:

- An entire manual
- An entire chapter of a manual
- The executable for a help system

I also need your content plan or outline, if I don't already have it. I may also need information about how the software works and, unless I have had a chance to complete a developmental edit on the material, how you think your audience intends to use the information module.

What I Do

I read your material at least twice. While I'm reading it, I evaluate:

FIGURE 2.1 Continued

- Whether the overall content is appropriate for the audience and purpose.
- Whether the information is organized into business tasks.
- Whether the tasks are properly designed.
- The order in which the tasks are presented.
- Whether the information within each task is chunked properly.
- Whether the information within each task is layered properly.

What You Get

You will receive your draft marked with changes and suggestions. The changes may be marked in red (only because red shows up well and I'm used to using it). The suggestions will be given to you in a separate memo.

- I may suggest alternative organization methods, such as rearranging chapters, combining or splitting chapters, or moving paragraphs.
- I may suggest altering the content of a task or restructuring the task.
- I may suggest adding or removing elements such as tasks, procedures, tables, graphics, or headings.

Feel free to question any of these suggestions. Some of them may be a result of my not understanding the product very well. If you think your readers will understand it better, then my suggestions will be moot.

Some of my suggestions may require additional writing. If you feel you cannot implement all of them, we will have a conference to discuss them, and I will be happy to prioritize them for you.

3. The Copyedit—Get Right Down to the Nitty-Gritty

The copyedit assumes that your structure and organization are well established and that neither they nor the contents of the information module will change substantially. A copyedit focuses on details—grammar, style, and syntax.

What I Need

An editor can copyedit anything from a single sentence to a complete library. To be efficient, however, I will need a complete information module, such as:

- An entire manual or help system
- An entire chapter of a book
- The executable for a help system

The larger the unit, the better the job I can do for you.

If possible, I also like to have a copy of the draft on which I performed the substantive edit, if in fact I did one!

FIGURE 2.1 Continued

What I Do
I read the material at least twice, checking for

- Conformance to the company style guide
- Adherence to the department's philosophy on software documentation
- Consistency of presentation
- Optimal word choice
- Proper syntax
- Proper grammar
- Proper/preferred spelling
- Correct cross-references
- Correct claims for graphics and tables
- Clarity

What You Get
You will receive your draft marked with changes and suggestions. The changes will be marked in red (I'm a creature of habit). The suggestions will be given to you in a separate memo.

Again, feel free to question any of the suggestions. However, be aware that correcting the problems identified at the copyedit stage is usually not negotiable. You will need a very good reason to avoid making some sort of change. You may choose a different solution from the one I have marked on your copy, but you will probably have to solve all of the problems I find.

4. The Proof and Production Edit—One Last Chance
The proof and production edit assumes that your product is complete and has one foot out the door. A proof and production edit focuses on mechanical details.

What I Need
I need a complete production unit, such as a complete manual. The unit should be finished and, in your opinion, ready to be given to a customer.

What I Do
I read the material once, for mechanical problems such as

- Typos
- Missing or incorrect production elements (such as part numbers)
- Improper capitalization
- Improper grammar
- Improper pagination
- Nonconformance to company style, including unmarked trademarks and other legal requirements

FIGURE 2.1 Continued

> **What You Get**
> You will receive your draft with all problems marked.
> Any problem that falls into one of the categories listed in "What I Do" must be corrected. Otherwise, your module cannot be submitted for production. But along the way, I sometimes find other problems that you may want to solve if you have time. Whether you solve these problems is completely up to you.
>
> *(Reprinted by permission of Jo Levy, San Jose, CA.)*

FIGURE 2.1 Continued

The "Levels of Edit Agreement" in Figure 2.2 takes a different approach. Each level of edit provides a list of tasks. Writers, managers, and editors agree which specific tasks will be completed and sign to indicate their agreement. Notice that the agreement also differentiates between preliminary documents (in first- or second-draft stage) and final documents, which are almost ready for release.

> Select **Light, Medium, Heavy** edit or **Sign-off**. You can also select individual options within other categories.
> The levels of edit are cumulative, progressing from light edit through the sign-off checklists. Sign-off is reserved for documents that have already received light, medium, or heavy editing.
>
> ☐ **Light Edit**
> ☐ Review and correct spelling, grammar, capitalization, and punctuation on a spell-checked file.
> ☐ Identify and suggest changes to incomprehensible sentences.
> ☐ Make general observations about organization. Point to specific examples.
> ☐ Check for consistent and appropriate typefaces and fonts.
> ☐ Check headers and footers for correct text, format, and consistent placement on the page.
> ☐ Check for consistent and appropriate presentation of numbers as numerals or words.
> ☐ Check for correct usage and presentation of abbreviations and acronyms, lists, notes, and cautions.
> ☐ Check format and page design, including size and placement of graphics.
> ☐ Check for legal issues, including appropriate identification and use of trademarks.

FIGURE 2.2 Levels of Edit Agreement at Cadence Design Systems, Inc.

- ☐ **Medium Edit.** Perform a light edit plus:
- ☐ Check cross-references to chapters against table of contents.
- ☐ Check that titles, chapter heads, and subheads are parallel, complete, and organized.
- ☐ Check appropriate placement of figures in text.
- ☐ Check passive constructions, tense, and other style issues. Suggest rewriting when appropriate.
- ☐ Check for correct syntax, appropriate word usage, and other language issues.
- ☐ Check tables for correct format and continuation onto consecutive pages.
- ☐ Check format against Cadence style.

- ☐ **Heavy Edit.** Perform a medium edit plus:
- ☐ Check for tone, coherence, and smooth transitions.
- ☐ Check overall integrity and quality of document including conformance to writing and editorial standards.

- ☐ **Sign-off Checklist for Preliminary Documents.** Use for documents that have at least a light edit.
- ☐ Check manual name and part number.
- ☐ For preliminary documents only, check that *Preliminary* notation appears across the title page and in the footer.
- ☐ For preliminary documents only, check legal notice (This material subject to change . . .) on title page.
- ☐ Check title of cover art on copyright page.
- ☐ Check for bad page breaks.
- ☐ Check for clean, camera-ready master. Note obvious errors and quality issues.
- ☐ Proof writer's final changes for accuracy.

- ☐ **Sign-off Checklist for Final Documents.** Use for documents that have at least a medium edit. Check the same issues as for preliminary documents plus:
- ☐ Check for an index. Final documents must have an index.
- ☐ Check a few page numbers in the table of contents and index.
- ☐ Check appropriate placement of index entries, especially *See* and *See also* references, which are automatically alphabetized with the "S" listings under each entry.

(Reprinted by permission of Cadence Design Systems, Inc.)

FIGURE 2.2 Continued

If you are editing your own writing, you can learn a systematic approach to editing by studying the "Levels of Edit Agreement" in Figure 2.2. In the next section of this chapter, you will see how these two different approaches have been considered in the development of a more generic "Checklist for Efficient Editing."

2.4 Planning an Editing Task

This chapter has described several ways you can approach levels of edit, but most companies, journals, magazines, and publishers will already have editing procedures in place. In that case, you can simply follow the suggested levels of edit.

Sometimes, however, you will be the only or the first writer or editor in a start-up company, or you will be working on your own as a freelance editor or writer preparing material for publication. You need a checklist that can help you plan your editing task, determine how much time and money the task will take, and allow you to negotiate with a client. The checklist that follows can be adapted to all types of business, scientific, and technical documents: journal articles, reports, proposals, online documentation, instructions, and manuals. The best way to use the checklist is to work through the entire document several times, each time checking on only one item. For example, look at all headings and subheadings at once; examine all figures and tables in sequence.

Some of the terms in this checklist may not be familiar to you. Check the glossary or the index of this book to find the section where that term is explained. The exercises in Part Two will teach you how to correct errors at the sentence level; the exercises in Part Three will teach you how to correct errors at the document level. In addition, Chapters 15 through 19 have individual checklists specific to that chapter's subject.

Checklist for Efficient Editing

Level One—Always ask these questions:
- ☐ 1. Is the distribution information (order number, document number, file number, bar code) correct?
- ☐ 2. Are the front and back cover, title page, and spine accurate? Title? Logo? Address? Phone numbers? Copyright information?
- ☐ 3. Are the preface and first page of text accurate, informative, and appropriate in tone?
- ☐ 4. Is the document complete (no missing pages)?

- ☐ 5. Are all pages complete (no missing sections)?
- ☐ 6. Are there any obvious typographical errors?
- ☐ 7. Are there any obvious policy, copyright, or political problems?
- ☐ 8. Can this document be released without a complete edit or rewrite?
- ☐ 9. Is there an index? Is it usable?

Level Two—For a more thorough edit, ask all the questions in Level One. In addition, ask:

- ☐ 1. Is the content technically accurate? Are calculations correct?
- ☐ 2. Are figures and tables technically accurate? Are references consistent? Is figure and table format consistent?
- ☐ 3. Is the format of text consistent?
- ☐ 4. Do the headings and subheadings show a clear hierarchy? Do they provide both key content words and task information? Are headings consistent?
- ☐ 5. Is the purpose clearly stated? Is the intended audience identified? Do the approach and tone match the purpose and needs of the intended audience?
- ☐ 6. Is the language correct, consistent, and clear?
  - subject-verb agreement
  - proper verb tenses
  - consistent word use
  - correct punctuation and capitalization
  - parallel construction
  - correct verb tenses
  - no extra words, undefined jargon, vague words

Level Three—For a still more thorough edit, apply all the questions from Level One and Level Two. In addition, ask:

- ☐ 1. Are paragraphs logically developed? Do transitions link sections and paragraphs?
- ☐ 2. Is the style effective?
  - Is passive voice used only for a reason?
  - Are sentences crafted so important information is in an emphatic position?
  - Do most sentences follow a known-to-new sequence?
- ☐ 3. Is back matter accurate and adequate?
  - Is the glossary complete and correct?
  - Is the reference list (bibliography) correct? Does it match in-text references?
  - Is the index complete? Accurate? Cross-referenced?

If you are a writer doing your own editing, apply the checklist as it is written to two or three documents and note those places where you need to adapt the checklist to your own needs. Then, using this checklist and those in Chapters 15 through 19, customize the checklist for the type of editing you do. If you are a freelance writer or editor, you might want to make your adaptation of the checklist into a formal agreement that you and your client can sign.

If you are a writer who sometimes edits for others, use a checklist to determine with the writer or manager how much editing you will take responsibility for. Such an agreement will allow you to ensure that the most important areas are covered without taking responsibility for everything in the document.

If you are the only writer or editor at a company or organization, a checklist will help you estimate your time and cost for editing. At one high-tech company, for example, experienced editors estimate that at the highest or third level of editing, they can edit six pages per hour on an early draft of the document. At the second level of editing, they can edit 10 pages per hour on a second draft. At the first level of editing, they can edit 15 pages per hour on a final draft. Keep track of the time it takes you for each level of editing, and soon you will be able to estimate your time and costs for editing documents at various levels of editing and various stages of the document's development.

EXERCISE 1

Learning about Local Levels of Editing

Interview a writer or editor at a local company, journal or magazine, or publishing house. Ask how levels of editing are determined and compare the response with the samples in this chapter. Report to the class what you have learned.

EXERCISE 2

Understanding the JPL Levels of Edit

Read the complete JPL Levels of Edit *document (see the reference list). How does this document clarify how much time and money are spent on editing?*

EXERCISE 3

Learning about Local Technical Review Procedures

Interview a technical person in your community. Ask how technical reviews of documents are handled; compare the response to the description of the process in this chapter. Report to the class what you have learned.

EXERCISE 4

Applying the Checklist for Efficient Editing

Apply the checklist at the end of this chapter to a workplace document and keep track of the amount of time it takes to complete each level of editing in the checklist.

References

Cadence Design Systems, Inc. 1996. *Levels of edit agreement.* Reprinted by permission of Cadence Design Systems, Inc.

Haugen, Diane. 1991. Editors, rules, and revision research. *Technical Communication* 38.1 (February).

Levy, Jo Carol. 1997. *Levels of edit.* Reprinted by permission of Jo Levy, San Jose, CA.

Nadziejka, David. 1996. Quality editing. *Intercom* 43.4 (April).

———. 1995. Needed: A revision of the lowest level of editing. *Technical Communication* 42.2 (May).

Shaver, Bob. 1995. Perfection. *Technical Communication* 42.3 (August).

Van Buren, Robert and Mary Fran Buehler. 1980. *The levels of edit.* 2nd ed. JPL Publication 80–1. Pasadena, CA: California Institute of Technology.

3 Editor-Author Relationships

Contents of this chapter:

- Authors and Their Responsibilities
- Editors and Their Responsibilities
- Making Editor-Author Relationships Work Well

Mention the word *editors* to some engineers who must write, and they mutter about "changing the meaning to make it sound better," or they call editors "nothing but comma guardians." Mention the words *scientist-author* to some editors, and they roll their eyes and complain, "That guy couldn't write at all. His sentences were 40 words long and incomprehensible, and he didn't want me to 'mess' with them." The fact is, there is sometimes conflict between authors and their editors—conflict that can range from an uneasy alliance to actual hostility.

On the other hand, as Bruce Speck, a professor of professional writing, says, ". . . both authors and editors could work more effectively if they openly recognized that text production is a collaborative effort in which different yet similar talents are used in the service of a common goal" (1991, 305).

Anjali Puri, an editor at a software company, clarifies these "different yet similar talents."

> In my experience, writing is a right-brained activity. When writing, it's important to write and not worry about mechanical, grammatical, and style issues. Editing is a left-brained activity. Here, logic, rules, definitions, grammar, organization are all being applied to the particular piece of writing. Objectivity exists since the editor did not write the documentation, and that distance lets the left brain function without attachments. Correcting your own writing is hard, and it leaves the door open for errors that could be avoided. (1991)

In addition, as editor C. E. Putnam notes, "What makes the editorial task really complex, challenging, and interesting is the fact that most issues of editorial concern—all those having to do with organization, length, style, point of view, choice of word, and turn of phrase (to say nothing about basic question of content and audience)—are legitimately open to discussion" (1985, 19).

Obviously then, there is—or should be—common ground between authors and editors. Both are concerned about producing a hardcopy document or online documentation that is understandable and usable, and by working together, both can produce a product that is better than either could produce alone. To achieve that "better" document, both author and editor need to learn their mutual responsibilities and the ways to build a positive working relationship.

3.1 Authors and Their Responsibilities

The general public often thinks of authors only in terms of novelists or nonfiction writers who write or ghostwrite sports and entertainment biographies and autobiographies. The truth, of course, is that those authors make up only a small percentage of the thousands of individuals who write professionally or who write as a part of their profession.

Many authors are *subject-matter experts* (SMEs). These SMEs may be scientists who write to share new discoveries in the field or laboratory; engineers who are experts on the way a piece of hardware functions or the way a software program interacts with another program; business executives who know about a manufacturing or marketing operation; attorneys; educators; government administrators. SMEs, in other words, include any person who has specialized knowledge that he or she wants or needs to share with others through the written word.

Other authors are *professional writers*. They may, indeed, be novelists or nonfiction authors who write as freelancers and work directly with publishing houses or magazines. They may also be technical writers, learning product engineers, information developers, or one of the other variant names of this profession. These writers are not usually experts in the technical field; instead, they are experts in language and communication. Their job is to get information from SMEs (those who *know*), organize it, and then write so that the information is understandable and usable to those who *don't know* but want or need to know the information or use the product. Other professional writers may work in government, corporate communication, marketing, or public relations.

Whether they are SMEs or professional writers, authors need to think of editors as does Jonathan Yardley, journalist, reviewer, and book author. Yardley says, "Writers need editors. . . . Or, perhaps more accurately, writers need good editors. There are almost as many bad editors as there are bad writers, which is to say plenty, and the writer who finds himself [sic] at the mercy of one is sure to come away bruised and bitter. . . . Perhaps it's best for writers to think of editors not as malign creatures wielding blue pencils but as professional readers—or, even better, as readers' representatives in the publishing process. We write, after all, to please and inform readers, so it stands to reason that if we fail to do so, we fail in our jobs. What the editor's function boils down to is to determine if we get across to readers and, if not, to help us figure out how to do so" (1990, B2).

Yardley's views were echoed by the scientists in one study of editor-author relationships at a major scientific laboratory. Carol Gerich, writing in 1994 about the revision process at the Lawrence Livermore National Laboratory, found that "about half the papers in this group are edited—an impressive number considering that the time spent editing is charged directly to the author's individual budget" (66). The two authors in Gerich's study estimated that they would use 75 percent of the editor's suggestions in final revisions. As one of the authors told his editor, "My attitude is to accept all these things [rhetorical fixes] if they don't change the meaning. It's your business to know this. You're the expert" (68).

Professional writers, for the most part, appreciate what good editors can do for them. For example, Tricia Lyons writes documentation and training materials at a company specializing in computer solutions for automotive and hardgoods companies. She talks about her editor in glowing terms: "Our editor is exemplary. She's charming, tactful, knowledgeable, and speedy. Her edits are easy to understand, and she's always available to explain or discuss a point. She's a joy to work with and makes us all shine as writers."

The question that arises, then, is this: What is the *author's* responsibility toward making the editor-author relationship work? Researchers who have studied this relationship agree that authors are responsible for following these five guidelines:

1. Following the company or publication style guide while writing drafts to reduce the editorial burden. Authors do not work in a vacuum; their creativity and personal style are always limited to at least some extent by the type of document they are writing and the constraints imposed by corporate or publication guidelines. The best way to ensure good editorial collaboration is to meet the guidelines whenever possible during the drafting process.

2. Agreeing with the editor on the goals of the project, expectations of the process, and the roles of each participant. This agreement should include a determination of the level of edit required and the deadlines for completion of each segment of the work. The best way to ensure a good working relationship is to have a written agreement, either a checklist or a memo of understanding that is the result of a meeting between the author and the editor. (See Chapter 2 for examples of these agreements.)

3. Meeting deadlines and returning drafts promptly. Editors—whether they are fellow writers or dedicated editors—are usually juggling several projects. To obtain the best kind of editing, authors need to respect schedules and deadlines.

4. Giving credit to the editor for his or her expertise. An editor is the language expert, but even more important—an editor is the reader's representative. If the message in the document is not "getting through"—whether the problem is organizational or syntactic—an author needs to listen to and respect the editor's response. Authors need to be open to suggestions, subdue their ego-involvement with their words, and learn to view their writing as a developing project, not a finished creation. Authors can learn from editors and apply one set of editorial suggestions to another document or the next part of the same document while it is being created.

5. Checking all the suggestions and changes made by the editor. With developmental or organizational problems, editors usually pose questions or suggest alternatives. Authors need to address these areas seriously and rethink the writing. At the sentence level, editors sometimes make the actual changes or corrections. When working online, they usually mark the changed passages with change bars. Again, authors need to check these corrections to ensure that diction, syntax, or even punctuation changes have not altered the meaning of the passage. Shirley Krestas, a former high-technology publications manager, explains: "As a manager, I knew that technical accuracy was vested with the writer. So if the writer said that a change would affect the accuracy of the information, I backed the writer."

3.2 Editors and Their Responsibilities

The editing half of the editor-author relationship can be carried out by any one or more of the following groups:

- subject-matter experts
- managers and other gatekeepers such as journal editors
- professional editors
- other writers functioning as editors

Each of these editorial groups may have a slightly different purpose, and often more than one kind of editor will work on a single document.

Subject-matter experts (SMEs) most often function as peer technical-content reviewers. Many companies routinely route documents to the relevant technical experts for content review. Gerich, in her study of the revision process at a national laboratory, found that the scientific authors were expected, though not required, to have articles reviewed by colleagues. According to Gerich, "Colleague reviewers enjoy a true partnership with authors in the revision process because their roles are frequently interchanged" (1994, 66). Because SMEs are technical experts, their job is to comment on technical accuracy, coverage of the topic, and logic of the argument. They are not expected to edit the language. In fact, Yvonne Kucher, an editor at a networking company, says, "[At this company,] editing and technical review happen concurrently, but we prefer that a document is edited before it goes for technical review. Once a technical person finds a grammatical error, he or she defocuses from the task."

Managers and other gatekeepers are those individuals who ultimately control whether or not a document is published. They may be managers in technical areas or in publications; they may be managing editors who accept articles for journal publication or acquisitions editors who offer contracts from a book-publishing house. Like SMEs, gatekeepers are primarily concerned with content, but they also evaluate the skill and effectiveness of the writing.

Professional editors are hired primarily to work with documents written by others. These editors may be highly experienced and able to work with authors in the broad areas of developing the documents: organization, choice of media for presentation, logic, and completeness. The same person (or a different professional editor) may copyedit at the sentence and paragraph level, fine-tuning the prose. Before publication, a professional editor may proofread the document. Editors working in a publications department may work with from 10 to 20 writers, and they often work on several projects at one time. Likewise, editors who work for journals or publishers often have several concurrent assignments.

Other writers are the people most often called upon to be developmental and copyeditors; as noted in Chapter 1, many companies employ no professional editors. Thus, while there may be provisions for technical review and management review, editing another writer's document for organization, logic, language, grammar, diction, syntax, and punctuation is likely to be part of a writer's responsibility. Most of the 50 writers I queried in preparing this book spend 20 to 30 percent of their time editing for others; some writers spend as much as 50 to 75 percent of their workday acting as an editor.

Like authors, editors have a responsibility to make the editor-author relationship work well. Editors have the following responsibilities:

1. Knowing the company or publication style and being able to articulate the reasons for editing decisions. Large companies, established journals, and publishing houses have publication style guides either in hardcopy or online. Those style guides may cover everything from accepted abbreviations to grammar imperatives to formats. Many organizations also use a more general style guide—such as *The Chicago Manual of Style* or the United States Government Printing Office *Style Manual*—as a backup. If you must edit in a startup situation or for a company that does not have a style guide, you will want to begin compiling a style sheet for the first documents you work on. As you make editing decisions, record the choices on the style sheet, and you will soon have the core of a style guide. See Chapter 7 for details on preparing a style sheet.

In addition to knowing the company style, you must be able to justify every editing decision you make, whether it pertains to grammar, punctuation, diction, syntax, organization, graphics, or document design. Editors must be language experts, not only knowing *what* to do, but also knowing *how to explain* each decision. As editor Sylvia Thompson notes, "The writer may say 'Who are you and why can you say that?' A writer will not respect you if you don't act like you know what you're doing. Therefore, an editor has to give the writer some justification for what he or she does."

You may be a very good writer. But if your knowledge of the rules of grammar and the guidelines influencing word choice and sentence construction is intuitive rather than explicit, you will need to review sentence-level editing. All the chapters in Part Two of this book will help you be a better editor. In addition, what you will learn about sentences will help you be an even better writer than you are now.

2. Agreeing with the author on the goals and expectations for the editing project and the level of edit desired. This agreement should be reached before you begin the project; you can use a checklist like that shown in Chapter 2, or you can write a memo of understanding after meeting with the author. It is important, whether you are working freelance or within a company, that the agreement and schedule are in writing.

3. Acting, while editing the document, as the reader's representative. In order to edit effectively, you must understand the purpose of the document and its context or relationship to other documents. You must also know who the intended reader is and the level of that reader's experience. The reader might be specifically identified: "This manual is for systems administrators who are experienced users of the previous version of this product." More commonly, reader identity is loosely defined: first-time users, technicians, experts, and so on. If the writer has not identified the intended reader, ask questions. Chapter 14 explains ways you can analyze readers and determine reader type, reader tasks, and reading style.

You must edit for readability: Does this make sense? Will the reader understand these words? Are new terms defined? Are the sentences logically constructed? Is the document free of punctuation and grammatical errors?

In many cases, you must also edit for usability: Can the reader perform this operation? Are all of the parts explained? Is the level of language accessible? One of the best ways to be the reader's representative is to ask questions, writing them down for the author either on Post-it® notes, in the margins, or in a separate memo. If the document is an instruction or procedural manual, you can literally user-test the instructions by performing the operation and noting problems and questions.

4. Giving positive feedback, giving direction, and avoiding changes to the author's style. Good editors recognize when authors have done something well, and they are as quick to compliment as they are to criticize. When they must criticize, good editors strive to do so in a non-threatening way. Judy Billingsley, who edits at a company building fault-tolerant computer systems, says, "Editing can cause hurt feelings and animosity. If you're abrasive, you won't get far." Publications manager Martha Cover agrees, saying: "Be careful not to be snide; you can't even be witty. Instead, ask questions to untangle the prose."

In addition to questions, authors appreciate direction. In other words, don't just say that something is wrong; suggest how it might be fixed. Editor Yvonne Kucher notes: "If something doesn't work, I give the writers options and explain what I'm looking for. It doesn't help to fix all the problems. Writers should learn and move on to other things."

Finally, when you edit, you must remember not to change the author's style just to make the document read the way you would write it. Each author has a style of writing that is unique to that person, and—if the author is at all competent—the style is also tailored to the needs of the specific assignment. Style is composed of many elements; among them are word choice, sentence length, types of sentences, sentence patterns, rhythm, sensory images, and tone. As an editor, you need to fix what is wrong, but you should preserve any stylistic qualities that work.

5. Meeting deadlines and fulfilling responsibilities. The burden is on both editors and authors to ensure that their relationship is good. Editors can build trust by returning their edited drafts on time and by completing whatever level of edit was agreed to at the beginning of the project. Editing can be time consuming, so beginning editors should keep track of the time spent on each project to develop some accuracy in forecasting how long an edit will take.

6. Working with the author after the edit. Editing is not finished when the last correction has been made on the screen or the paper. Some authors want to discuss the edit; others want all the editor's comments in writing so they can internalize them before any discussion. But usually, the writing in a document improves the most when editors and authors can talk to each other. Editor Yvonne Kucher feels strongly that "Editors need interpersonal skills. The workplace is becoming fragmented with communication by e-mail, phone, fax, and print vendors on disk. As an editor, I need to communicate with writers, and when I talk to people, I need to see the expressions on their faces."

You perform an added service in editing when you initiate an editor-author conference. During the conference, you need to establish rapport, use tact and diplomacy, and be willing to admit your own mistakes. Editor Judy Billingsley says, "I listen when my writers take exception to my edits. If the writer has a valid reason for changing the standard, I will make an exception or change the standard. I won't change what the writer has done if I can't add value."

Editors and authors work best when they work as a team, sitting down and working together to produce an effective document. Editing and writing are not two separate and parallel paths to that end. Editing and writing should converge on one path.

3.3 Making Editor-Author Relationships Work Well

Good editor-author relationships are clearly the result of an attitude of professional respect from both participants. Editors need to respect authors for their subject-matter expertise and for their efforts at taming the material and making it understandable to the reader. Authors need to respect editors for their language expertise and for their ability to evaluate the document in context and to analyze the parts of the document ensuring that each part communicates its message effectively.

Writers who sometimes act as editors are in a good position to recognize this need for professional respect on both sides of the relationship. They want respect for their writing *and* for their editing.

Part of that professional respect should come from realizing that editors and authors are members of a team with the same goal: producing a document or documentation that best meets the reader's needs. As a team, an author and an editor can produce writing that is better than either individual could have produced alone. Editor-writer Marjorie Hermansen says that this collaboration "produces an increase of ideas, an expanded sense of creative energy—SYNERGY. A synergistic environment, writer + editor = collaboration, where the whole is greater than the sum of all its parts, represents the writer/editor relationship at its best" (1990, WE15).

EXERCISE 1

Understanding Your Attitude Toward Being Edited

As a writer, think on paper about your experiences being edited.

- Have you participated in classroom small-group editing sessions? If so, what help did you receive? Was it a positive experience? Why or why not?
- Have your writing instructors edited your documents in draft? How? Did their comments help you produce a better document? Was it a positive experience? Why or why not?
- Have your documents been edited by other writers or by an editor at a company? How did they help you? What problems occurred?

Briefly write down your responses in all the applicable areas. Then summarize your present attitude toward being edited.

EXERCISE 2

Understanding Your Strengths and Weaknesses as an Editor

List five strengths you can bring to the editing process. What are you particularly good at? List your five editing weaknesses. How can you improve in those areas where you are weak? Summarize your present attitude toward acting as an editor.

EXERCISE 3

Analyzing Your Editing Experiences

Write about your previous experiences editing someone else's work. What were the results? How did the other writer respond to criticism? What would you do differently in another editing context?

EXERCISE 4

Volunteering as an Editor

If you have never edited for someone else, volunteer your editing services to another student, to a coworker, or to a nonprofit agency. Agree with your client what level of edit you will perform (see Chapter 2). After the edit, write a brief report on your client's responses to suggestions and changes. Discuss the responses in class.

References

Gerich, Carol. 1994. How technical editors enrich the revision process. *Technical Communication* 41.1 (February).

Hermansen, Marjorie S. 1990. The writer/editor relationship: Collaboration—a step beyond compromise. *Proceedings.* Washington: 37th International Technical Communications Conference.

Puri, Anjali. 1991. Collaborative writing and editing. Unpublished document. San Jose, CA: San Jose State University.

Putnam, C. E. 1985. Myths about editing. *Technical Communication* 32.2 (April).

Speck, Bruce W. 1991. Editorial authority in the author-editor relationship. *Technical Communication* 38.3 (August).

Yardley, Jonathan. 1990. The eclipse of editing. *Washington Post,* August 6.

4 Working as an Editor

Contents of this chapter:

- Understanding the Abilities and Skills of an Editor
- Clarifying the Duties of an Editor
- Enhancing the Editing Job
- Flourishing Professionally

"To be a good editor, one must be a good *writer*, with strong, versatile, and flexible language skills." Writers who must act as editors should be encouraged by this comment from Lola Zook (1994, 2), long-time editor and one of the pioneers in the movement for professionalizing editing. Writers are, almost by definition, people who excel in the use of language and who enjoy the challenge of creating meaning by using the building blocks of words. Editors must also excel in the use of language and enjoy the challenge of clarifying the meaning of documents.

Nevertheless, editing is different from writing. Editors must be objective and dispassionate, always taking the broader view and the reader's perspective. Therefore, writers who must also be editors of their own or someone else's work need to learn how to shift roles. This shift is especially difficult when writers must edit their own work. Shirley Krestas, who places editors and writers in industry, says, "If you're going to be both the writer and the editor, you have to be able to step out of your writing mode, reread your style guide and reset your focus before you begin to edit." If possible, you should let the document rest for as much as two days so you can approach it from a fresh perspective.

4.1 Understanding the Abilities and Skills of an Editor

Editors also need specific abilities beyond writing skill to succeed in their work. For ease of discussion, those abilities can be grouped into four categories: organizational, analytical, interpersonal, and communication.

Organizational abilities involve looking at the big picture and being able to evaluate how the parts contribute to the whole or can be made to do so. Editors need good pattern-recognition skills, whether they are examining the subpoints in a section to evaluate equal importance or the parts of a sentence to ensure parallelism. They need logical thought to maintain consistency and to see if one point leads to another. In addition to the organization of text, editors need a good eye for document design, and they need to know the basic guidelines for good page or screen design. Finally, a part of organizational ability is time and project management: the skill to handle a big project and perhaps coordinate the work of others.

Analytical abilities enable an editor to take a document apart, examine the pieces, and evaluate how well the pieces work together. Editors must pay attention to detail, have a high boredom threshold, and be able to remember the details of material read 20 pages or screens ago. They need to be able to assume the reader's perspective and examine how well the document works from that perspective. They need to be adept at making decisions and evaluating potential ethical problems. Because they are usually working in a technical or scientific environment, they need to be interested in technology and willing to learn a technical field. And finally, in an area that overlaps with communication skills, editors need to be able to find and correct errors in grammar, diction, syntax, and punctuation.

Interpersonal skills are essential for success in editing. Editors need to understand the writer's point of view and be able to keep their own ego out of editing decisions. They need to establish collegial relationships with authors and with others involved in the publication process: managers, graphic artists, designers, production staff. They need persistence and follow through, but they also need flexibility. Technical editor Doug MacBeth says, "Teamwork is very important in today's workplace. I've seen editors with excellent language skills fail in their editing careers because they weren't able to get along with others."

Communication skills are, of course, at the center of an editor's needed abilities. Editors need an ear and an eye for language—to know what works for the reader and what is accepted usage in the profession; to know what is needed to communicate with an international reader; to know the rules and conventions of written English. In addition, editors must know how to communicate with their writers, giving constructive criticism, negotiating areas of conflict, and justifying editorial decisions. Finally, in this era of rapid technological change, editors need to keep current, learning how to use new publishing and editing tools as they become available.

4.2 Clarifying the Duties of an Editor

Because editing costs both time and money, anybody acting as an editor needs to "add value" to the document and—by extension—to the product or the company. One way you can ensure that you are adding value is by clearly understanding the major duties of an editor. Those duties will vary from industry to industry and even from company to company within an industry, but editing duties can be usefully grouped into six general categories, four dealing directly with editing and two requiring organizational and interpersonal skills. Editing duties include: editing for applicability, readability, usability, and ethical and legal exposures; managing projects; and maintaining a collegial and collaborative environment.

Editing for Applicability

As an editor, you must always ask two questions when you first encounter a document:

1. What is the purpose of this document?
2. Who is the intended reader?

Together, these two questions dictate the *applicability* of the document; that is, how well the document accomplishes a "fit" between its intention and its achievement.

Frequently, the purpose is explicitly stated in an introduction or overview. If the purpose is not explicit, you must examine the document plan, search-read the document to determine the purpose, or ask the writer what it is. If finding the purpose of the document is difficult, the document may have a problem at the outset.

The type of intended reader may also be explicitly stated in a title, an overview, or a preface. Sometimes there is more than one intended reader; a manual, for example, may have both a tutorial for first-time users and a quick-start summary for experienced users. Documents like research reports may be written to a specific, named individual, or they may be intended for publication in a journal whose readers have a common level of interest and expertise. When you edit, you always need to ascertain who the intended reader is and then examine the document to see how well it meets that reader's needs. See Chapter 14 for help in analyzing readers.

Editing for Readability

Once you are satisfied that you know the purpose and intended reader, you need to examine the document's *readability*. Professor Thomas Huckin calls readability the extent to which a document's meaning "can be easily and quickly comprehended for an intended purpose by an intended reader" (1983, 91). The key words in this definition are *easily, quickly,* and *comprehended.* To comprehend is to understand, and readers of technical, scientific, and other workplace documents are busy people who are reading—not for pleasure—to learn what something means, how to do something, or what action they should take. All elements of the document must contribute to that understanding, including words, sentences, paragraphs, organization, graphics, and design elements.

The chapters in Part Two of this book will sharpen your skill in working at the sentence level, while the chapters in Part Three will cover principles of organization, graphics, design, and accessibility. Evaluating readability and suggesting ways to improve it are two of the most important services you can perform as an editor.

Editing for Usability

Except for memos and letters, workplace writers produce more step-by-step instructions than any other kind of document (Redish 1989). Those instructions and manuals can benefit greatly from a third kind of editing: editing for usability.

Usability means that the instructions are not only readable and understandable, but that they can be successfully carried out to accomplish a goal. Usability can "*verify* that a manual is accurate, . . . *validate* that people can use the manual to accomplish certain tasks, or . . . *diagnose* specific problems in the product and the manual—and . . . suggest solutions" (Redish 1989, 68).

Large, established organizations often have elaborate usability laboratories with video cameras, one-way viewing windows, and psychology experts to observe typical users in action and evaluate how well the instructions work. But small companies and startups usually lack the money and the time for such formal testing. In those situations, an editor can perform a valuable service by being a first usability tester. That means following the instructions in a hands-on situation: installing the thermostat or the software; unpacking and assembling the printer, the machine, or the toy; replacing the broken element; testing the new component. Good usability testers assume nothing but follow the instructions as they are written and note problem areas either by talking into a recorder or by making notes as they proceed. If you edit for usability, ask the following questions:

- Can you understand what is written without having to read the material again and again?
- Are the instructions in a logical sequence?
- Are graphics provided when they are needed?
- Is the information correct? Is what is supposed to happen actually happening?
- Is the table of contents easy to use?
- Can you find what you want in an index?

In organizations that have usability labs, editors can work with usability experts to fine-tune the documents—even, sometimes, the products. The questions above apply to usability testing of the document, and that is an editor's primary responsibility. However, editors can also recommend product enhancement based on their hands-on experience with the product.

Editing for Ethical and Legal Exposures

Editing for *ethical and legal exposures* (potential problems) can be an important part of an editor's job. While large companies will have legal departments that are empowered to examine documents for legal exposures, small companies and startups may rely entirely on writers and editors for both legal and ethical integrity.

As an editor, you may need to examine the following items:

1. Instructions to determine if the proper danger, warning, and caution notices are inserted before the applicable steps.
2. Trademarks and copyrights of your own company and any others named in the document to ensure that trademarks are protected by proper usage and that copyrights of written material are acknowledged.
3. Information about a new product, service, or method that might be patentable. In the United States, a company has only one year to file a patent after an invention is publicly announced. In some other countries, any public disclosure forfeits the company's right to apply for a patent in that country (Good 1993).

In addition, you may need to examine the writing itself, looking for imprecise wording that can mislead a reader, ambiguity that can confuse an issue, false implicature that will lead a reader to infer something that was never stated, and understatement of the negative, which can fail to alert readers to possible effects of an action. Herbert Michaelson, an engineer and long-time technical communicator, tells writers, "Because you can easily overlook [pitfalls of practical ethics] in your own writing, ask a colleague to help seek out the improprieties. An independent review of the initial draft could also show where important information was omitted" (1990, 58).

Managing Projects

Writers working as part-time editors seldom are responsible for managing an entire publication project, but they still need *management skills*. For example, if your primary job is writing online documentation and you are tapped to assist another writer with a copyedit, you need to manage both projects and finish them within deadlines.

To do this, you need some idea of the time you require to carry out editing tasks. A 1995 research project on the technical editing process yields some statistics on editing productivity that can serve as a beginning point. Education Professor Thomas A. Duffy found that editors "can edit 28.8 pages per day if they are doing a high-level edit focusing on comprehensibility, coherence, etc. If they are doing a copyedit, productivity increases to an average of 38.4 pages per day. The estimates [from 27 editors] for a high-level edit ranged from 8 to 60 pages per day, whereas, with one exception, the estimates for copyediting ranged from 10 to 80 pages per day" (267).

Recognize that these are experienced full-time editors reporting, and notice too that the range is very broad. When *you* edit, you need to keep track of the time you spend so that you can develop some accurate estimates of your own productivity for both kinds of editing. Duffy's study does not specify the number of hours in that "day," so you are better advised to keep track of your editing productivity by the hour.

When you have some experience, or if you work as a full-time editor, you might be asked to coordinate the efforts on a writing-editing task or a larger writing-editing project; you might even be assigned to coordinate the production process, working with other departments and vendors of services and supplies.

Maintaining a Collegial and Collaborative Environment

When you are an editor, one of your duties is to establish and maintain a collegial and collaborative environment. *Collegial* means that all participants bear collective responsibility for the project; *collaborative* means that the participants work together. You achieve such a mutually beneficial environment by fostering good editor-author relationships, as detailed in Chapter 3. Frequently, the editor sets the tone for the editor-author relationship, and you will be most successful as an editor if that tone is cordial.

4.3 Enhancing the Editing Job

True professionals are those who perform their daily assigned tasks to the best of their ability but also go beyond those daily tasks. They seek ways to make the job easier, to help others improve their performance, to provide guidelines and sources of information. As an editor, you can enhance your performance and the performance of your fellow writers and editors by carrying out whichever of the following suggested actions are appropriate.

Learn the Style and Format Guidelines for Your Company or Field

Your primary source of information for editing should be your company's or organization's style guide or style manual. Most established companies have a style guide either in hardcopy or online; writers should know and follow its suggestions, and editors should use it as a reference when making editorial or format decisions. If there is no style guide, you should begin compiling one (see Chapter 7) as soon as you begin your first editing assignment.

You should also be familiar with the general style guides for your field. *The Chicago Manual of Style* is the most frequently cited source for general style in the workplace. Other style guides frequently used include:

- The United States Government Printing Office *Style Manual*
- *The New York Public Library Writer's Guide to Style and Usage*
- The Journal of the American Medical Association *Manual for Authors and Editors*
- The Associated Press *Stylebook and Libel Manual*
- The Council of Biology Editors *CBE Style Manual: A Guide for Authors, Editors, and Publishers in the Biological Sciences*
- The American Psychological Association *Publication Manual of the American Psychological Association*

If you work in an organization whose writers frequently publish in scientific or technical journals, you can also collect authors' instructions and style guides from the editorial boards of those journals and make them available to the writers.

Collect Reference Materials

When you are editing, you will have questions. Few of us can remember all the details of hyphenation, the distinctions between words that mean almost the same thing, the fine points of graphic design. If you don't have questions, the writers you work with will, and all of you need access to authoritative information. You can enhance your editing performance if you collect books or computer files that provide reference material.

Dictionaries

All writers and editors need access to dictionaries. At a minimum, you need a good college dictionary like the *Random House College Dictionary* or the *American Heritage Dictionary* in hardcopy or on disk. For best results, everyone in the department should use the same version. You also need access to an unabridged dictionary—*Webster's Third New International Dictionary*—and whatever specialized dictionaries apply to your field. In the computer industry, for example, several editors I interviewed used Alan Freedman's *The*

Computer Glossary and T. D. Pardoe and R. P. Wenig's *Data Communication and Networking Dictionary*.

Grammar

For information about grammar that goes beyond the scope of this book, a good source is Toby Fulwiler, Alan Hayakawa, and Cheryl Kupper's *The College Writers' Reference*, which treats grammar and punctuation as editing choices. A somewhat more traditional grammar book is Lynn Quitman Troyka's *Simon and Schuster Handbook for Writers*. A short, succinct treatment of grammar is found in Margaret Shertzer's *The Elements of Grammar*.

Language, Usage, and Rhetoric

Classic sources about language and usage include S. I. Hayakawa's *Language in Thought and Action* and H. W. Fowler's *Modern English Usage*, newly out in a third edition. Other useful resources in this area are William Strunk and E. B. White's *The Elements of Style*, Joseph Williams's *Style: Ten Lessons in Clarity and Grace*, William Zinsser's *On Writing Well*, Richard Lanham's *Revising Business Prose*, and Scott Rice's *Right Words, Right Places*.

Graphics and Document Design

Key resources in this area include Jan White's *Editing by Design*, Edward Tufte's *Envisioning Information*, and William Horton's *Illustrating Computer Documentation*.

These references are only places to begin; you will want to add others as you encounter good sources.

Become an Expert in Writing and Editing Indexes

Indexes are access aids—ways of helping readers find the information they need. An excellent index can make a mediocre book shine, and a bad index can ruin an otherwise well-written book. But indexes are hard to write, and because they are the last section written, they are usually completed under severe time pressure.

Fortunately, excellent indexing software is available that will compile, set levels, alphabetize, and continually update index entries. Nevertheless, indexing software is only a tool. It doesn't do the work for you. How well you use this tool affects the outcome of the index, but writing the index is a real writing job that demands thought, analysis, and attention to detail. You can become a valued colleague if you are a good indexer. See Chapter 17 for details on writing and editing indexes.

Save and File Generic Portions of Written Documents

Editor Laurel Grove suggests that editors should save those portions of written documents that writers can reuse by making only slight modifications. Such material, called "boilerplate," might include the following modules:

- for proposals, résumés of key personnel and descriptions of past work
- for statements of work, safety and quality assurance material
- for environmental reports, general descriptions of the area
- for field reports, a description of the field methods (1990)

Compile Abbreviation and Acronym Lists

Well-established companies and organizations probably already have reference lists of common abbreviations and acronyms in a style guide or glossary. But technology changes rapidly, and someone needs to update these lists. In new or small organizations, quite possibly no one has compiled a definitive list, which provides an opportunity for an enterprising editor.

Establish a Glossary of Technical Terms

Antonia Van Becker, publications manager at a semiconductor company, says, "In every product, we discuss definition of terms at some point. Often we do a page at the beginning of the project that defines the new or changed terms. If terms are not defined at the beginning, the issue comes up during a content review when the technical experts disagree about the meaning of a sentence."

An editor should ensure that terms are defined and that the SMEs and the writers agree on the definitions. Beyond that, an editor can establish and circulate a glossary that will keep all participants up to date.

4.4 Flourishing Professionally

Whether you are working as an editor or writer, you are a *communicator*, and the communication field is continually changing and expanding. As tools like page-layout programs, software editors, and graphics programs improve, the job of a communicator involves more responsibilities. If you want to succeed—even flourish—in this field, you need to keep up professionally. Here are five ways to perfect your skills and expand your knowledge and your opportunities.

Improve Your Communication Skills

Attend in-house seminars on communication whenever they are offered, or enroll in college or continuing education classes to improve your communication skills. Colleges across the United States have recognized the importance of and the need for good communicators, and course offerings have rapidly expanded. Look for classes in technical communication, business communication, marketing, public relations, international communication, and instructional technology. Ask to see course syllabi and talk to instructors to ensure that the courses will meet your needs. For more narrowly focused courses, investigate college continuing education offerings; you may find courses as specific as Advanced Editing, Writing Online Documentation, and Designing Web Pages.

Become a better interviewer and listener. Take a course in interviewing skills or check out a book on the subject. Make use of a tape recorder when you undertake a new assignment to interview an author or an information source. Take notes and then fill in the details by listening to the tape.

Become a better speaker. Take a class in public speaking, join a Toastmasters Club, seek out small supportive groups where you can practice speaking. Videotape your presentations and evaluate what works and what doesn't.

Increase Your Knowledge of the Industry

Whether your field is book publishing, software documentation, environmental science, government, or medical writing, learn more about the industry. Subscribe to newsletters, magazines, and journals. Join professional societies like the Society for Technical Communication or the American Medical Writers Association; attend meetings, volunteer for leadership positions, and participate in society competitions. Network with others in the organization.

Expand Your Professional Skills

The professional skills you need will vary from field to field, but everybody who is an editor or writer needs computer skills. Technical publications manager Cheryl Herfurth says that you may need to learn new *platforms* (hardware and operating systems); *developmental tools* (page layout and word processing programs); *graphics tools* (draw programs); *technical skills* (networking, databases, C or C++ programming); *specialized skills* (online documentation, hypertext, information mapping, usability testing) (Brown 1994). You might also investigate video production, UNIX, multimedia, and networking.

Learn Management Skills

Attend seminars or classes in project management, meeting management, and time management. Ask your manager questions. Volunteer to run a meeting and ask for a critique of your performance afterward. Talk to production editors and find out what their job entails. Be active on a project team and observe how other team members function. Learn how to deal with print vendors. Durthy Washington, who manages her own business in editing and writing, says, "I wish I had learned [in school] the role of editors in the real world: that excellent writing and editing skills are only the beginning. People skills, management skills, and keeping up with technology make the job interesting."

Keep Your Résumé and Portfolio Current

However much you enjoy your present position, you should be prepared to apply for another job. Layoffs among writers and editors are common, or—more happily—better opportunities may present themselves. Keep your résumé information up to date and be prepared to supply both a formatted résumé and a scannable résumé.

Collect editing samples and house them in a portfolio that can be read and assimilated quickly by an interviewer. Shirley Krestas, who places editors and writers in industry, says, "As an editor you should save your 'befores' and 'afters.' If you don't, it's difficult to prove what you can do or have done. If you can only show the final product, many people may have worked on it." Try to have examples of a variety of editing tasks to show your versatility; for each one, explain briefly in writing your assignment, the document's purpose and intended reader, and the level of editing you carried out.

You might also prepare short "success stories" showing how you used interpersonal skills like collaborating, coordinating, negotiating, problem solving, and scheduling. Anne Bers, a senior technical writer, says that otherwise these skills remain hidden because they don't appear on résumés or job descriptions (Brown 1994).

Editing can be a full-time job and a lifetime career. It can also be the springboard to related professional careers like writing, managing, training, and teaching. The chapters in Part Two of this book will help you become an excellent editor at the sentence level: a copyeditor, a proofreader. The chapters in Part Three will help you learn how to edit for organization and for international readers, and how to edit graphics, screens, and access aids. Editing is both detail oriented and broad in scope. As senior editor Mary Scroggins, explains, ". . . editors are the readers' advocates. As such, we search for words, phrases, and stylistic techniques that allow readers to understand exactly, not partially, what the author intended. Editing is not an exact science; it is an art guided by instinct and enhanced by training and the tools of the trade" (1988, 7).

EXERCISE 1

Interviewing a Communicator

Conduct an informational interview with an editor, a writer, a usability expert, or a person working in international documentation. Prepare about 20 questions before you go to the interview. Ask about the writing and editing process, the skills needed by editors, the computer software and hardware used, the kinds of writing done, the levels of editing followed. Ask for information that will help you prepare for the profession.

Remember: You are not asking for a job; you are seeking information. Write an account of the interview or report to your class what you have learned.

EXERCISE 2

Learning about Laws Governing Communications

Research copyright, trademark, or liability laws and report your findings to the class.

EXERCISE 3

Analyzing Style Manuals

Examine one of the general style manuals listed in section 4.3 to see what it covers and what it emphasizes. Write a brief review or report orally to the class.

EXERCISE 4

Analyzing Local Style Guides

Examine the style guide from a company or organization in your community to see what it covers and what it emphasizes. Compare it to one of the general style guides listed in section 4.3.

EXERCISE 5

Examining Dictionaries, Grammars, and Other Reference Books

Examine one of the reference books listed in section 4.3 for its approach and its applicabilty to editing. If you are in a class, each class member should report on a different book.

EXERCISE 6

Researching Communications Seminars and Classes

Investigate what college, continuing education, and professional seminars or classes are available in your community that will enhance your skills.

EXERCISE 7

Learning about Professional Organizations

Find out what professional organizations are available in your community, when they meet, and how students or newcomers can join.

References

Brown, Dennise. 1994. Perspectives on change. *Connection*, (April).
Duffy, Thomas. 1995. Designing tools to aid technical editors: A needs analysis. *Technical Communication* 42.2 (May).
Good, Charles. 1993. Publish and perish! *Intercom*, (March).
Grove, Laurel. 1990. The editor as ally. *Technical Communication* 37.3 (August).
Huckin, Thomas. 1983. A cognitive approach to readability. In Paul V. Anderson et al., eds. *New Essays in Scientific Communication: Research, Theory, Practice.* Farmingdale: Baywood.
Michaelson, Herbert. 1990. How an author can avoid the pitfalls of practical ethics. *IEEE Transactions on Professional Communication* 33.2.
Redish, Janice and David Schell. 1989. Writing and testing instructions for usability. In Bertie Fearing and W. Keats Sparrow, eds. *Technical Writing: Theory and Practice.* New York: MLA.
Scroggins, Mary. 1988. In search of editorial absolutes. *Intercom*, (August/September).
Zook, Lola. 1994. The defining effort: An introduction. In Charles Kenmitz, ed. *Technical Editing: Basic Theory and Practice.* Arlington, VA: Society for Technical Communication.

5 Style, Sentence-Level Editing, and Editing Marks

Contents of this chapter:

- Style and Sentence-Level Editing
- Definitions of Copyediting and Proofreading
- Editing Marks

Professionals in the workplace often speak about "company style," "government style," "scientific style," and "personal style." Nevertheless, many people are not clear what the term "style" means or how style affects communication. Before you tackle the concepts that undergird sentence-level editing, you should know this: When you edit sentences, you are affecting the style of the document.

5.1 Style and Sentence-Level Editing

What is style? As a writer, you might think of style as a way of creating the text. According to business professors Fielden, Fielden, and Dulek, "The message is *what* is said. Style is the *way* the message is said. Organization is *when* (early or late in a communication) or in what order your points will be made" (1984, 29). They also provide this more complete definition: "Style: That choice of words, sentences, and paragraphing, which, by virtue of being appropriate to the message situation and to the relative power position of writer and reader, will produce the desired reaction and result" (19). As business communicators, Fielden and his colleagues delineate six styles that derive from three sets of opposites:

> forceful style or passive style
> personal style or impersonal style
> colorful style or colorless style

Each of the six styles is effective in specific situations and for specific types of readers.

In another approach to defining style, you might consider the degree of formality needed for the specific occasion. Documents can be written in formal style, informal style, or semiformal style.

Formal style is writing that is highly structured, impersonal, and deliberate and thus distances the reader from the writer. Formal style is characterized by nouns and third-person pronouns like *it*, use of the passive voice, and long, complex sentences. Some scientific reports are written in formal style.

Informal style is writing that is casual and conversational. Because it projects a friendly tone, informal style can be effective in one-to-one situations like memos and letters. Informal style uses personal pronouns like *I* and *you*, contractions like *can't* and *we'll*, personal names, active voice, and short sentences.

In *semiformal style*, the writer also uses the pronouns *I* and *you*, uses active voice and short sentences whenever possible, but structures the document more tightly, and follows the conventions of Standard Written English (Rew 1993). This book is written in semiformal style, as is most current workplace writing—whether technical, scientific, business, or government. For writers, style is the overall effect created by word choice and sentence and paragraph structures.

As an editor, you might think of improving style as a way of improving the text. When you edit, your aim should be to help solve what Joseph Williams calls "the single most serious problem that *mature* writers face: a wordy, tangled too-complex prose style." Williams, a linguistics professor and author of the classic book *Style: Ten Lessons in Clarity and Grace*, explains the importance of editing in these words: "The serious part of writing is rewriting. Samuel Johnson said it about as well as anyone: 'What is written without effort is in general read without pleasure.' The effort is more in the editing than in the writing" (1989 Preface).

Thomas R. Williams, a technical communications professor, ties a definition of style to the task of sentence-level editing in this way: "Style is the cumulative effect of grammatical, punctuation, spelling, usage, diction, voice, and sentence structure choices (to name a few) made at the sentence level. Our style is more constrained in Standard English, which is what we use in technical communication, with the effect that the meaning carried by discourse following the conventions of Standard English is less ambiguous—more likely to be interpreted or decoded consistent with the author's intent" (1997). When you edit your own work or someone else's at the sentence level, you should be aiming to improve the style, making the document easier to understand.

Sentence-level editing, then, is the process of examining and correcting a document at the individual sentence and word level. Many other terms are used to describe this process: microediting, line editing, language clarification, line-by-line editing, literary editing, correctness editing, copyediting, and proofreading. The two terms most commonly applied to sentence-level editing are copyediting and proofreading.

5.2 Definitions of Copyediting and Proofreading

In the days when original drafts were typed on a typewriter, the distinction between copyediting and proofreading was clear. *Copyediting* was the review and correction process an editor performed on doublespaced draft copy: the typed version. Once the copyedited draft was corrected by the author, a typesetter set the document into type. This new version of the document was returned to the author and editor in *galley proof*, long sheets of paper single-spaced in the intended type size and style but not yet in the final page format. These galleys were then *proofread* by the proofreader. The process consisted of comparing the corrected original draft to the typeset version and fixing any errors introduced by the typesetter.

With the advent of word processing, computer networking, and desktop publishing, the distinctions between copyediting and proofreading are no longer so clear. The author's original draft is now typically lodged in a computer file. An editor may read and correct the draft either on hardcopy (a version printed out on paper from the file) or online (a version displayed on the screen). Changes and corrections can be made to the file by both author and editor, and that file will most often produce the final version of the document—even if the document is typeset and printed as a book or glossy four-color annual report.

Editors working in industry are therefore as likely to talk about the *number* of their editing passes through a document as they are to talk about copyediting and proofreading. Likewise, writers in the workplace often talk simply of drafts and *edits* of drafts. A writer might say, "This document is in early draft stage; it needs a developmental edit." Or a writer might say to a writing colleague or a professional editor, "Before the installation manual goes to print, could you give it a final edit?"

Writers and editors also talk about *alpha*, *beta*, and *final* drafts. Alpha drafts are first drafts that are ready for developmental editing (an examination of purpose, audience, content, and organization). Beta drafts are second drafts that are ready for technical review and copyediting (an edit of sentences, words, and punctuation). Final drafts are just that: The document is in final form and ready for proofreading (the last edit) and then release to the public. See Chapter 2 for explanations of the various levels of editing.

The distinctions are also sometimes blurred between sentence-level editing and document editing. Of the 13 experienced editors I interviewed while preparing this book, one told me that she "did it all at once," and several others had only two passes at a document. With tight schedules and rapid product cycles, "Nobody has time to edit separately for content, organization, clarity, and correctness," said former technical editor Ruth Chase, who now manages a writing group at a software company. "At a certain stage, you have to put on blinders in order to get the product out," said Jo Levy, a professional freelance editor. She tries to have four editing passes at a document but can't always get them.

Thus, it is probably more accurate to speak of *initial* editing and *final* editing at the sentence level, but because the terms copyediting and proofreading are so well known, I will continue to use them in this book.

The first sentence-level edit, commonly called copyediting, is the process of analyzing and correcting a document to ensure that it is well written, consistent in style, and error free. The editor works on a manuscript *before* it is set into its final version; usually copyediting is done on the author's last clean draft. After editing (and any necessary changes by the author) the manuscript is ready to be printed. For details on the process of copyediting, see Chapter 7.2.

The final sentence-level edit, called proofreading, is the last editing pass a document or screen receives before it is released to the public. For many documents in the workplace, the final edit consists of comparing the last edited version to the new "clean" version to ensure that the changes have actually been made. In publishing, proofreaders check to see if errors have been introduced during typesetting. Editors, whether they are professional editors or colleague writers functioning as editors, usually do not have time to read the entire document again. Some critical documents in industry—marketing materials and annual reports, for example—may be read in their entirety at a final edit to ensure that they are absolutely error-free. In book or journal publishing, both authors and professional proofreaders will usually read the entire document once page proofs are available. But even students who must be their own editors need to take one final editing pass at a document before they turn it in. For details on the proofreading process, see Chapter 6.3.

One effect of the computer revolution is that in the workplace the secretary has almost disappeared. No longer can you, as the writer, scratch out a rough draft and expect a secretary to type it, insert the punctuation, and correct your errors of diction and syntax. Instead, you will probably "keyboard" your own document, and you will be responsible for its accuracy and readability. When it goes for review, the writing (and the errors) will be yours. That puts an additional burden on you as a professional—not only to be a good technical worker and an accurate and knowledgeable writer—but also to be an excellent editor and proofreader.

5.3 Editing Marks

Copyediting and proofreading marks are shorthand symbols that are useful whether you are a writer correcting your own work or an editor suggesting changes for a colleague. Fortunately, both copyeditors and proofreaders use the same kinds of marks, though proofreaders in publishing have a few additional marks that copyeditors don't need. The most common marks are listed in Figure 5.1 with their meanings and examples of their use. As you work through the exercises in this book, you will have many opportunities to use these marks.

See Chapter 4 for useful references for editors.

CHAPTER 5 STYLE, SENTENCE-LEVEL EDITING, AND EDITING MARKS 57

Mark	Meaning	Example
℘ or ⌒	Delete	The t̶h̶e̶ solar cell
⌒	Close up, no space	The out jet
℘	Delete and close up	The photo voltaic cell
#	Insert space	the flatplate
∧	Insert word or letter	col̬ector
∾	Transpose	photo voltiac cell
⋏	Insert comma	solar cell˛a device
⊙	Insert period	the device⊙On a scientific
⋎	Insert apostrophe	suns energy
(:) or :̂	Insert colon	Connect the following:̂
⋏;	Insert semicolon	the collector;however
=	Insert hyphen	flat=plate collector
⋎⋎	Insert quotation marks	crystals are ⋎plugged into⋎
⊢m⊣	Insert em dash	The flat plate—its cells in place—is mounted
⊢n⊣	Insert en dash	pp. 218–224
stet	Restore to original	Two types of c̶r̶y̶s̶t̶a̶l̶ material (stet)
¶	Start new paragraph, or no paragraph intended if preceded by "no"	top of the roof. ¶The collectors
No ¶		
≡	Capitalize	the photovoltaic cell
/	Lower case	The P̸hotovoltaic C̸ell
( )	Parentheses	the two parts (cells and modules) are
[ ]	Brackets	the principle [sic] use
◯	Spell out	has (approx.) five requirements
‖	Align	‖cell ‖ module ‖ collector
⌐	Move left	⌐module

FIGURE 5.1 Copyediting and Proofreading Marks

Mark	Meaning	Example
⊐	Move right	module ⊐
⊓	Move up	⊓ module The
⊔	Move down	The ⊔ module
∨	Superscript	x2
∧	Subscript	√2
___	Italicize	The <u>collector</u>
∿∿∿	Make boldface	the photovoltaic cell
⌒	Run in	finish the line ⌒ by adding information

FIGURE 5.1 Continued

Copyediting Sample

Copy is generally doublespaced, giving you room to insert any needed changes or corrections. Copyediting marks are placed directly where the changes are to be made as shown in Figure 5.2. (If the copy is singlespaced, make changes as you would in proofreading. See the proofreading sample.)

¶ What is the difference between silicon, silicone, and silica? This article will tell you. ¶ Silicon is the second most abundant chemical element in the earth's crust and the seventh most abundant in the universe. Silicon has the ability to act as an electrical semiconductor, which makes it indispensable to the microelectronics industry. Purified silicon is a gray solid or a brownish powder. But it tends to combine with other elements and is found in nature only in compound form in molecules with other kinds of atoms. ¶ Silicones are polymers—compounds formed of long-repeating molecules—that contain silicon along with oxygen, carbon, and other elements. You may encounter them while preparing for a back packing trip or while working on your car.

FIGURE 5.2 Copyedited Document

CHAPTER 5   STYLE, SENTENCE-LEVEL EDITING, AND EDITING MARKS   **57**

Mark	Meaning	Example
℘ or ⌒	Delete	The the solar cell
⌒	Close up, no space	The out let
⌢	Delete and close up	The photo voltaic cell
#	Insert space	the flatplate
∧	Insert word or letter	colector
∩	Transpose	photo voltiac cell
⌃	Insert comma	solar cell a device
⊙	Insert period	the device On a scientific
∨	Insert apostrophe	suns energy
⊙ or ⌃	Insert colon	Connect the following
⌃	Insert semicolon	the collector however
=	Insert hyphen	flatplate collector
∨∨	Insert quotation marks	crystals are plugged into
1/m	Insert em dash	The flat plate its cells in place is mounted
1/n	Insert en dash	pp. 218 224
stet	Restore to original	Two types of crystal material (stet)
¶	Start new paragraph, or no paragraph intended if preceded by "no"	top of the roof. The collectors
No ¶	preceded by "no"	
≡	Capitalize	the photovoltaic cell
/	Lower case	The Photovoltaic Cell
⊰ ⊱	Parentheses	the two parts cells and modules are
[ ]	Brackets	the principle sic use
◯	Spell out	has approx. five requirements

‖	Align	cell
		  module
		  collector

⌐	Move left	module

FIGURE 5.1 Copyediting and Proofreading Marks

Mark	Meaning	Example
⌐⌐	Move right	module ⌐⌐
⌐⌐	Move up	⌐⌐ module The
⌐⌐	Move down	The ⌐⌐ module
∨	Superscript	x2
∧	Subscript	y2
___	Italicize	The <u>collector</u>
∿∿∿	Make boldface	the photovoltaic cell
⌒	Run in	finish the line ⌒ ⌒ by adding information

FIGURE 5.1 Continued

Copyediting Sample

Copy is generally doublespaced, giving you room to insert any needed changes or corrections. Copyediting marks are placed directly where the changes are to be made as shown in Figure 5.2. (If the copy is singlespaced, make changes as you would in proofreading. See the proofreading sample.)

¶ What is the difference between silicon, silicone, and silica? This article will tell you. ¶ Silicon is the second most abundant chemical element in the earth's crust and the seventh most abundant in the universe. Silicon has the ability to act as an electrical semiconductor, which makes it indispensable to the microelectronics industry. Purified silicon is a gray solid or a brownish powder. But it tends to combine with other elements and is found in nature only in compound form in molecules with other kinds of atoms. ¶ Silicones are polymers, compounds formed of long-repeating molecules—that contain silicon along with oxygen, carbon, and other elements. You may encounter them while preparing for a backpacking trip or while working on your car.

FIGURE 5.2 Copyedited Document

Their water repellency, low surface tension, and other qualities make them excellent as water proofers, brake fluids, lubricants, adhesives, and rust preventers. They are also used in surgical implants and prostheses. The toy bouncing putty that comes inside a plastic egg is a form of silicone. Silica, or silicon dioxide, is a compound of silicon and oxygen atoms. Most of the rocks in the earth's crust contain silica, it takes a number of forms depending on its level of purity. Sand, sandstone, quartz, flint, amethyst, jasper, onyx, agate, and opal are all forms of silica. Silica is an excellent drying agent and dehumidifier, which is why delicate instruments like cameras are often packaged with a small envelope of silica gel.

FIGURE 5.2 Continued

Proofreading Sample

Proof is nearly always singlespaced, and there should be few errors. The sample in Figure 5.3 has more errors than are typical on proof so that you can see a variety of corrections. When you correct proof, first divide the page into two parts with an imaginary line down the middle. Then identify changes for the left side in the left margin, and changes for the right side in the right margin. This helps typesetters find the changes quickly without reading the whole document. Insert a caret (∧) where the change is needed. If you are both deleting and inserting, use this mark (/)), which means "delete and insert." Use a slash mark to separate each correction in a line. Circle any words written in the margin that are instructions to the author or typesetter.

What is the difference between silicon, silicone, and silica? This article will tell you.
Silicon is the second most abundant chemical element in the earth's crust, and the seventh most abundant in the universe. Silicon has the ability to act as an electrical semiconductor, which makes it indispensable to the microelectronics industry. Purified silicon is a gray solid or a brownish powder. But it tends to combine with other elements and is found in nature only in compound form, in molecules with other kinds of atoms.

Silicones are polymers—compounds formed of long repeating molecules—that contain silicon along with oxygen, carbon, and other elements. You may encounter them while preparing for a backpacking trip or while

FIGURE 5.3 Document That Has Been Proofread

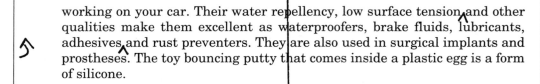

working on your car. Their water repellency, low surface tension, and other qualities make them excellent as waterproofers, brake fluids, lubricants, adhesives, and rust preventers. They are also used in surgical implants and prostheses. The toy bouncing putty that comes inside a plastic egg is a form of silicone.

Silica, or silicon dioxide, is a compound of silicon and oxygen atoms. Most of the rocks in the earth's crust contain silica, which takes a number of forms depending on its level of purity. Sand, sandstone, quartz, flint, amethyst, jasper, onyx, agate, and opal are all forms of silica. Silica is an excellent drying agent and dehumidifier, which is why delicate instruments like cameras are often packaged with a small envelope of silica gel.

FIGURE 5.3 Continued

References

Fielden, John S., Jean D. Fielden, and Ronald E. Dulek. 1984. *The business writing style book.* Englewood Cliffs, NJ: Prentice-Hall.

Rew, Lois Johnson. 1993. *Introduction to technical writing: Process and practice.* 2nd ed. New York: St. Martin's.

Williams, Joseph M. 1989. *Style: Ten lessons in clarity and grace.* 3rd ed. Glenview, IL: Scott, Foresman.

Williams, Thomas R. 1997. Prepublication review. April.

6 Final Editing or Proofreading

Contents of this chapter:

- Final Editing: The Art of Proofreading
- The Importance of Final Editing
- The Process of Final Editing

One principal goal of this book is to help you become a better writer by sharpening your sense of the English language and by improving your language skills. The best way to begin that process of language awareness is to learn, when editing, to be a different kind of reader.

6.1 Final Editing: The Art of Proofreading

In the normal reading process, you anticipate and predict meaning based on a few word groups that you see as your eye moves quickly across the line. You do not read each word, nor do you consciously notice punctuation. If you are the author of the text before you, you may even read what is *not* there: those words that were in your mind as you composed but perhaps never made it on to the screen or paper.

When you perform a final editing pass on a document, you need to change your reading style from *reading* to *proofreading*. Proofreading involves attending to the text in detail—"looking specifically at each word and mark of punctuation, carefully noting not only what is there but also what is not there" (Harris 1987, 464).

To succeed at this new kind of reading, you need to see what is actually on the page rather than what you expect to see. Thus, proofreading is by no means a mechanical process; instead, it requires thinking, concentration, and attention to detail.

You might think that proofreading skills are no longer needed now that we have software spelling, grammar, and style checkers. It is certainly true that spelling checkers help eliminate many spelling and typing errors, and a document should always be run through a spelling checker before it is proofread. Likewise, style or grammar checkers may note such things as the missing half of a set of parentheses. However, all of the professional writers and editors I queried said that while they always used spell checkers, they did not use style checkers because they were time consuming and relatively unhelpful. See Chapter 1 for specific comments on style checkers.

6.2 The Importance of Final Editing

How important is final editing or proofreading? The three examples that follow show how lapses in proofreading can have expensive, embarrassing, and even legal consequences.

The marketing editor at a computer company told me recently that the marketing department was forced to destroy 5,000 glossy, four-color marketing brochures when someone discovered (just before they were to be distributed) that the field office phone numbers listed on the brochure cover were incorrect. As readers, we often skim over numbers like addresses, zip codes, and phone numbers unless we have a specific need for that information. Unfortunately, at the final editing stage, the editor read the document but did not proofread it and therefore did not catch the errors. It was an expensive mistake.

Sometimes mistakes are both expensive and potentially embarrassing. While recently reading the annual report of a large U.S. company that transports, stores, processes, and sells agricultural products, I discovered the last-minute fix of an error that had been missed in final editing. In the president's letter to stockholders, I noted two words that appeared in white type on an orange background—the only such words in two pages of standard black type. (The president's and chief executive's signatures were also in orange.) I scraped at the words with my fingernail and discovered that the orange rectangle was a sticker. The words in orange, *carbon monoxide* were pasted over the words in the text, *carbon dioxide*. The sentence read as follows:

> . . . [our company's] "expanded ethanol production . . . is already on stream and committed, much of it for export, where it drastically reduces **carbon monoxide** in the air of Sao Paulo, Brazil and other cities."

You can see that using the wrong word in this sentence drastically changed the meaning, yet it is an error that a spell checker will not flag, nor will an editor—unless that editor is proofreading with great care and for meaning. The company's solution of pasting in a new word on each annual report was expensive, but probably not as expensive as reprinting all the annual reports—more than 34,000 of them.

In government documents, even punctuation marks can be critical. For example, editing consultant Carolyn Matalene reports: "Getting [state water regulations] right requires proofreading, again and again, because a misplaced comma can change the law. Defining a public water system as '15 or more taps or 25 or more people served 60 days out of the year,' means something different from '15 or more taps or 25 or more people, served 60 days out of the year.' These writers know that not getting it right will have consequences" (1995, 45).

How can you change your reading style to perform proofreading? Editing experts suggest three methods:

1. **Use a pointer such as a pencil to point to each word as you proofread or use an index card or ruler to cover text until you expose each line.** By preventing your eyes from leaping ahead, this method forces you to read line by line or word by word.

2. **Read the document aloud.** Again, this method forces attention to individual words and sentences.
3. **If you are the writer, set the document aside for a time before you edit.** The elapsed time will help you see the document as a reader, not as a writer.

In publishing, and sometimes in industry, proofreading involves comparing the typeset or printed version to the last corrected draft. In this case, the proofreader may look only at changes made to the manuscript to ensure that those changes were also made to the typeset version. However, this kind of editing might well miss problems like the carbon dioxide/carbon monoxide error discussed above, so if time permits, the entire manuscript should be proofread. Names and numbers should be checked for accuracy against the *original* document.

According to editing experts, the best way to learn editing skills is to begin as a proofreader. Karen Judd, author of *Copyediting: A Practical Guide*, says: "To get started in copyediting, be prepared to spend several months or even a couple of years as a freelance proofreader" (1990, 263). Writing in *The Editorial Eye*, Editorial Experts, Inc., say: "The neophyte technical editor should begin developing a familiarity with one or two scientific disciplines through proofreading; wise supervising editors assign such tasks to newcomers to the field" (1986, 1).

6.3 The Process of Final Editing

Final document copy is almost always singlespaced, which leaves little room between the lines to insert corrections or changes. Therefore, editors usually divide the text in half with an imaginary vertical line down the center. Corrections to the left half of the text are placed in the left margin; corrections to the right half are placed in the right margin. A caret (∧) is inserted in the text where the actual correction is to be made. If more than one correction appears on the line, the marks are separated by slash marks. See the example of the use of proofreader's marks on text in Chapter 5.3.

When proofreading a document, follow this process:

1. **Choose a proofreading method—covering the text with a card, using a pointer, or reading aloud—and follow it throughout the document.** Concentrate on details: individual words, numbers, punctuation marks.

2. **For your first efforts, make a copy of the document and mark on it, later transferring your marks to the original.** Alternatively, edit in pencil at first, so you can erase if necessary. Keep a copy of your editing marks for later reference.
3. **Use a color that will show up on the manuscript.** Many editors use red ink. Technical editor Ruth Chase, however, does all her editing in green ink. "It shows up on the paper," she says, "yet it doesn't have the negative implication of the schoolroom invoked by red ink."
4. **Keep a sheet of paper on which you can write questions, or attach small adhesive notes to the text with your queries.** When you are adept at proofreading, you can write each query in the margin with a circle around it.
5. **Group the items to be checked, and make several passes through the document.** Check the items in one group at each pass. Suggested groups are as follows:
   - **Numbers and names.** Check for accurate use of numbers, including addresses, phone numbers, and part numbers. Also check for correct numbering of lists, accurate alphabetizing, correct hierarchies in outlines, and correct use of numbers in bibliographic citations. Check all mathematical operations. Check proper names and acronyms against the original because spell checkers do not catch errors in nonstandard words.
   - **Accuracy of words.** Look for spelling, wrong word, and typing errors. Pay special attention to long words, breaking them into syllables as you read.
   - **Punctuation.** Look for errors in punctuation, capitalization, and hyphenation of syllables. Many word processing programs will automatically break words into syllables correctly, but the final edit should correct any problems.
   - **Crosschecks.** Crosscheck that the entries in the table of contents are the same as the headings and that the page references are accurate. Check running headers and footers. Ensure that in-text bibliographic references appear in the reference list. If time permits, check the accuracy of a sample of index entries by looking at the referenced page to see if the information appears. See Chapter 17 for details on editing access aids.
6. **If you are comparing the last edited draft to the final version, make sure that all marked changes have been made.** In addition, correct any other errors that you find. If time permits, read the entire document.
7. **Make all your marks clearly and neatly.** Another person must decipher your marks, and that person cannot read your mind.

Remember that in the final edit you essentially prepare the document for publication by eliminating distracting surface errors from the manuscript. This is not the time to make major changes or tinker with the style, even if you are the author. The only exception to this guideline is that if you find a major error, you must fix it.

Plan for plenty of time to complete final editing or proofreading passes. Experienced editors plan to complete 8 to 10 pages per hour; as a beginner, you may well take much longer, especially while you are learning the editing marks. The exercises in this book will give you much practice in editing, and you will increase your pace as you proceed. Develop the habit of keeping a log of the time you spend on each editing project.

EXERCISE 1

Proofreading by Comparing a Proof Copy to the Original Master

To sharpen your editorial eye, edit the following brief passage by comparing the master version to the proof copy. Since this passage is in Latin, you will not be able to read the word groups for meaning. Use the editing marks explained in Chapter 5. Selected answers appear in the Appendix.

Master Version

Cujus genas ac faciem, omnemque undique totius venerandi capitis superficiem, pannus subtilissimus operiendo obtegit, qui ita omnibus menbris subpositis districtissima sollicitudinis arte cohaesit, quasi caesariei, pelli, temporibus, ac barbae, conglutinatus sit. Qui ex nulla parte, alicujus arte, altius aliquantulum a cute vel carne elevari, divelli vel subrigi, potuit.

Reginaldi monachi Dunelmensis libellus de Admirandis beati Cuthberti Virtutibus quae novellis patrate sunt temporibus. Ed. J. Raine. London: Sartees Society, 1835, p. 41.

Proof Copy To Be Corrected

Cujus genas ac faciem, omnenque undique totius venarendi captitis superficiem, pannus subtilissimus operienda obtegit, qui ita omnibus membris subpositis districtissima sollictudinis arte coheasit, quasi casariei, pelli temporibus, ac barbae, conglutinatus sit. Qui ex nula parte, alicujis arte, altius aliquantulm a cute vel carne elaveri, divelli, vel subrigi, potuit.

CHAPTER 6 FINAL EDITING OR PROOFREADING **67**

EXERCISE 2

Proofreading for Accuracy

While your primary responsibility in proofreading is to ensure that the final copy agrees with the original, you will be a better editor if you also pay attention to the meaning of the content—eliminating inaccuracies that may have escaped other editing passes. Find and correct the errors in the following sentences.

1. The contract called for reports four times a year: March 31, June 31, September 30, and December 31.

2. Armco maintains sales offices in the following 10 cities: Boston Fort worth, Colorado Springs, Chicago, Atlanta, Jacksonville, Raleigh, Buffalo, Nashville, Los Angeles, and Newark.

3. Send your response directly to Senator Ardis McManus at the state capital building in Hartford, CN.

4. Polar bears are larger than all other bears: the adult male can be 11 feet (3 meters) long, compared with the 9–10 feet (2.7–3 meters) of the grizzly bear.

5. Because Winnipeg is adjacent to the Wisconsin border, the machine parts can be transferred to barges at Duluth.

6. Every medication is capable of producing side effects. Most patients experience little or no problems while taking this medication.

7. This license is issued solely as a license to drive a motor vehicle in this state; it does not establish eligibility for employment, voter registration, or public benefits.

8. Despite the relative influence of this area, one out of three people in the county uses the services of a United Way agency.

9. The price of a safety seat should be mainly a secondary consideration.

10. As the microwaves bounce off the metal walls of the cooking cavity, the microwave oven does not get hot.

11. The deficits were absolutely unexceptable.

EXERCISE 3

Proofreading the Final Copy

Using the editing marks explained in Chapter 5, correct all the errors you find in the following document. Write questions for the author in the margins, placing a circle around each question. This document is a student's summary of an article from a professional journal (Wing 1987). To provide practice, errors not in the original document have been introduced. Also remember that this is a student paper. You may find stylistic items that you would like to improve, but the final edit is not the time to do it. Correct what is wrong but do not rewrite.

```
Topic: THE CORPORATE LIFE
Source: Power Ruth M. "Who Needs a Technical Editor? IEEE
Transactions on Professional Communication PC-24 NO. 3.
(September 1981): 139-140.

       Ensuring quality control in written communication is the
job of a technical editor. With a better understanding of
the responsibilities of a technical editor, an engineering
manager might want to hire either a full-time or part-time
technical editor. The 1991 article "Who Needs a Technical
Editor?" provides information to eningeering managers about
the responsibilities, qualifications for technical editors.

       Power begins her article with a scenario in an engin-
neing office. Old and young engineers believe that they do
not need technical editors. The engineers believe that their
writing and editing skills are adequate. However, when the
manager receives poorly written reports from his engineers,
he must spend hours to correct simple grammatical and typo-
graphical errors. The managers conclusion: Engineers can't
write. However, Powers says that engineers are not the only
ones who have probelems writing correctly and clearly. Gov-
ernment and lawyers also use too much jargon in their writ-
ing. Power says that a technical editor can solve many prob-
lems with clear and correct writing.
```

CHAPTER 6 FINAL EDITING OR PROOFREADING   **69**

According to Power, a technical editor should be able to perform a variety of job functions such as organizing data, checking references, writing copy, organizing layouts, copy editing, proofreading, and making sure that the whole project runs smoothly. Power concludes that the technical editors job is one of quality control for written communication.

After providing her definition of a technical editor, Power breaks the rest of her article into three topics: Editor's functions, Qualifications, and Hiring options (for employers).

A technical editor's function should start at the early stages of development on a written document. Although he or she is not actually writing the document, the editor can help authors start writinng the document by giving suggestions on style and organization. The editor can provide continuing guidance as the project develops. Power says that asking an editor to come in at the last minute and "fix-up" a paper would be frustrating to the editor and the writer. To give engineering managers a better idea of what technical editors do all day, Power lists 21 jobs (see table 1) that editors are performing in industry today. The author also gives several scenarios where technical editors can be helpful. The Topics of the scenarios include technical editing of a journal article, assistance in preparation of a technical article, individual coaching, assistance in preparing a speech, and assistance with translations.

When listing her qualifications of technical editors, Power uses the Society for Technical Communication as a source of information. She lists characteristics of technical editors that include gender, age, education, work experience, average salary, and technical background.

In her final topic, hiring options, Power lists several places where employers can start their search for a technical editor. Power cautions the engineering mangers that technical editors may not be the solution to all their written communication problems. If the manger, engineers, and technical editor can not work as a team, the editor will be very little help. The editor may even cause more controversy because he might have an opinion totally different than the manager and engineers. Power concludes by answering the title of her article--"Who Needs a Technical Editor?" The answer to the question is "anyone who is responsible for communicating technical information and who needs help in getting projects organized, following up on details, coordinating groups of writers, untangling convoluted sentences, suggesting just the right word, settling quesitons of grammar, or finding that one last typo.

Although this article was intended for engineering managers, I found this document very informative and useful for technical writers and editors. Brown says that in the last 10 years industry has recognized the need for better quality control in written documents. She presents many strong arguments to managers to hire technical editors. The author gives a very clear definition of a technical editor. However, there was other information I found valuable to me as a writer and editor.

If the engineering manager decides he needs a technical editor, Power lists several places (telephone directories, colleges, or the Society for Technical Communication) where the manager can begin his search for an editor. I found this information very valuable. When I begin looking for a job, I will have some idea of where managers are looking for potential employees.

When I begin looking for a job, I want to be sure my qualifications are what the manager is expecting. Power lists qualifications and probable backgrounds of technical editors. I would like to know what my employer expects from me on the job. Also, it would be nice to know what type of background your competition (other people that are applying for the job) might have.

After I find a job, a salary must be determined. Brown lists several different salaries with corresponding amounts of experience. This information is nice to have also. When I find a job, I would like to know what my peers are earning, so I have some idea of how much the job might pay.

The author stresses one point about technical editors to managers that I like. She reminds managers that editors will not solve all the problems of projects. Editors may cause even more controversy between themseleves, engineers, and managers during a project. She says that Managers, engineers, and editors must work together as a team for a project to be successful.

The most valuable part of Power's article was the list manager's might use to determine if they need a technical editor. After reading this list, I have a better idea of what managers are going to expect from me in the "real world". This is very valuable information. What a student learns in the classroom is sometimes very different from what management wants in the real world.

EXERCISE 4

Comparing Proofread Versions

Compare your corrections with the marked document that follows. If you missed errors, resolve to proofread more closely next time. Since you have not yet reviewed grammar and punctuation, don't be discouraged. You will develop a sharper eye as you proceed with the exercises in this book.

```
Topic: THE CORPORATE LIFE
Source: Power, Ruth M. "Who Needs a Technical Editor?" IEEE
Transactions on Professional Communication PC-24 NO. 3.
(September 1981): 139-140.

        Ensuring quality control in written communication is the
job of a technical editor. With a better understanding of
the responsibilities of a technical editor, an engineering
manager might want to hire either a full-time or part-time
technical editor. The 1991 article "Who Needs a Technical
Editor?" provides information to enfineering managers about
the responsibilities of qualifications for technical editors.

        Power begins her article with a scenario in an engin-
eering office. Old and young engineers believe that they do
not need technical editors. The engineers believe that their
writing and editing skills are adequate. However, when the
manager receives poorly written reports from his engineers,
he must spend hours to correct simple grammatical and typo-
graphical errors. The managers conclusion: Engineers can't
```

write. However, Power says that engineers are not the only ones who have problems writing correctly and clearly. Government and lawyers also use too much jargon in their writing. Power says that a technical editor can solve many problems with clear and correct writing.

According to Power, a technical editor should be able to perform a variety of job functions such as organizing data, checking references, writing copy, organizing layouts, copy editing, proofreading, and making sure that the whole project runs smoothly. Power concludes that the technical editors job is one of quality control for written communication.

After providing her definition of a technical editor, Power breaks the rest of her article into three topics: Editors functions, Qualifications, and Hiring options (for employers).

A technical editor's function should start at the early stages of development on a written document. Although he or she is not actually writing the document, the editor can help authors start writing the document by giving suggestions on style and organization. The editor can provide continuing guidance as the project develops. Power says that asking an editor to come in at the last minute and "fix-up" a paper would be frustrating to the editor and the writer. To give engineering managers a better idea of what technical editors do all day, Power lists 21 jobs (see table 1) that editors are performing in industry today. The author also gives several scenarios where technical editors can be helpful. The Topics of the scenarios include technical editing of a journal article, assistance in preparation of a technical article, individual coaching, assistance in preparing a speech, and assistance with translations.

When listing her qualifications of technical editors, Power uses the Society for Technical Communication as a source of information. She lists characteristics of technical editors that include gender, age, education, work experience, average salary, and technical background.

In her final topic, hiring options, Power lists several places where employers can start their search for a technical editor. Power cautions the engineering mangers that technical editors may not be the solution to all their written communication problems. If the manger, engineers, and technical editor cannot work as a team, the editor will be very little help. The editor may even cause more controversy because he or she might have an opinion totally different from the manager and engineers. Power concludes by answering the title of her article, "Who Needs a Technical Editor?" The answer to the question is "anyone who is responsible for communicating technical information and who needs help in getting projects organized, following up on details, coordinating groups of writers, untangling convoluted sentences, suggesting just the right word, settling questions of grammar, or finding that one last typo."

Although this article was intended for engineering managers, I found this document very informative and useful for technical writers and editors. Power says that in the last 10 years industry has recognized the need for better quality control in written documents. She presents many strong arguments to managers to hire technical editors. The author gives a very clear definition of a technical editor. However, there was other information I found valuable to me as a writer and editor.

If the engineering manager decides he or she needs a technical editor, Power lists several places (telephone directories, colleges, or the Society for Technical Communication) where the manager can begin searching for an editor. I found this information very valuable. When I begin looking for a job, I will have some idea of where managers are looking for potential employees.

When I begin looking for a job, I want to be sure my qualifications are what the manager is expecting. Power lists qualifications and probable backgrounds of technical editors. I would like to know what my employer expects from me on the job. Also, it would be nice to know what type of background my competition (other people who are applying for the job) might have.

After I find a job, a salary must be determined. Power lists several different salaries with corresponding amounts of experience. This information is nice to have also. When I find a job, I would like to know what my peers are earning, so I have some idea of how much the job might pay.

The author stresses one point about technical editors ~~to managers~~ that I like. She reminds managers that editors will not solve all the problems of projects. Editors may cause even more controversy ~~between~~ among themselves, engineers, and managers during a project. She says that Managers, engineers, and editors must work together as a team for a project to be successful.

The most valuable part of Power's article was the list manager's might use to determine if they need a technical editor. After reading this list, I have a better idea of what managers are going to expect from me in the "real world." This is very valuable information. What a student learns in the classroom is sometimes very different from what management wants in the real world.

EXERCISE 5

Proofreading a Typeset Document

Using the marked document in exercise 4 as the final corrected draft, proofread the typeset document that follows and mark any errors that were not corrected in the final version. Be sure to compare the two documents.

Topic: THE CORPORATE LIFE
Source: Power, Ruth M. "Who Needs a Technical Editor?" <u>IEEE Transactions on Professional Communication</u> PC-24 NO. 3.
(September 1981): 139–140.

Ensuring quality control in written communication is the job of a technical editor. With a better understanding of the responsibilities of a technical editor, an engineering manager might want to hire either a full-time or part-time technical editor. The 1991 article "Who Needs a Technical Editor?" provides information to engineering managers about the responsibilities qualifications for technical editors.

Power begins her article with a scenario in an enginneing office. Old and young engineers believe that they do not need technical editors. The engineers believe that their writing and editing skills are adequate. However, when the manager receives poorly written reports from engineers, he or she must spend hours to correct simple grammatical and typographical errors. The manager's conclusion: Engineers can't write. However, Power says that engineers are not the only ones who have problems writing correctly and clearly. Government workers and lawyers also use too much jargon in their writing. Power says that a technical editor can solve many problems with unclear and incorrect writing.

According to Power, a technical editor should be able to perform a variety of job functions such as organizing data, checking references, writing copy, organizing layouts, copy editing, proofreading, and making sure that the whole project runs smoothly. Power concludes that the technical editor's job is one of quality control for written communication.

After providing her definition of a technical editor, Power breaks the rest of her article into three topics: editor's functions, qualifications, and Hiring options (for employers).

A technical editor's function should start at the early stages of development on a written document. Although he or she is not actually writing the document, the editor can help authors start writing the document by giving suggestions on style and organization. The editor can provide continuing guidance as the project develops. Power says that asking an editor to come in at the last minute and "fix-up" a paper would be frustrating to the editor and the writer. To give engineering managers a better idea of what technical editors do all day, Power lists 21 jobs (see Table 1) that editors are performing in industry today. The author also gives several scenarios where technical editors can be helpful. The topics of the scenarios include technical editing of a journal article, assistance in preparation of a technical article, individual coaching, assistance in preparing a speech, and assistance with translations.

When listing the qualifications of technical editors, Power uses the Society for Technical Communication as a source of information. She lists characteristics of technical editors that include gender, age, education, work experience, average salary, and technical background.

In her final topic, hiring options, Power lists several places where employers can start their search for a technical editor. Power cautions the engineering mangers that technical editors may not be the solution to all their written communication problems. If the manager, engineers, and technical editor cannot work as a team, the editor will be very little help. The editor may even cause more controversy because he or she might have an opinion totally different than the manager and engineers. Power concludes by answering the title of her article--"Who Needs a Technical Editor?" The answer to the question is "anyone who is responsible for communicating technical information and who needs help in getting projects organized, following up on details, coordinating groups of writers, untangling convoluted sentences, suggesting just the right word, settling questions of grammar, or finding that one last typo.

Although this article was intended for engineering managers, I found this document very informative and useful for technical writers and editors. Power says that in the last 10 years industry has recognized the need for better quality control in written documents. She presents many strong arguments to managers to hire technical editors. The author gives a very clear definition of a technical editor. However, there was other information I found valuable to me as a writer and editor.

If the engineering manager decides he or she needs a technical editor, Power lists several places (telephone directories, colleges, or the Society for

Technical Communication) where the manager can begin searching for an editor. I found this information very valuable. When I begin looking for a job, I will have some idea of where managers are looking for potential employees.

When I begin looking for a job, I want to be sure my qualifications are what the manager is expecting. Power lists qualifications and probable backgrounds of technical editors. I would like to know what my employer expects from me on the job. Also, it would be nice to know what type of background your competition (other people that are applying for the job) might have.

After I find a job, a salary must be determined. Power lists several different salaries with corresponding amounts of experience. This information is nice to have also. When I find a job, I would like to know what my peers are earning, so I have some idea of how much the job might pay.

The author stresses one point about technical editors to managers that I like. She reminds managers that editors will not solve all the problems of projects. Editors may cause even more controversy between themselves, engineers, and managers during a project. She says that managers, engineers, and editors must work together as a team for a project to be successful.

The most valuable part of Power's article was the list manager's might use to determine if they need a technical editor. After reading this list, I have a better idea of what managers are going to expect from me in the "real world." This is very valuable information. What a student learns in the classroom is sometimes very different from what management wants in the real world.

TABLE 1
Do You Need a Technical Editor?
Use This Checklist to Write a Job Description

Prepare or supervise preparation of reports, manuals, letters, proposals, memorandums, forms, procedures, scripts, articles, charts, and tables.
Work with engineers to achieve effective written and oral communications.
Designand present in-house writing courses.
Coach individuals in effective writing techniques.
Edit and proofread prepared copy at the request of an author.
Provide authors with outlining and editing assistance.
Write or help authors write abstracts, introductions, sumaries, and articles for publication.
Review, edit, and rewrite as necessary, in cooperation with the authors, all technical copy.
Prepare articles for publication according to the editorial requirements of specific journals.
Prepare and maintain a style manual specifically designed for the needs of the organization.
Maintain a reference library to help referee differences of opinion on grammar, spelling, and punctuation.
Help authors prepare speeches, select visual aids, and rehearse presentations
Review and edit all technical documents for general readability, consistency of style, logical flow, and correctness of spelling, punctuation, and grammar.
Advise on graphics, printing methods, and binding practices.
Supervise filling of technical documents.
Write procedures for a word processing area.
Supervise and manage a word processing area.
Maintain overall quality of materials processed through a word processing area, editing and proofreading as required.
Assist in preparation of materials for a source documentation library.
Prepare a publications guide to ensure uniformity of layout, typing, and printing.
Provide timely and effective services as required to save engineering time.

© 1981 IEEE

EXERCISE 6

Comparing Proof Corrections to Final Copy

Check your marks against the corrected document that appears in the Appendix.

References

Editorial Experts, Inc. 1986. On becoming a technical editor. *The Editorial Eye*. 130 (June).

Harris, Jeanette. 1987. Proofreading: A reading/writing skill. *College Composition and Communication* 34:4 (December).

Judd, Karen. 1990. *Copyediting: A practical guide*. 2nd ed. Los Altos, CA: Crisp.

Matalene, Carolyn. 1995. Of the people, by the people, for the people: Texts in public contexts. In John Frederick Reynolds et al. *Professional Writing in Context: Lessons from Teaching and Consulting in Worlds of Work*. Hillsdale, NJ: Erlbaum.

Power, Ruth. 1981. Who needs a technical editor? *IEEE Transactions on Professional Communication* 24.3 (September) 139–140. Table reprinted by permission of IEEE.

Wing, Wesley. 1987. The corporate life. Unpublished paper. San Jose State University. Reprinted by permission.

7 Editing Copy and Constructing Style Sheets

Contents of this chapter:

- Developing Language Awareness
- Learning a Process for Editing Copy
- Constructing a Style Sheet
- Working Toward a Style Guide

One of the ways to become a better writer in the workplace—whether in business, publishing, science, or industry—is to develop language awareness. In other words, you need a solid understanding of the grammar, conventions, and usage of written English: how the language is handled effectively.

7.1 Developing Language Awareness

Language awareness means that you can do all of the following tasks:

1. **Notice or detect a problem.** Those who are expert writers and editors, according to Carnegie-Mellon researchers Linda Flower, John Hayes, and others, have in their mind a "correct template" for how the text should read. When the words on a page or screen do not match this template, experts can recognize that a problem exists (Flower 1986).

2. **Determine or diagnose the problem.** Noticing a problem does not mean the editor or author knows exactly what the problem is. You have to ask "What is the problem?" To diagnose, you need to know the guidelines and conventions of good writing and how to analyze the document until you find the problem. You need to be able to move from "That doesn't sound right" to "The sentence needs to be reordered." You need to go from "This list is not clear" to "The list items are not parallel in structure."

3. **Select a method of fixing the problem.** Sometimes the method of fixing the problem is to rewrite: to add more information or change the way the information is arranged. More often, the method is to revise at the sentence level. For example, you may have to make subjects and verbs agree (a grammatical problem), to change words or eliminate jargon (a diction problem), or to change the order of the sentence to make it easier for the reader to understand (a syntax problem).

4. **Understand and be able to explain, or justify, the changes you have made.** You will be a better writer when you are confident of your editing skills at the sentence level because you will know what you are doing. You will be a better editor when you can explain to a writer why something needs to be fixed. Writers tend to be defensive about the words they have built into text, but if you can justify your editing changes with a knowledgeable and coherent explanation, they will accept those changes.

Just as proofreading is different from regular reading, so editing at the sentence level is different from writing. But because surface distractions and errors at the sentence level keep readers from understanding the larger message of a document, good editing is critical to effective writing.

Editors have a variety of explanations for what they do at this level of editing. Ruth Chase, senior editor at a software company, calls this "polishing a book." Sylvia Thompson, senior editor at a division specializing in electronic mail systems, calls it "improving the communication by removing distractions." At companies with tight schedules, copyediting takes place at the first editing pass; at those with a slightly longer development cycle, copyediting occurs at the *beta* (or second) draft level. The beta draft may circulate concurrently for technical content review by subject matter experts (SMEs) and for copyediting review by editors or colleague writers.

7.2 Learning a Process for Editing Copy

The first *sentence-level* edit, commonly called copyediting, is the process of ensuring that a document is well written, consistent in style, and error free. The editor works on a manuscript *before* it is set into its final version; usually copyediting is done on the author's last clean draft. After editing (and any necessary substantial changes by the author), the manuscript is ready to set into print. Good copyediting requires:

- an eye for detail and a willingness to check for omissions and discrepancies
- a thorough understanding of the grammar and punctuation conventions in standard written English
- the ability to use words accurately and to sense and correct poor sentence construction

You can edit your own writing, but you first need to distance yourself from it. Type a clean copy and put it aside for a day or more. By then you'll no longer be so emotionally tied to your document, and you can see the actual words instead of the words you intended.

Most workplace editors work on hardcopy or paper documents rather than on softcopy or on screen. Editors say that it is easier to read paper documents, and they are more likely to find errors on hardcopy. Editor Sylvia Thompson says, "I think it's too much to ask an editor to look at details on a small screen. I print out even online help systems for editing, although they will be used on a screen." Freelance editor Jo Levy edits on hardcopy but looks at screens to check the appearance of a panel on the screen.

In editing copy, whether you are working on your own or someone else's draft, you should follow these steps:

1. Read the entire piece to get a sense of the whole.
2. Read all headings and subheadings in sequence, checking for accuracy, parallelism, and consistent format.
3. Read all tables and figures in sequence, checking for correct format, completeness, and consistency.
4. Read the paragraphs for consistent development from paragraph to paragraph and within each paragraph. Make sure that transitions signal intended movement from idea to idea.
5. Read individual sentences to ensure correctness of
    grammar
    word choice
    sentence construction
    punctuation
    spelling
    capitalization
6. Doublecheck numbers, equations, and formulas for accuracy and consistency.

On doublespaced hardcopy, editing marks are made *within* the manuscript exactly where the change must be made. On singlespaced hardcopy, editing marks are made in the margins, as they are in proofreading. Comments and questions are written in the margins, surrounded by a circle to distinguish comments from corrections. See Chapter 5 for a sample of doublespaced copyedited text. See Chapter 6 for a sample of edited text that is singlespaced. The sample in Chapter 6 is for proofreading, but the technique is the same in copyediting.

On softcopy, as an editor you may have the capability of inserting editing marks on screen, of adding editing comments, or of making actual changes and marking them in color or with change bars. Whatever your company's procedures, remember that you must communicate with the writer. Ethically, you cannot change the document without indicating what the changes are and where they take place. For this reason, most editors work on hardcopy and let the writers make the actual changes to the file.

7.3 Constructing a Style Sheet

Part of the challenge of editing copy effectively is simply remembering the stylistic choices you have made as you work through a document. Did you decide to use *copyedit* (one word), *copy-edit* (one word hyphenated), or *copy edit* (two words)? Is it *online, on-line,* or *on line*? Do you spell it *grey* or *gray*? Is the abbreviation *p., pg.,* or *pp.*?

For each editing project, it pays to construct a simple style sheet that you can fill out the first time you make each editing decision. On the style sheet, list those items that you have decided to treat in a certain way. For example:

- your preferred spelling of a word (traveled, not travelled)
- abbreviations (when to use periods, what abbreviations are acceptable in this document)
- punctuation (if you will use a comma after a short introductory phrase like "In this way, . . .")
- typographic style (use of quotation marks or boldface or italics for highlighting words)
- numbers (when you will spell out a number and when you will use a numeral)
- footnotes and reference style
- capitalization

If you are working, your company may already have a style guide either online or in hardcopy. A style guide lists the company's preferred way of handling items like those above. You should read that style guide carefully and note on your personal style sheet the items pertaining to your project. A style sheet will help you remember decisions you have made and will ensure a consistently edited document.

How To Create a Style Sheet

At the beginning of an editing project, create a simple style sheet by following these three steps:

1. Take five blank sheets of paper, 8-1/2 by 11 inches. Treat page 1 as a cover, listing on it whatever is appropriate. For example:

   Document Title   Author
   Document Number Editor
   Job Number

2. Divide page 2 into four equal sections and label each section with a category that fits this document: for example, Punctuation, Tables, References, and Typographic Style. See Figure 7.1.
3. Divide pages 3 to 5 into four equal sections and label each section with two letters of the alphabet—AB, CD, and so on. On these pages you'll list preferred spellings, hyphenations, abbreviations, and other stylistic choices.

As you work through the manuscript, you will fill out the style sheet and use it for a reference. Thus, if you treat a definition a certain way on page 3, when you come to page 13, you'll be able to look at your style sheet and treat the next definition in the same way. For more information on using a style sheet, consult Karen Judd's *Copyediting: A Practical Guide*.

Example of a Style Sheet

The examples of an individual style sheet in Figure 7.1 show the suggested method for construction. Be sure to consult your company's style guide for approved company style decisions. Figure 7.2 shows some typical types of entries.

7.4 Working Toward a Style Guide

You will want to construct a style sheet for every document you edit. If you work for a large or established organization, you will probably have access to a company style guide, either online or on hardcopy. That style guide will help you make most of your editing decisions and will influence the entries on your style sheet.

If your organization does not have its own style guide, it may rely on one of the standard style guides published in book form. The most widely recommended style guide is *The Chicago Manual of Style;* for government documents, its counterpart is the United States Government Printing Office *Style Manual*. See Chapter 4.3 for the titles of other style guides for specific disciplines.

However, you may find yourself working in a start-up company or as a freelance editor. While you can rely on the general style guides for advice, you may also want to work toward developing a style guide that meets the needs of your particular company or discipline. Durthy Washington, who owns her own editorial consulting firm, says:

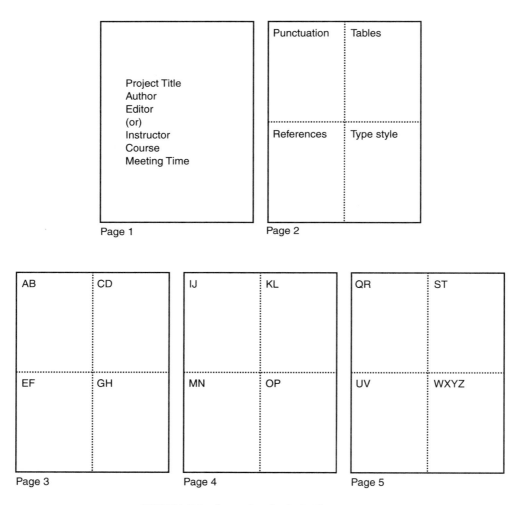

FIGURE 7.1 Example of a Style Sheet Format

> A style guide creates and defines the standards for corporate documents and helps establish a consistency among these documents that reflects a company's commitment to quality and style. Its primary function is to streamline the document development process and thus significantly reduce the time required to create new information products. But an effective style guide can also serve as a training tool for new writers, editors, and documentation managers and enable companies to realize significant cost and time savings. (1993, 505)

Washington goes on to suggest a process for conducting a requirements analysis, developing the style guide, and promoting it within the organization.

AB	CD
antimagnetic ball-like busing	cathode ray tube (no hyphen) check valve (two words)
EF	GH
east (direction) the East (part of the country) egg-shaped Export-Import Bank	gate-crasher gatepost (one word) Hz hemostat

FIGURE 7.2 Typical Entries in a Sample Style Sheet

If you are working as a writer or an editor, the thought of *also* putting together a large style guide might be daunting. The general style guides run to several hundred pages, and even some company style guides are very long and crammed with detail. But you need not begin with a long document. Bill Sullivan, writing in *Intercom* (the magazine of the Society for Technical Communication), suggests starting small. "Start by writing individual articles one page or so long, like online help topics. Circulate them one at a time by e-mail and suggest that people keep them in a folder, or circulate them in hard copy in a three-ring binder. Put dates on the articles in case you decide to revise them. . . . Just as it is easier to read and comprehend one page than a whole book, it's easier to write that way. Unless you work in a very unusual place, other responsibilities are likely to take priority. If you haven't got time to write a whole book well, you may still be able to do a good job on one page at a time" (1996, 25).

EXERCISE 1

Constructing a Style Sheet Template

Following the examples in Figure 7.1, construct a style sheet template. Make several copies.

EXERCISE 2

Recording Style Decisions

If you are in an editing class, use one style sheet template to record style decisions made by the class as you work through various documents. For example, will the word be in-house *or* in house? *Save the other copies to use when you begin editing whole documents.*

EXERCISE 3

Using a Style Sheet

If you are learning editing on your own, use a style sheet template for each document you edit, carefully recording your editing decisions as you make them.

References

Flower, Linda, John R. Hayes, Linda Carey, Karen Schriver, and James Stratman. 1986. Detection, diagnosis, and the study of revision. *College Composition and Communication* 37 (February), 16–55.

Judd, Karen. 1990. *Copyediting: A practical guide.* 2nd ed. Los Altos, CA: Crisp.

Sullivan, Bill. 1996. How do I develop a style guide? *Intercom* 43.8 (October).

Washington, Durthy. 1993. Creating the corporate style guide: Process and product. *Technical Communication* 40.3 (August).

8 Capitalization, Spelling, Numbers, and Document Style

Contents of this chapter:

- Capitalization
- Spelling
- Numbers
- Document Style

Choices in capitalization, spelling, number usage, and document style are usually specified in a company or publication style guide. General style guides, like *The Chicago Manual of Style* and the United States Government Printing Office *Style Guide,* are often used as backup resources. Specific technical professions also have style guides. For example, the Council of Biology Editors, the American Mathematical Society, and the American Chemical Society all publish style guides governing usage in publications of those professions.

While you are learning editing skills, you may wish to examine some of these style guides to familiarize yourself with their organization and requirements. In the meantime, this book will give you enough essential guidelines so that (1) you can practice editing capitalization, spelling, numbers, and document style and (2) set up your own style sheet when you copyedit a document.

8.1 Capitalization

The first rule in capitalization is to follow your company style guide, but if no style guide exists, follow these recommendations or *The Chicago Manual of Style* in setting up your own style sheet for capitalization.

General Rules

Capitalize the following items:

Names of
- people
- organizations
- structures
- vehicles
- months

- days of the week
- holidays
  (but not seasons like winter)
- geographic areas
  (but not directions like north)

Titles of
- people
- books
- articles

- chapters
- courses or studies (if they are official titles)

88

Specific Rules

- In titles and headings, capitalize all major words. Capitalize a preposition, article, or conjunction if it is the first or last word of the title or heading.

    Through the Ages: A Look at Political Movements
    The Department of Health and Human Services

    Reference lists in technical articles or books follow special rules of capitalization for titles depending on the chosen documentation system. For examples see Chapter 17.6, *The Chicago Manual of Style,* or Lois Johnson Rew's *Introduction to Technical Writing: Process and Practice.*

- Capitalize trade names, even those trade names that have taken on common usage.

    Kleenex   Teflon
    Plexiglas   Formica
    Scotch tape   Styrofoam
    Xerox

    Companies have very specific rules about trade name usage, so be sure to check the company style guide or legal department whenever you must include trade names in a document.

- Capitalize references to figures, tables, and chapters when they are followed by a number or letter. Do not capitalize them when you make a general reference.

    See Table 14.
    All the tables that follow are from the *Statistical Abstract of the United States.*

- In lists, capitalize the first word in the list if each item is a complete sentence.

    A hurricane starts when the following conditions occur:

    1. Water vapor evaporates from the surface of a warm tropical ocean.

    2. Thunderclouds form and release heat.

    3. The thunderclouds form a spiral.

    In lists that complete a sentence, do not capitalize the list items.

    Hurricanes are classified in categories by destructive potential. The categories specify damage that is

    1. minimal   4. extreme
    2. moderate   5. catastrophic
    3. extensive

Avoid using all caps for blocks of text; research has shown that text in both capital and lower case letters is easier to read because the shapes of the letters are significantly different.

8.2 Spelling

For a technical writer, one of the advantages of a word processor is its spell checker. Spell checkers generally have a large vocabulary of common words and will flag misspellings and typos so they can be corrected. Editors in industry always ask writers to run a spell checker on a document before it is submitted for review. However, spell checkers do not always work properly, nor will they solve all spelling problems. Therefore, as a writer or an editor, you need to be very aware of the individual words that appear in the text and ensure the following:

1. **The words are correct for the context in which they are used.** A spell checker will not flag real words even if they are incorrectly used in context.

    . . . a currant hot software product
    . . . you must fined your mistakes
    This revue highlights the problem of . . .

2. **The words are consistent throughout the document.** Some spell checkers will accept more than one spelling. For example,

    gray and grey
    theater or theatre
    disk drive or disc drive

    In this case, a company style guide will usually specify usage. If you are creating your own style sheet, choose one variant and use it consistently.

The following guidelines will help you become a better speller whether you are a writer or an editor.

1. **Run a spell checker before you edit a document.**
2. **Buy a good dictionary and look up every word you're not sure of.** At work, ask your company to choose *one* dictionary as a standard and buy one for each employee. Use the dictionary all the time.
3. **Keep a list of your own problem words, either on three-by-five inch cards, on the wall by your desk, or on your computer terminal.** Use it as a quick reference. Few people have more than 20 problem spelling words, and a list will save you time.

4. **When you edit a document, check for spelling by reading slowly, one sentence at a time.** Use a pointer or cover the text with a card. This forces your attention on individual words.

5. **Ask a colleague to read your documents to check for spelling.** Offer to read his or her work in return.

EXERCISE 1

Editing for Capitalization and Spelling

Edit the following sentences for capitalization and spelling errors. Consult a dictionary. Use standard editing marks to make any necessary changes. Selected answers appear in the Appendix.

lowercase __/__ capitalize ___ insert __∧__ delete __ℓ__

1. After chapter 10 was reviewed for technical accuracy, the Technical Writing Manager asked senior editor P. E. Barazza to copyedit the document.

2. The entire program benefitted from the dialog between the laboratory personal and the customers.

3. The Northeast suffered its worst storm of the season, and the Government Offices canceled their office hours for two days.

4. The program error defys correction; we think the programmers deliberetely introduced a bug.

5. The gage was in error, but an accurate reading was not necessary to the anologue readout.

6. He recieved a grant from the Rockerfeller foundation, but it was not adequate to support his work.

7. The principle agency involved is the department of agriculture; however, the budget bureau is also peripherilly concerned.

8. Because sulphur was stored in bldg. no. 8, fire chief R. A. Wu ordered a review of Fire Policys and Proceedures.

9. The anologue to digital converter was a good use of the currant technologie's capability.

10. Because three words were mispelled on the applicant's Résumé, Margaret Black, the publications manager, discarded the application.

8.3 Numbers

As a technical writer or editor, you will use or review numbers of all kinds—counting numbers, ordering numbers, fractions, measurements, and percentages. You need to establish guidelines for consistent use, so you will know when to use numerical symbols (2, 36, and 580) as opposed to word equivalents (two, thirty-six, and five hundred eighty).

Because no single standard has been accepted in industry or publishing, you must establish guidelines for number use. When you must write about numbers:

1. **Determine your company's accepted style and follow it.**
2. **If your company has no accepted style or you are working on your own, follow these guidelines, which have been adapted from the Government Printing Office (GPO)** *Style Manual.* These guidelines assume that while words may make smoother sentences, numerals are easier for readers to understand.

Numerals

Use numerals in the following situations:

- In counting 10 items or more.

    18 disk drives

    1,712 stocks and 514 bonds

- In ordering items from the 10th on.

    the 12th floor

    125th Street

- In sentences containing 2 or more numbers, when 1 number is 10 or more.

Last month the shipping and receiving department processed 452 orders with a work force of only 7 people.

- In units of time, measurement, or money. Although days, weeks, and years are technically units of time, use words for designations up to nine.

   The meeting is scheduled for 2:30 P.M.

   Clay pipe with inside diameter greater than 18 inches must be tested for "D" load in conformance with ASTM C-301.

   The bid was for $32,000.

   We anticipate a 10 percent increase in the unit cost over the next three years.

   Use one sheet of filter paper (No. 595, 5.5 cm).

   If a sentence contains both numbers and numbers used as units, follow the general rule of spelling out up to nine for the numbers and the rule for numerals for the units.

   He ordered five 3.5-inch bolts.

- Combined with words for very large numbers. In this case, use a number followed by a word.

   2 gigabytes

   an $8 million budget

- Combined with words if two numerals appear together. In this case, spell out the lower number and use numerals for the higher number.

   12 two-foot sections of pipe

   eight 20-pound boxes of detergent

Words

Use the word equivalents of numerals:

- In counting up to nine items.

   four books

   nine surveys

- In ordering items up to the ninth.

   the second building on First Street

   the ninth screen shot

   Seventh Avenue

- To begin a sentence.

  Twenty-seven species have been studied thus far.

  Three tables provide details.

  If the numeral is more than two words, it is better to reorder the sentence.

  NOT   Two hundred forty-seven surveys were returned by customers.

  BUT   Customers returned 247 surveys.

- To indicate approximations.

  during the nineties

  But not when the approximation is used with a modifier.

  almost 20 officers

  over 200 responses

- Combined with numbers if two numerals occur together. In this case, spell out the lower number and use numerals for the higher number.

  two 3-gallon tanks

  12 six-ounce packages

- To indicate fractions, unless the fraction is used as a modifier.

  one-half of the total distance

  BUT   2 1/2 yards of fabric

  When possible, change the fraction to a decimal.

  NOT   3 1/2 acres

  BUT   3.5 acres

Number Use for International Readers

Many U.S. companies routinely do more than 50 percent of their business in other countries, and the written documents dealing with that business must be readable and usable to residents of those countries. Documents are often translated from English into 15 to 20 different languages. Writers and editors of workplace documents can improve communication by using numbers in ways that can be easily understood by translators or by English-as-a-second-language (ESL) readers. In editing number use, follow the company style guide. If there is no style guide, follow these general guidelines:

1. **Use the metric system if possible: meters, liters, kilometers, and so on.** The standard system in science and engineering is known as the MKS system, for meter, kilogram, and second. If the document requires measurements in pounds, inches, and miles, provide the metric equivalents as well, if you can.

2. **Avoid writing dates using numbers only** (for example, 4-11-99). Instead, always write out the name of the month. Consider writing the date military style (11 April 1999).

3. **Be consistent and clear in the way you write telephone numbers.** In the United States, we often use a combination of parentheses and en dashes to separate the area code and the local number: (805) 394-3434. Other countries, however, indicate telephone numbers in other ways. Some European countries separate groups of telephone numbers by spaces only; other countries may use decimal points. A recent trend in the United States uses the decimal point to separate number groups: 1.800.623.6000.

4. **Clearly identify the number system you are using, whether it is monetary, metric, or information on addresses, dates, or time.**

See Chapter 19 for more information on international issues.

EXERCISE 2

Revising Numbers for Clarity and Consistency

Evaluate each of the following sentences to see if it conforms to the rules for number use. Make necessary changes, using standard editing marks.

insert ⋀   delete ℓ   lower case /   spell out ◯

1. In the northeast, water temperatures near the surface during the spring and summer range from 50 degrees F to 70 degrees F. Bottom temperatures at 100 feet range from 8.9 degrees C to 12.2 degrees C.

2. Currents are primarily tidal in origin and do not exceed five knots.

3. Visibility in Great Lakes waters ranges from 100 feet in Lake Superior to less than one foot in Lake Erie.

4. The five Great Lakes have a total area of 95 thousand square miles, and the two-thirds of these lakes that lie within U.S. boundaries represent almost fifty percent of the fresh water acreage in the country.

5. The singular hose regulator may freeze in the free-flow position after twenty to thirty minutes exposure in polar waters.

6. In temperatures of 50 degrees F or less, use one-quarter to 3/8 inch wet suits or variable volume wet suits.

7. The phylum *Coelenterata* includes about ten thousand species. Only 70 species have been involved in injuries to humans.

8. In a popular model of the Diver Propulsion Vehicle (DPV), two twelve-volt batteries (in series) provide about one hour of operation at full power.

9. The operation with the atmospheric diving system began at 8:00 A.M. and lasted for five hours and 59 minutes below the ice with minimal discomfort for the operator. By two in the afternoon, the operator was back on the surface. With conventional diving methods, the decompression obligation would have been more than 8 days.

8.4 Document Style

The style or "look" of a document is determined by many things, including the capabilities of your word processor and printer. One decision that should be set at the copyediting stage, however, is the way in which words will be differentiated or "set off" from other words.

Notice that in the paragraph above, I chose to call attention to the words *look* and *set off* by enclosing those words in quotation marks. In this paragraph, however, I have italicized them. I could also have set those words in bold type, underlined them, or set them in all capital letters.

These type style choices are usually set by a company or profession's style guide. However, if you are setting the style yourself, choose one system and follow it consistently throughout the document. As you read technical and scientific literature, you will notice that within a profession, style choices are usually consistent. In this section, I have followed the convention of setting off words with quotation marks when I mean "so called" and with italics when I refer to a word as a word. Italics are also conventionally used for book titles referred to within text. Words are usually set in bold type for emphasis or for headings. Sometimes words are bolded if they also appear defined in a glossary. If you do not have the capability to italicize or bold, you can substitute underlining and capitalization.

A second style choice to make early in copyediting is the way to display paragraphing and lists. Paragraphing is a conventional way of breaking information into "chunks" that are small enough to be handled by the reader's short-term memory. The chunk of information that constitutes a paragraph is visually defined by white space—either indentation before the first word or extra white space between the lines of type. This is a stylistic choice.

Lists are also chunks of information defined by the white space surrounding them. Before copyediting, you must determine the way that lists will be displayed. *Numbered lists* help the reader understand sequence (in instructions), importance in a hierarchy (in outlines), or the number of items or operations. Numbers can be bolded for more emphasis. *Bulleted lists* are effective when sequence or the number of items is not important. Bullets can be solid or open circles, squares, or diamonds; again, this is a style decision. See Chapter 16 for more on design of lists.

EXERCISE 3

A Comprehensive Exercise

Edit the following document, using standard editing marks. Write any questions in the margin.

"Manufacturing Scuba Cylinders" Adapted from an article appearing in *Skin Diver;* March, 1967. P. 46

A scuba cylindet starts as a 5 ftx10' sheet of steal plate. This plate is heated to 1300 deg. in an enormus Spheroidized annealing furnace that heats up to 3,000,000 b.t.u.'s and it is kept their for five days. The Steel molecules in the sheat are re-arranged into a uniforme mass of moleculer spheres. This anealing process' turns the hard steel into soft, workable metal.

The plyable sheet is allowed to cool, then it is cut into strips. These strips are run thru a huge punch press which stamps out blank disks a quarter in thick & 23" in dia. Each disc is then conveyed to a hydraulic press weighing 400 tons. It is pressed into a large 'cup' shape in this giant machine. Next it is moved to another presss to be "drawn". This is done by a hydraulic-ram tha pushes the "cup' thru a die; narrowing and elongating it.

EXERCISE 4

A Comprehensive Exercise

Edit the following document, using standard editing marks. Write any questions in the margin.

README.1

This is the README file from DISK 1 of a 2 disk set.

This is a shorttext file that describes FLODRAW, lists the files on DISK 1, and tells you how to use the 10 minute tutoiral.

FLODRAW is a 'what-you-see-is-what-you-get' graphics editor that produce black and white full page diagrams on an IBM PC or compatable compter. FLODRAW has been designed to handle symbols quickly and easily. Its ideal for documents such as flocharts, organization charts, system diagrams, and other symbol oriented documents. FLODRAW is is keyboard controlled only.

FLODRAW combines text and graphics on the document. The program is equiped with basic graphics editing for lines and circles, in addition to pixleediting.Automatic generation of lines and arrows supports diagrams lie flocharts.

FLODRAW comes with symble libraries for flow charts, HIPO charts, and electricc diagrams. You can desingn your own symbols, save them, and combine them into new libaries, or add them to existing libraries.

9 A Review of Basic Grammar

Contents of this chapter:

- Definitions of Major Sentence Parts
- Recognizing Sentence Parts
- Definitions of Other Important Sentence Parts
- Recognizing Subjects and Verbs in Complicated Sentences

Every professional editor I interviewed in preparing this book listed a basic understanding of traditional English grammar as a critical skill for writers and editors. *Grammar,* according to linguistics professor Martha Kolln, is "what the language 'can and will do.'" Explaining further, Kolln says ". . . grammar refers to the underlying system of rules that enable people to speak their native language" (1991, 1, 2). When you edit, whether the document is your own or someone else's, you need to understand both how the language works and what common terms people use in talking about the language.

Native speakers of English often learned these rules in about the eighth grade, and both writers and editors have told me anecdotes about a Mr. Schwartz or a Miss Rudolph, who insisted that their students learn traditional grammar. These lucky writers and editors also voice their appreciation to their grammar teachers. If you didn't learn the basics of grammar in your earlier years, or if English is your second language, this chapter and those that follow will fill that gap. This chapter covers essentials: definitions and examples of the parts of the basic sentence followed by exercises to help you recognize those parts. Neither this chapter nor Chapters 10 and 11 cover everything there is to know about grammar and punctuation. If you want more details, consult one of the references listed in Chapter 4.3. However, this book provides enough basic information for you to edit competently at the sentence level.

If you are confident of your understanding of grammar, skip this chapter. If you want to test your competence, take and score the following quiz. See the Appendix for the answers.

Diagnostic Quiz on Grammar

For the sample sentence, answer each of the questions below.

SAMPLE SENTENCE

Those eight companies, which require writers to use sophisticated software publishing tools, are leaders in the industry.

1. Is this a sentence? How do you know?
2. What is the simple subject?
3. What is the complete subject?
4. What is the verb? Is there more than one verb?
5. What kind of verb is it?

6. What is the predicate?
7. What is the word *leaders* called?
8. List and identify any phrases in this sentence.
9. List and identify any clauses in this sentence.
10. List and identify any verbals in this sentence.

9.1 Definitions of Major Sentence Parts

Writers and editors cannot talk to each other about sentence patterns or word order unless they share some terms to describe the parts of a sentence. Like auto mechanics and airline pilots, writers and editors have a stock of jargon—technical words to describe what they are doing. Following are some terms and definitions you should know in order to understand and explain how sentences work. More definitions appear in section 9.3.

- A *sentence* is an independent unit of words that contains a subject and a verb and closes with a mark of punctuation.
- A *subject* is what or who is talked about in a sentence. The subject may be a single word or a group of words. Subjects are usually nouns (name words) or pronouns (replacements for name words).

The simple subject is the main noun or pronoun.

<u>She</u> worked.

The complete subject is the main noun or pronoun plus all its associated words.

<u>The manager of the publications department</u> worked.

Subjects can be multiple.

<u>The software and the hardware</u> were released yesterday.

- A *verb* is a word or words that show action, occurrence, state of being, or condition. Verbs can be singular or plural.

The engineer <u>opened</u> the window. (action)

Profits <u>have decreased</u>. (occurrence)

The writer <u>was</u> an expert. (condition) a "linking" verb

The meeting <u>is</u> in the conference room. (condition)

Verbs can be multiple.

Scientists <u>investigate</u>, <u>analyze</u>, and <u>report</u>.

Verbs can contain more than one word, usually to convey information about tense, mood, or voice.

>   The results <u>will be repeated</u> every six months.

- The *predicate* includes the verb plus additional words that complete the meaning of the sentence. The predicate may be a single word or a group of words.

    >   Editors [subject] evaluate [predicate].

    >   Editors [subject] evaluate text, graphics, and design [predicate].

- The *direct object* (also called the object of the verb) is the thing or person that receives the action of the verb. It answers this question: What or who receives the action?

    >   The technician installed the system [direct object].

- The *indirect object* is the thing or person that indirectly receives the action of the verb. An indirect object answers the question *to what? for what?* or *to whom? for whom?*

    >   The editor sent the manager [indirect object] an e-mail response [direct object].

- The *subject complement* is a word following a linking or condition verb that renames or describes the subject.

    >   The expert was an *attorney*. [renames]

    >   The index is *incomplete*. [describes]

- *Linking or condition verbs* include is, was, and other forms of *be;* appear, become, get, look, remain, seem, feel, taste, smell.

- A *phrase* is a group of related words that does not contain a subject and verb. A phrase is not an independent unit. A phrase functions as a part of speech. Common types of phrases are prepositional phrases, verb phrases, infinitive phrases, and gerund phrases.

    >   under review [prepositional]

    >   to understand the sentence [infinitive]

    >   to stand for reelection [infinitive and prepositional]

- A *clause* is a group of related words containing a subject and a verb. Some clauses are independent units, while others depend on the rest of the sentence for completion.

    >   because the building was unoccupied

    >   which was not covered

    >   the committee met three times

9.2 Recognizing Sentence Parts

Besides knowing what the major sentence parts are, you also need to be able to recognize them in a sentence. Once you can do that, as a writer you'll be better able to understand why a sentence doesn't "sound right" or "look right." As an editor, you'll be able to explain to writers why a change is needed in a particular sentence.

The following exercises will give you practice in recognizing subjects and verbs and other major sentence parts like direct objects and complements. Selected answers appear in the Appendix.

EXERCISE 1

Recognizing Subjects and Verbs

Underline the simple subject once and the verb twice. Draw a slash mark between the subject half and the predicate half of the sentence.

1. Hydraulic mining was invented by Edward Mattison at a gold mine in Nevada County, California, in 1853.
2. Mattison brought water to his claim by a hose with a tapered nozzle.
3. The strong water jet washed gold-bearing gravel into his sluices.
4. The method was very effective, and soon many California gold miners built hydraulic mining systems.
5. Sixty million gallons of water were used daily at the Malakoff mine alone.
6. The company built a 45-mile long canal to supply water for the Malakoff mine.
7. The canal and reservoir cost $750,000.
8. The mine tailings polluted the rivers of California with sand and gravel.
9. Floods occurred downstream and destroyed both crops and fields.
10. Hydraulic mining ended in California in 1884; angry farmers were granted an injunction against discharge of tailings into streams and rivers.

104 PART TWO SENTENCE-LEVEL EDITING

EXERCISE 2

Recognizing Subjects and Verbs

Underline the simple subject once and the verb twice. Note that some sentences have more than one verb. Draw a slash mark between the complete subject and the predicate.

1. The belt sander is the best all-around sander for a woodworker.
2. The straight-line action can give a surprisingly fine finish.
3. However, belt sanders can be heavy and awkward on edges or small pieces.
4. An inexpensive stand for the sander allows the woodworker's hands complete freedom.
5. The stand accepts 3-inch and 4-inch belt sanders.
6. They mount and dismount in a matter of seconds with two U-bolts and four wing nuts.
7. The stand can be used with the belt running vertically.
8. In addition, the stand can be flipped 90 degrees for horizontal work.
9. The stand's worktable is ribbed for easy collection of sawdust between the ribs.
10. The entire unit can be bolted to a workbench permanently or clamped to it temporarily.

EXERCISE 3

Recognizing Subjects, Verbs, Direct Objects, and Subject Complements

Underline the simple subject once and the verb twice. Mark D.O. over any direct objects and S.C. over any subject complements.

1. In 1851, Henry Burden built the largest water wheel in America.
2. This overshot wheel was 60 feet in diameter and 22 feet wide.
3. It produced 278 horsepower; in theory, it could produce 1000 horsepower.
4. Burden's wheel was a "bicycle-wheel" pattern.
5. The iron spokes were radially laced; they were not crossed.
6. Only tension supported the wheel's rim and the water buckets.
7. A segmental gear at the rim supplied the drive.
8. The wheel powered the roll-trains at Burden Iron Works.
9. Burden's machinery was a wonder of the age.
10. Burden also invented a horseshoe-making machine; it could produce 3600 horseshoes an hour.

9.3 Definitions of Other Important Sentence Parts

- A *modifier* is a word, phrase, or clause that limits or describes another word.

    Adjectives modify nouns and pronouns.

    A *fine* spray of *distilled* water was used to clean the surface.

    Adverbs modify verbs, adjectives, and other adverbs.

    He *carefully* stacked the cartons in the truck.

- *Prepositions* are words that join with a noun or pronoun to form a prepositional phrase. The noun or pronoun in the prepositional phrase is called the object of the preposition. Prepositional phrases then *modify* another word in a sentence.

    An oil well was successfully drilled (*in 1859*) (*near Titusville*) (*in northwestern Pennsylvania*).

To be a preposition, a word must have an object; that is, it must make a prepositional phrase. Sometimes the same words can be used as simple adverbs.

    The mosquito landed (*on his arm*).   prepositional phrase

    The speaker droned *on*.   adverb

The object of a preposition cannot also be the subject of a sentence. Thus, if you mark off prepositional phrases first, you will find it easier to locate the simple subject or direct object in a sentence.

    (On page 274) <u>is</u> the <u>answer</u> (to your question).

- *Conjunctions* are words that join two or more words, phrases, or clauses.

    the choke valve *and* the air valve

    Position the diaphragm *and* tighten the attaching screw.

- A *verbal* is a word derived from a verb but not functioning as a verb, though a verbal can be part of a verb phrase.

    Verbals can be

infinitives	*to pump*
present participles	*pumping*
past participles	*pumped*
gerunds	*pumping*

    Verbals can function as subjects, direct objects, modifiers, objects of prepositions, and parts of verb phrases.

    The mechanic plans *to pump the tank dry*. (direct object)

    Pumping procedures are explained in the service manual. (modifier)

    The *pumped* oil was stored in an underground tank. (modifier)

    Pumping iron is good exercise. (subject)

CHAPTER 9 A REVIEW OF BASIC GRAMMAR **107**

EXERCISE 4

Recognizing Modifiers, Prepositional Phrases, and Verbals

First, enclose prepositional phrases in parentheses. Then, underline the simple subject once and the verb twice. Mark direct objects and subject complements.

1. Minimizing space stress is a task of the Aerospace Human Factors Research Division of NASA.

2. Fifty specialists are studying possible mental pressures on astronauts during prolonged tours in the U.S. space station.

3. Previous shuttle missions have spent about a week in orbit, but astronauts at the space station can expect a stay of 90 days.

4. After 30 days, symptoms of isolation and confinement appear and begin to cause problems.

5. Soviet cosmonauts have experienced long missions in space, and their stress problems have been studied intensively by scientists.

6. Two cosmonauts had a fist fight in the space capsule after 90 days in space.

7. Space stress caused problems during a Gemini mission in 1965 and on Skylab 4 in 1973.

8. Stress reducers include private sleeping quarters, room for movement, windows, and plants for greenery and food.

9. The favorite activity of astronauts in space is viewing Earth through windows.

10. Skylab astronauts missed fragrances and the sounds of birds, thunder, and rain.

11. Before flights, NASA psychiatrists suggest training in compatibility and cohesiveness.

12. Cosmonauts have been abandoned in desert and ocean storms for stress testing.

9.4 Recognizing Subjects and Verbs in Complicated Sentences

Students often have trouble recognizing subjects in more complicated sentences. Some examples are listed below.

- **The complete subject is a long phrase including prepositional phrases.** One easy way to find the simple subject is to enclose each prepositional phrase in parentheses first; since the object of a preposition can't be a simple subject, you can narrow the choices.

    The principal source (of the selenium) (in the San Luis Drain and the Kesterson National Wildlife Refuge) is the Panoche fan area.

- **The subject is out of usual order.** Some sentences begin with words like *it is, there is,* or *there are. There* and *it* used in this way are called *expletives* and, though they appear in the subject position, the real subject follows the verb.

    There are some 18 irrigation entities in the Grasslands.

    Other sentences are simply out of usual subject-verb-object order.

    Between the two highways is a national park.

- **The subject is a verbal.** A verbal is a form of a verb that can be used as a noun. Some verbals used as subjects are infinitives; others are gerunds.

    Infinitives: to be, to run, to study

    To run the Boston Marathon/ has been his goal from the beginning.

    Gerunds: studying, typing, working

    Swimming/replaced jogging for the summer activity.

    Drenching the soil in irrigation/washes salts and trace elements into the drain areas.

EXERCISE 5

Recognizing Subjects and Verbs in Complicated Sentences

Underline subjects once and verbs twice. Put parentheses around each prepositional phrase.

1. For light-skinned people, tanning is the body's way of protecting itself from harmful ultraviolet rays.
2. The energy in ultraviolet (UV) light stimulates chemical changes in the skin.
3. Among these is the beneficial formation of vitamin D.
4. The pale skins of people in northern climes are transparent to UV light.
5. Creating a protective umbrella of melanin is the skin's way of protecting itself from UV rays.
6. Some tanning occurs within hours of exposure to sun.
7. To protect against harmful UV rays is good preventive medicine.
8. Using a good sunscreen and wearing protective hats and visors are the recommended measures of dermatologists.
9. There are good reasons for protecting the skin: maintaining skin elasticity, avoiding skin cancer, and allowing the body to produce antibodies against disease.
10. Protecting your skin makes good sense.

Reference

Martha Kolln. 1991. *Rhetorical Grammar.* New York: Macmillan.

10 Punctuating Sentence Types

Contents of this chapter:

- Sentences, Fragments, and Run-ons
- Compound Sentences
- Complex Sentences with Subordinators
- Complex Sentences with Relative Pronouns

Central to increasing your *language awareness* is a clear understanding of the types of sentences used in written English and the way writers identify those types of sentences for readers. One use of punctuation marks is to show readers the types of sentences they are reading. This chapter reviews the use of punctuation in sentence types, reminding you how to do the following: create complete sentences, end sentences effectively, and separate the parts of compound and complex sentences.

10.1 Sentences, Fragments, and Run-ons

A *sentence* is an independent unit of words that contains a subject and a verb and closes with a mark of punctuation. Those end marks help readers understand the purpose of the sentence. A sentence that makes a statement ends with a *period*. A question ends with a *question mark*, and a strong statement ends with an *exclamation mark*. (For definitions of major sentence parts, see Chapter 9.)

The two sentence problems called "fragments" and "run-ons" can be considered punctuation problems. In both cases, the writer has punctuated a word group as a sentence when it is either not a sentence at all (a fragment) or is two or more sentences (a run-on). Other names for run-ons are comma-spliced or fused sentences.

If you are a native speaker of English, you probably can remember high school or middle school English teachers complaining about fragments and run-ons so much that you learned to avoid them most of the time. Now, as an adult, you are often faced with advertising that deliberately uses fragments for a hard-hitting message. For example, look at the sentences below from high-technology advertising.

> Forget manual coloring and overlays. VST can draw your VLSI, gate arrays, and other complex IC designs. **In minutes. In big E-size formats. And with hundreds of bright, vibrant shading colors to define data paths, junctions, or layers.**
>
> By finding out what's hot, we've created the broadest range of communications circuits around. **For voice, data, and picture transmission systems.**
>
> Work faster. **No more juggling of hardware adapters.** Everything, from programming algorithms and electrical parameters to pin-signal assignments, is stored on floppies. **Masters and backups included.**

Fragments like these, while acceptable in advertising, would be unacceptable in a workplace document.

What do you need to remember, then, as a writer or editor? Primarily, you must remember always to suit your writing to the intended *reader*. To avoid problems, *always* write complete, properly punctuated sentences.

How do you write complete sentences? You keep before you the definition of a sentence: a group of related words *with a subject and a verb* that is an independent unit. In addition, you learn and observe the rules for punctuating compound and complex sentences.

Most *fragments* are phrases that are broken away from the preceding sentence. They simply need to be joined to that sentence with a comma, with a colon, or occasionally with a dash. Most *run-ons* are composed of two or more sentences; they run on because they lack the punctuation that tells the reader when to stop.

As a writer or editor, you need to think of using punctuation marks as though you were putting out traffic signs and signals. When you drive, you almost unconsciously obey a variety of signs—stopping, yielding, slowing, merging, turning right or left. Readers do the same thing; they rely—almost unconsciously—on punctuation marks to tell them when ideas are linked equally, linked unequally, expanded, and ended.

10.2 Compound Sentences

When writers wish to combine two or more ideas of *equal* importance, they often choose to write a *compound* sentence. For example, look at the two sentences below.

> K. Meyerbock wrote the code for the new release.
>
> He now heads the project team.

If you want to combine these two related ideas in one sentence and give equal importance to each idea, you might say:

> K. Meyerbock wrote the code for the new release; he now heads the project team.

What does this combination of ideas say to the reader? A simple chronology: Meyerbock did this and now he does that. You might say instead:

> K. Meyerbock wrote the code for the new release, and he now heads the project team.

But remember that sentences convey meaning. Is there more meaning behind these two ideas? Is the new job a result of writing the code? You might say:

> K. Meyerbock wrote the code for the new release, so he now heads the project team.

You could also say:

> K. Meyerbock wrote the code for the new release; therefore, he now heads the project team.

Each of these four sentences correctly combines the two ideas into a compound sentence that gives equal importance to each idea. But the four sentences have slightly different meanings, differences caused by the choice of conjunction and the punctuation.

Definitions

Before you can learn the punctuation rule for compound sentences, you need to review the meanings of four key terms:

- A *clause* is a group of related words that contains a subject and a verb. The subject may be understood. A short sentence is called a clause when it becomes part of a larger sentence.

    now she is eligible

    fasten the plate to the right side [you fasten]

    who ran the meeting

- An *independent clause* is a clause that can stand alone as a complete sentence.

    shortwave transmissions have unique qualities

- *Coordinators* are connecting words. They are also called coordinating conjunctions. Coordinators join items of equal value or importance. Coordinators can join words, phrases, or clauses. The common coordinators include:

    and   or   yet   for
    but   nor   so

- *Adverbial connectives* are adverbs that can be used, with a semicolon, to join independent clauses. They are also called conjunctive adverbs. They include the following words:

accordingly	furthermore	moreover	then
also	hence	namely	therefore
anyhow	however	nevertheless	thus
anyway	indeed	now	too
besides	instead	otherwise	
consequently	meanwhile	still	

Phrases can be used in the same way as adverbial connectives.

as a result	for example	in fact	in addition

Punctuation Rule for Compound Sentences

A compound sentence can be punctuated in one of three ways:

1. independent clause ⟨, coordinator⟩ independent clause

    The replacement parts arrived this morning, and we can proceed with the repairs.

2. independent clause ⟨;⟩ independent clause

    The replacement parts arrived this morning; we can proceed with the repairs.

3. independent clause ⟨; adv. connective,⟩ independent clause

    The replacement parts arrived this morning; thus, we can proceed with the repairs.

The adverbial connective often begins the second clause, but it can also appear elsewhere in that clause.

> In double-spaced text, paragraphs are usually indented; in single-spaced text, however, they may be set off by a line of white space.

As the writer, *you* have the flexibility to influence the reader by the method of compounding you choose. How do the sample sentences that illustrate the punctuation rule differ in meaning? Be prepared to discuss in class.

Compound Sentences with Colons

In most compound sentences, the two clauses have parallel ideas that add information (using *and* as a connector) or that contrast information (using *but* or *however*). Occasionally, however, the first clause of a compound sentence introduces a second clause that summarizes or explains the first clause. In this case, the two clauses may be joined by a colon.

> Classes are application oriented: People can walk away and use the knowledge at their jobs the next day.
>
> This expert system has a specific function: It helps the company create a personnel policy manual of up to 500 pages.

In this kind of compound sentence, the second clause usually begins with a capital letter, but this is a stylistic choice.

EXERCISE 1

Sentence Combining: Writing Compound Sentences

To review your understanding of key terms, answer each of the following questions without looking at the definitions. Then check your answers against the written definitions.

1. What is a clause?
2. What is an independent clause?
3. How would you define a compound sentence?
4. When should you write a compound sentence?
5. What are the three ways most compound sentences may be punctuated?
6. When do you use a colon in a compound sentence?

Following are pairs of simple sentences that are related in some way. Rewrite each pair into three compound sentences, using each of the three ways of joining a compound sentence. Be sure to write a compound sentence, even though you could, perhaps, write a simple sentence. Mark the most effective one with a star and be prepared to defend your choice. Selected answers appear in the Appendix.

1. Barbed-wire fences are common throughout the United States.
   Most city dwellers have never looked at a barbed-wire fence.

2. A barbed-wire fence consists of posts in the ground with wire between them. The barbs are spaced at 6-inch intervals on the wire.

3. The barbs must be firmly fastened to the wire. They must not break off or slide along the wire.

4. Four hundred patents for barbed wire have been recorded. Only two attempts were commercially successful.

5. Jacob Haish invented one method. Joseph Glidden, a neighbor of Haish in northern Illinois, more successfully marketed his wire.

6. Glidden's first wire was made in 1873. Within the next five years, sales amounted to more than 80 million pounds a year.

7. In Glidden's method barbs are clasped around the wire. A second wire then twists around the first and grips the barbs firmly.

8. A good barbed-wire fence will last 60 to 70 years. Stringing one is more complex than most people realize.

9. Some fence builders insist on cedar, locust, or Osage-orange wood for fence posts. These woods will last in the ground for many years.

10. Cottonwood or pine posts are more readily available. They will rot quickly unless creosoted.

EXERCISE 2

Recognizing and Punctuating Compound Sentences

The following sentences are unpunctuated. Underline subjects once and verbs twice. Then, using standard editing marks, insert the correct punctuation to make a good compound sentence. Defend your choice in the space below by citing that portion of the rule you used.

Insert a comma ⋀ Insert a semicolon ⋀

1. In software technology, the word *object-oriented* can be used to refer to computer interfaces it can also be used to refer to computer programming.

2. A user can manipulate a graphical image on the screen and control the computer.

3. For example, a picture or icon can represent a report and the user can print the report by dragging the icon and dropping it on an icon of a printer.

4. *Object-oriented* in this sense simply means using objects in fact it means little more than that the system has a good graphical-user interface.

5. In computer programming, *object-oriented* however refers to a particular way of organizing the internal elements of a program.

6. Conventional programs are designed like a super-fast robot running around a warehouse and reorganizing the stock.

7. The stock represents the computer's data and the robot represents the operative force manipulating the data.

8. The robot blindly executes the sequence of instructions the robot's instructions need to be altered if the warehouse is ever remodelled.

9. The robot's tasks are complex so it's difficult to revise all the instructions to accommodate a new situation.

10. Object-oriented programs resemble an army of robots each robot specializes in a particular kind of stock in its warehouse.

Based on Gary Hendrix. 1992. An object-oriented introduction. *Symantec*. 3.1 (Winter).

EXERCISE 3

Recognizing and Punctuating Compound Sentences

The following sentences are unpunctuated. Underline subjects once and verbs twice. Then, using standard editing marks, insert the correct punctuation to make a good compound sentence. Defend your choice in the space below by citing that portion of the rule you used.

Insert a comma ⋀   Insert a semicolon ⋀

1. Choosing the most economical plastic for a part by comparing the cost per pound of various plastics is a mistake some plastics weigh twice as much per cubic inch as others and could cost twice as much to ship.

2. A more meaningful comparison is cost per cubic inch but cost/strength values should be analyzed as well.

3. Selecting the best qualified material is not a simple task of comparing the numbers on published data sheets such values can be grossly misleading.

4. One could decide on the material with the highest ASTM test values however this process seldom results in the best choice.

5. The choice of any material should be based on the best combination of properties not on only one property.

6. The best material therefore usually represents a trade-off among satisfactory properties ease of processing and cost it is seldom the plastic with the highest values in any single category.

7. The figures in published data sheets represent laboratory tests nevertheless they do not duplicate real-life molding conditions.

8. The molding variables are beyond their control so material suppliers have chosen to make simple standardized ASTM laboratory tests.

9. The tests are not material *performance* tests for predicting real-life results over a period of time instead they are material *quality* tests.

10. The ASTM tests were never intended to compare one material with another on the basis of strength after molding different configurations and different molding processes can change the values significantly.

11. ASTM tests are essentially short-term tests with all variables fixed except the one being measured for example testing for impact strength is often done at a standard rate of loading.

12. A short-time tensile test may show a high strength value yet with reduced loading the tensile strength can drop to as little as 20 percent of the short-term strength.

EXERCISE 4

Correcting Fragments and Run-Ons

Underline subjects once and verbs twice. Then, using standard editing marks, make the changes necessary to have a properly punctuated, complete sentence.

1. Recording/playing time is the major difference between the two formats. A maximum of 8 hours for VHS compared with 5 for Beta.

2. Select plants to be used as females, these will be primary and secondary flowers that are beginning to unfold.

3. Please send me a technical report about the EXXXTRA ski. A report dealing with the design of your ski, the materials used in building the ski, and how your skis compare to your competitors' skis.

4. There are three tracks as shown in Fig. 2. The audio or sound, video or picture, and control or synchronization.

5. However, the IRS has caused problems with the law-abiding citizen by auditing a percentage of tax returns every year, this creates problems of finding records and taking time off to correct any problems found, in general the IRS invades an individual's privacy.

6. Today the principal idea behind microwave cooking is the same. To save money, time, and offer the convenience needed when preparing meals.

7. All recorders are equipped with a carrying handle. Which lies flat against the unit for listening or storage purposes and swings up for easy transport.

8. The same could hold true for breaking a mirror someone could get cut by the broken glass.

10.3 Complex Sentences with Subordinators

Sometimes when you combine two or more ideas in a sentence, you want to make each idea equally important. Then you write a *compound* sentence. At other times, you want to combine ideas, but you want to make one idea dominant and the other subordinate. Then you write a *complex* sentence. Occasionally, you will want to do both, and then you will write a *compound-complex* sentence.

Consider the two sentences below.

> Analytical chemistry deals with precise numbers and equations. Organic chemistry deals with reactions and synthesis.

Depending on the point you wish to make, as a writer you will choose one of these two ideas as dominant.

Suppose the document is about analytical chemistry. Then, that sentence is the main or *independent* clause; the sentence about organic chemistry is a subordinate or *dependent* clause. You can form a dependent clause by beginning it with a *subordinate conjunction*.

> *While* organic chemistry deals with reactions and synthesis, analytical chemistry deals with precise numbers and equations.

The independent clause can come at the beginning or the end of the sentence.

> Analytical chemistry deals with precise numbers and equations *while* organic chemistry deals with reactions and synthesis.

However, if the document is primarily about organic chemistry, you might begin the analytical chemistry clause with a subordinate conjunction.

> *While* analytical chemistry deals with precise numbers and equations, organic chemistry deals with reactions and synthesis.

Again, the independent clause can come at the beginning or the end of the sentence.

> Organic chemistry deals with reactions and synthesis *while* analytical chemistry deals with precise numbers and equations.

Do you see how the meaning has changed? You changed it by choosing the main clause. You could also change the meaning significantly by choosing a different subordinate conjunction. Suppose you use *although*:

> *Although* analytical chemistry deals with precise numbers and equations, organic chemistry deals with reactions and synthesis.
>
> Organic chemistry deals with reactions and synthesis, *although* analytical chemistry deals with precise numbers and equations.

If you are the writer, you make these decisions subconsciously as you write. But when you edit, you need to rethink the motivation for making one idea more important than another. If you are editing someone else's sentences, you need to think like the reader. What relationship of clauses would help the reader to understand?

Definitions

- A *clause* is a group of related words that contains a subject and a verb. The subject may be understood. A short sentence is called a clause when it becomes part of a larger sentence.

    > [you] copy the file to the C disk
    >
    > Anderson will act as supervisor

- An *independent clause* is a clause that can stand alone as a complete sentence.

    many companies have supported the program

- A *dependent clause* is a clause that cannot stand alone as a complete sentence. It is subordinate to an independent clause.

    until a replacement for the manager is found

- A *complex sentence* is a sentence containing one independent clause and one or more dependent clauses.

    After Rosa Hsue was promoted, she assumed management of the project.

- *Subordinators* include the words below. Subordinators are also called subordinate conjunctions. Subordinators can begin dependent clauses.

after	if, even if	when
although	in order that	whenever
as	once	where
as if	since	whereas
as though	though	wherever
because	unless	while
before	until, till	why

Punctuation Rule

To punctuate a complex sentence when the dependent clause begins with a subordinate conjunction, do the following.

- If the dependent clause stands first, it must be followed by a comma.

    Once a file variable has been declared, a programmer must associate the variable with the file on disk.

- If the dependent clause stands after the independent clause, no punctuation is required unless the subordinator is *though*, *although*, or *whereas*.

    A programmer must associate the variable with the file on disk once a file variable has been declared.

    More than 14,000 people apply to the program, although fewer than 1,400 are accepted.

EXERCISE 5

Writing Complex Sentences with Subordinators Showing Cause or Logic

Subordinate conjunctions can be divided into two classifications: temporal/spatial and causal/logical. Using each causal/logical subordinator, form a dependent clause. Then attach that dependent clause to an independent clause, punctuating correctly. Place some of the dependent clauses at the beginning and some at the end of the sentence.

1. although, though
2. because
3. if, even if
4. since
5. unless
6. whereas

EXERCISE 6

Sentence Combining: Writing Complex Sentences with Subordinators Showing Cause or Logic

To review your understanding of key terms, answer each of the following questions. Write the answers without looking at the definitions; then, check your answers against the written definitions.

1. What is a dependent clause?
2. What is a subordinate conjunction?
3. When should you subordinate a clause?
4. Write the punctuation rule for a complex sentence with a subordinate conjunction.

The common causal/logical subordinators are because, if, even if, since, although, though, unless, whereas.
   Using a causal/logical subordinator, join the following independent clauses to make a complex sentence. Be sure to write a complex sentence that shows cause or logic, even though you could write a simple or compound sentence. Use a different conjunction for each one, and place the dependent clause

in different positions. Punctuate correctly. Be prepared to defend your choices. Use standard editing marks.

insert___∧___ delete___⌒___ make lowercase___/___ capitalize___=___

1. Chemically, Silly®Putty is a liquid. It resembles a solid.

2. The molecular structure stretches. Silly®Putty can be slowly pulled.

3. Silly®Putty will shatter. It is struck with a hammer.

4. A rubber ball has a 50 percent rebound capacity. Silly®Putty's rebound capacity is 75 to 80 percent.

5. The early Silly®Putty was sticky. In 1960 it was reformulated to be non-sticky.

6. It is nontoxic and harmless to children and furniture. Some people have called Silly®Putty the ultimate plaything.

EXERCISE 7

Writing Complex Sentences with Subordinators Showing Time or Place

Following are some of the common temporal/spatial subordinators. Using each one, write a good complex sentence. Be sure to write a complex sentence, even though you might be able to write a simple or compound sentence.

Determine whether the dependent clause is more effective at the beginning or at the end of the sentence. Punctuate correctly.

1. after
2. before
3. once
4. since
5. till, until
6. when, whenever
7. where, wherever
8. while

EXERCISE 8

Sentence Combining: Writing Complex Sentences with Subordinators Showing Time or Place

The common temporal/spatial subordinators are where, wherever, when, whenever, after, before, since, once, until, till, while. *They are used in the same way as the causal/logical subordinators.*

Combine each set of the following sentences into a single complex sentence by using one of the temporal/spatial subordinators. Be sure your dependent clause indicates time or place, and be sure to punctuate correctly. Use standard editing marks.

insert ∧   delete ⌒   make lowercase /   capitalize ≡

1. Silly®Putty was discovered by accident in 1945. James Wright dropped boric acid into a test tube containing silicone oil.

2. The experiments took place at the New Haven research laboratories. General Electric hoped to use silicone to make synthetic rubber.

3. Wright examined the resultant compound. He discovered its bouncing characteristics.

4. GE's mystery goo was gaining popularity at New Haven cocktail parties. A toy store named the Block Shop was developing a catalog.

5. The catalog contained a page of adult toys including Silly®Putty. Silly® Putty's popularity grew.

6. A writer bought some and played with it. He wrote almost a page about Silly®Putty for the *New Yorker*.

7. Peter Hodgson borrowed $147 to market the first Silly®Putty. Lack of funds forced him to use plastic eggs to hold the product.

EXERCISE 9

Recognizing and Punctuating Dependent and Independent Clauses

Underline subjects once and verbs twice. Circle the subordinator. Using standard editing marks, punctuate the sentence properly and, in the space below, justify your punctuation by citing the rule you followed.

Insert comma  Insert semicolon 

1. Although the dean of an engineering college once denied the very existence of technical writing many of us are confident of its reality.

2. The important distinction is that sequential contexts call for inflexible lines of thought and impersonal forms of expression whereas associative contexts permit random and diverse patterns of thought.

3. If imaginative literature attempts to control men's souls functional English should control their minds.

4. When we enter the world of pure symbol the difference between the two kinds of communication—scientific and aesthetic—becomes more pronounced.

5. Because technical writing endeavors to convey just one meaning its success, unlike that of imaginative literature, is measurable.

6. Objection may be raised to this distinction between the two kinds of writing because it makes for such large and broad divisions.

W. Earl Britton

7. If the rising tide of antipathy to technology is to ebb we tech writers must relate things in a human way to man.

John Frye

8. Even if explicit judgments are kept out of one's writing implied judgments based on selective perception will get in.

9. When the conclusions are carefully excluded however and observed facts are given instead there is never any trouble about the length of papers in fact they tend to become too long since inexperienced writers when told to give facts often give far more than are necessary because they lack discrimination between the important and the trivial.

S. I. Hayakawa

10.4 Complex Sentences with Relative Pronouns

Dependent Clauses Used as Adjectives

Complex sentences consist of dependent and independent clauses. Some dependent clauses are formed by adding subordinating conjunctions to clauses and thus reducing their independence. Other dependent clauses are formed by adding *relative pronouns* to clauses to reduce their independence. These clauses may be used either as adjectives or as nouns. A clause used as an adjective stands after the word it modifies and is called a *relative clause*.

Common relative pronouns that can begin clauses include

| that | who | whose |
| which | whom | what |

For example:

that	The paper *that she edited* was published last week.
which	The American Medical Writers Association, *which has over 3,000 members,* serves communicators in the biomedical and health sciences.
who	The operator *who entered the data* made several errors.
whom	The representative *whom we elected* served three terms.
whose	The engineer *whose invention was patented* received a company award.

Each relative clause above is used as an *adjective*; that is, it modifies or explains the noun preceding it, answering the questions "What?" or "Who?"

Dependent Clauses Used as Nouns

Dependent clauses can also be used as *nouns* in the sentence. The clauses may look the same, but they are not punctuated. They are called *noun clauses*.

For example:

that we lost	(subj.) *That we lost the contract* was a contributing factor to the layoff.
what caused the leak.	(dir. obj.) Everyone understands *what caused the leak*.
whom it might concern	(obj. of prep.) He wrote the letter to *whom it might concern*.

Punctuation Guidelines

Dependent clauses used as *nouns* are generally not separated from the rest of the sentence by commas. Dependent clauses used as *adjectives* sometimes must be set off by commas. Follow these guidelines to decide if the dependent clause should be set off by commas.

1. If the clause is necessary to establish the specific identity of the noun it follows, it is a *primary identifier*. That is, it restricts or limits the meaning of the word it describes. A primary identifier is called *restrictive*, and no commas surround it. In terms of meaning, some rhetoricians say that a restrictive modifier *defines*.

> The man *who was arrested yesterday* is free on bail. (restrictive)
>
> The printer *that he ordered two weeks ago* has now arrived. (restrictive)

2. If the clause merely adds extra information to the noun it describes, it is a *secondary identifier*. Such a secondary identifier is called *nonrestrictive*, and it should be surrounded by commas. In terms of meaning, a nonrestrictive modifier *comments*, and it could be dropped from the sentence without materially affecting the meaning.

> Joe Patkin, *who was arrested yesterday*, is free on bail. (nonrestrictive)
>
> The laser printer, *which he ordered two weeks ago*, has now arrived. (nonrestrictive)

3. Generally, *that* begins restrictive clauses (without commas), and *which* begins nonrestrictive clauses (with commas). However, many writers confuse the use of *that* and *which*, so you cannot rely on the words to tell you if the clause is restrictive or not. However, if you follow the guidelines in this book when you write and edit, you will help your reader understand the relationship of clauses.

> The memo that was circulated yesterday contained an error. (restrictive)
>
> The workforce-reduction memo, which was circulated yesterday, contained an error. (nonrestrictive)

EXERCISE 10

Writing and Punctuating Complex Sentences with Relative Clauses

Write one sentence using each clause as a restrictive or "primary" adjective modifier. Then write another sentence using the same clause as a nonrestrictive or "secondary" adjective modifier. Punctuate correctly.

1. who was a member of the committee
2. that   raised the temperature
   which
3. whom she respected
4. whose membership had lapsed
5. that   started the program
   which
6. who was promoted in 1998
7. that   was printed last month
   which

EXERCISE 11

Recognizing Relative Clauses and Relative Pronouns

Underline the relative clauses in each sentence below. Circle the relative pronouns. Label any noun clauses.

1. Scientists searching for the event that killed the dinosaurs think that they have found it.

2. The new report, which was published in the journal *Science*, provides evidence of the crash of a comet or asteroid.

3. The researchers who worked on the study included geologists and paleontologists.

4. The impact, which occurred 65 million years ago, coincided with the extinction of the dinosaurs.

5. Using sophisticated new tools, the scientists examined a crater that was formed in Mexico's Yucatan Peninsula.

6. The crater, which measured 110 miles in diameter, had been discovered in the 1950s by geologists who were looking for oil.

7. A team from NASA, to whom the information was given, used satellite images to map the crater.

8. The evidence that supports the dinosaurs' demise is based on grains of glassy material from deep within the crater.

9. Geologist Walter Alvarez, whose theory seems to be supported by the crater, speculates that the dust and smoke from the impact blocked the sun and led to climate changes.

10. That the crater exists is accepted by most scientists; geologist Charles Officer, however, completely rejects the theory.

132 PART TWO SENTENCE-LEVEL EDITING

EXERCISE 12

Recognizing and Punctuating Restrictive and Nonrestrictive Clauses

Underline the relative clause in each sentence. In the space below, mark it restrictive or nonrestrictive. Label noun clauses. Punctuate correctly. Change that/which if necessary.

Insert comma

1. Dyslexia is a condition that is characterized by difficulties in writing, reading, and organizing information.

2. Students who are dyslexic may need to write four drafts of a paper while nondyslexic students produce a single draft.

3. Dyslexic students often must read assignments three or four times to discern meaning in difficult text or text which the author has not clearly organized.

4. Thus, most college students who suffer from dyslexia are also dedicated and self-aware.

5. Those whose schedules are tight know how to set priorities and budget time.

6. Researchers who have studied dyslexia over the past 20 years have found that the brains of dyslexics differ in both structure and function from the brains of nondyslexics, especially in the area of the brain involved with language.

7. During an exam, dyslexic students who are expected to process a great deal of written information which is often subtle and complex may need to be given more time.

8. Intellectual skills which call for the ability to synthesize information, generate hypotheses, and analyze situations are not affected.

9. Students at Brown University who wrote a booklet about how dyslexics perceive things differently from other people performed a service to the university community.

10. One student who is enrolled at Brown described dyslexia as "the monster that eats time."

EXERCISE 13

Recognizing and Punctuating Restrictive and Nonrestrictive Clauses

Underline the relative clause in each sentence below and mark it restrictive or nonrestrictive. Label any noun clauses. Punctuate correctly and change that/which if necessary.

1. Word processing is available with computer systems which range in size from personal microcomputers to huge commercial installations.

2. Even in large installations, users may work at terminals which seem to be their own personal word processors.

3. Word processing provides a mechanism to prepare material that would otherwise be typewritten.

4. Word processing also provides features which facilitate the preparation of the first draft and any later drafts of a document.

5. Information entered on the keyboard is not put on the paper but is collected and put into memory, the word processing component that holds that particular data.

6. To alter text in memory, the user may strike special keys to produce signals which are not part of the text but are commands to alter or delete.

7. Once a text has been corrected the user can strike a function key which directs the word processor to print that text.

134 PART TWO SENTENCE-LEVEL EDITING

8. The user needs to convey two kinds of information to the word processor: first, data that will eventually be printed and second, commands to the machine itself which are not to be printed.

9. To show the characters that are to be removed, a cursor is supplied on the screen.

EXERCISE 14

Putting It All Together: Recognizing and Punctuating Dependent Clauses

The following paragraphs contain a number of dependent clauses, some beginning with subordinate conjunctions and others with relative pronouns. Underline each dependent clause and label the subordinate conjunction or relative pronoun. The sentences are unpunctuated, so drawing on all that you have learned, punctuate correctly. Change that/which if necessary, and be prepared to justify your decisions.

Each human eyeball is equipped with six extrinsic muscles which hold it in position in its orbit and rotate it to follow moving objects. Because the eyes work together they are directed to the same object and converge for near objects. Besides the extrinsic eye muscles, there are also muscles within the eyeball. The iris is an annular muscle that forms the pupil and allows light to pass to the lens lying immediately behind. This muscle contracts to reduce the aperture of the lens in bright light although another muscle controls the focusing of the lens.

People often think that the lens bends the incoming rays of light to form the image. However, the region where light is bent the most in the human eye to form the image is not the lens instead it is the front surface of the cornea. Although the lens is rather unimportant for forming the image it is important for *accommodation*. This is done by the shape of the lens. When the radius of curvature of the lens is reduced for near vision the lens becomes increasingly powerful and so adds more to the primary bending accomplished by the cornea. The lens is built up of thin layers like an onion and is suspended by a membrane, the zonula, which holds it under tension. Accommodation works in this way. When near vision is required tension is reduced on the zonula allowing the lens to spring to a more convex form—the tension being released by contraction of the ciliary muscle.

Adapted from R. L. Gregory
Eye And Brain: The Psychology Of Seeing

EXERCISE 15

Putting It All Together: Recognizing and Punctuating Sentence Types

Punctuate the following sentences correctly. Below each sentence identify the types of clauses and justify your punctuation. Use standard editing marks.

1. Whenever you make a decision about buying individual stocks or stock market mutual funds there is generally one overwhelming question in your mind How fast will my investment increase or decrease in value?

2. While no one can answer that question with complete certainty you can get some idea of how fast a stock will tend to go up or down by looking at something called a beta indicator.

3. Sometimes known as the beta coefficient, a beta is a measurement of stock volatility and it will give you an indication of how fast a stock will rise or fall in value in relation to a standard stock market index such as the Standard & Poor's 500.

4. When you use the beta as an investment tool keep in mind the risk and reward trade-off the more risk you take, the higher the returns, and vice versa.

5. Generally speaking, blue chip companies like IBM will have low betas and will be slow steady earners. The same is true for income funds and balanced funds although you should keep in mind that mutual funds

generally have lower betas because they can adjust their portfolios to market changes.

6. If you're looking for more speculative investments and you think the stock market is going up look for a stock with a higher beta and keep your eye on it at all times.

7. If you think that the market is going down you could look for a stock or a mutual fund with a negative beta that is a stock that tends to go up when the market declines.

8. Stocks with negative betas are rare though so don't hold your breath.

9. Keep in mind that a low beta does not mean that you won't lose money on a stock or a stock fund a low beta simply means that the stock or stock fund will tend to rise and fall more slowly than the average comparable stock.

10. If you want to find the beta of an investment the best way is to look it up determining an accurate beta can be difficult for non-economists.

11. The beta indicator is somewhat controversial and some analysts may not use it.

Adapted from Wm. Donoghue

EXERCISE 16

Putting It All Together: Recognizing and Punctuating Sentence Types

For each sentence, underline the simple subject once and the verb twice. Punctuate correctly. Do not rewrite the sentences, but change that *to* which *if necessary. In a numbered list, identify the sentence type and justify your punctuation choices by citing the appropriate rule.*

[1]A computer virus is the segment of a software program code that attaches itself to a computer file and then migrates from one file to another. [2]Typically, a virus has these two

distinct functions it spreads itself from one file to another and it implements the damage which was planned by the perpetrator. ³The damage may include erasing of a disk, corrupting of applications, or creating havoc on the computer.

⁴*Benign* viruses that usually conceal themselves until some predetermined date or time do nothing more than display some sort of message to the user. ⁵Although they are not intended to be malicious benign viruses can cause system crashes, printing problems, and application errors.

⁶*Malignant* viruses inflict malicious damage consequently they are more dangerous. ⁷Applications which are infected do not work as they are designed. ⁸Infected applications might terminate abnormally or they might write incorrect information into documents. ⁹The viral code might also alter the directory information on one of the volumes therefore preventing it from mounting on the desktop. ¹⁰In this case, users will be either unable to launch one or more applications or applications will be unable to locate the documents that need to be opened.

¹¹When a system is infected with a virus the infected file can be replaced or repaired. ¹²The way that is guaranteed is to delete the infected file and replace it with a clean, original copy. ¹³However, repairing a file can be done with an antivirus system. ¹⁴Because most viruses spread infections by adding code the repair process involves these steps. ¹⁵(1) Use the antivirus system to locate the virus code then remove it from the infected file. ¹⁶(2) After you complete the repair scan the file again to ensure that one virus was not hiding another. ¹⁷(3) Replace the repaired file if it doesn't perform in the same way that it did before the repair.

Adapted from "What's In a Virus." 1992. *Symantec*. 3.1 (Winter).

11 Punctuating within Sentences

Contents of this chapter:

- Abbreviations and Acronyms
- Apostrophes
- Colons
- Commas with Introductory and Interrupter Words and Phrases
- Commas and Semicolons in a Series
- Dashes, Commas, Parentheses, and Brackets
- Hyphens
- Quotation Marks and Ellipses

Many students think punctuation was invented by English teachers primarily to cause trouble for students. In fact, punctuation has evolved as the language has evolved. Those marks function like traffic signals, guiding a reader through a written piece smoothly and effortlessly. When you drive, you respond to traffic signals almost automatically—stopping, yielding, and slowing without much conscious thought. In the same way, good punctuation can guide a reader through a maze of words almost automatically, telling the reader when to stop, when to slow, when to speed up, and what words belong together.

Good editing clarifies writing by ensuring that punctuation marks meet the reader's subconscious expectations. In many ways, punctuation is conventional: Readers expect a mark to signal the end of an idea, to indicate a question, to separate the items in a series. If you edit to meet those expectations, readers can concentrate on the content, and in workplace writing, the content is the important consideration. In other ways, punctuation is flexible: Writers use punctuation marks to emphasize important material and to control the pace of the writing. In either case, writers and editors must know how punctuation is usually used (the so-called "rules") and what options are available. You already know most of the conventions or rules that govern punctuation, but this chapter will help you review them quickly and give you practice in applying them.

11.1 Abbreviations and Acronyms

An abbreviation is a shortened form of a word; like a contraction, an abbreviation is formed by omitting some of the letters of a word. In many abbreviations, a period replaces the omitted letters, but in special types of abbreviations, no periods are used. The trend in workplace writing is to limit the use of periods in abbreviations.

Common Abbreviations That Usually Require a Period

- A person's initials. In initials preceding a person's name, a space follows the period. Occasionally, we refer to a famous person by initials only, and no periods are used.

    L. C. Wu
    Dorothy K. Simpson
    FDR and JFK

- Designation of streets, avenues, and buildings in addresses.

    1734 Constitution Ave.
    Bldg. D

- A title preceding or following a person's name.

    Ms. Anne Trelawny
    Col. Avery Flushing
    Dr. William Blake
    Tyler Lange, Jr.

- An academic degree.

    B.S.   Ph.D.
    M.D.   M.B.A.

- A company name. Always use the abbreviation if it is part of the official company name; otherwise, use the abbreviated form only in addresses, lists, and bibliographies.

    General Service Co.
    Environmental Education Assn.
    Pyramid Supply, Inc.

- A date or a time.

    Oct. 12, 1999
    6:00 P.M.
    32 B.C. (or 32 B.C.E.)

- A part of a book in a cross reference, table, or figure.

    Ch. 24
    p. 76

- A country name when used as an adjective or a noun. However, the initials USA are usually used without periods.

    U.S. Dept. of Agriculture
    U.K. death duties

- Measurement notations that could be confused with words.

    in.   =   inch
    tan.   =   tangent
    at.wt. =   atomic weight

Abbreviations That Do Not Require a Period

- Measurements. Use the abbreviation with a unit of measure only, and only when it cannot be confused with an existing word.

    3000 kc   175 Hz   450 mm

- Postal abbreviations for states.

    NY   LA   AZ

- SI (Système International) units. These are international units of measurement.

m	=	meter	cd	=	candela
kg	=	kilogram	min	=	minute
s	=	second	d	=	day
A	=	ampere	h	=	hour
K	=	kelvin	L	=	liter
mol	=	mole			

- Acronyms and initialisms. An acronym is formed from the first letter or letters of several words. Acronyms are pronounced as words. Initialisms are also formed from the first letter of each word, but the letters are pronounced individually.

    Acronyms

radar	=	radio detecting and ranging
laser	=	light amplification by stimulated emission of radiation
scuba	=	self-contained underwater breathing apparatus
NATO	=	North Atlantic Treaty Organization
RAM	=	random access memory

    Initialisms

EST	=	Eastern Standard Time
UFW	=	United Farm Workers
MIT	=	Massachusetts Institute of Technology
NFL	=	National Football League

    The first time you use an acronym or initialism, put the identifying words first, then write the acronym in parentheses. Always do this unless you are sure that *all* your readers will recognize the shortened term.

EXERCISE 1

Abbreviations, Acronyms, and Initialisms

Using standard editing marks, edit the following sentences for errors in abbreviations, acronyms, and initialisms. Selected answers appear in the Appendix.

Insert period ⊙   Delete period ⌀
Insert space #   Delete space ⌒

1. Alternative fuels for energy are making a "surprisingly significant" contribution to US energy needs, according to R.S. Martin.

2. Martin, deputy asst. sec. for renewable energy at the US Energy Dept, notes the advantages of alternate fuels: little damage to the environment, safety, and limitless supplies.

3. However S Freedman of the Gas Research Institute, speaking at an energy conference at M.I.T., pointed out the relatively high cost and lack of proven large-scale capacity of alternative fuels.

4. At Luz International Ltd in the CA. Mohave Desert, a single solar plant produces more than 90 percent of all electricity drawn from the sun.

5. In 1981, it cost the first Luz solar plant 24 cents to produce 1 kilowatt hour (K.w.h.) of electricity. By 2000, experts anticipate costs will be down to 6 cents per Kwh.

6. Luz International, Ltd. has 365 acres of glass mirrors with a generating capacity of 200mw (megawatts or million watts).

7. The mirrors focus the sun's rays on oil-filled glass tubes, heating the oil to 735 deg and turning water into steam.

8. Another alternative energy source is wind. At U.S. Windpower, 23-ft.-tall wind turbines generate enough electricity to power 90,000 homes.

9. There are more than 17,000 wind turbines in the U.S.A. with a total capacity of 1,500 m.w., as much as a large nuclear plant.

10. The U.S. uses more than 83 quads of energy a year. (A quad is a unit of energy equalling 1,000,000 B.t.u.s—what it takes to raise one pound of water one degree.)

11.2 Apostrophes

The apostrophe is used for three reasons:

1. to show possession
2. to show the omission of letters or numbers
3. to prevent misreading of numbers or figures used as words.

To Show Possession

- Add *'s* to singular nouns and pronouns to show possession.

    the program's usability

    Mary White's memo

    nobody's fault

- Add *'s* to singular nouns that end in *s* to show possession. This is never wrong. If the pronunciation is awkward, however, you may use the apostrophe only.

    Ross's summary

    this class's schedule

    Yeats' biography

    The United States Government Printing Office (GPO) *Style Manual* requires only the apostrophe; follow your company style guide on this point or determine now what your personal style will be and note it on your style sheet.

- Add only the apostrophe to plural words ending in *s*.

    three members' reports

    all networks' programs

- Add *'s* to plurals that do not end in *s*.

    men's statistics

    women's responsibilities

    Personal pronouns are already possessive and do not need apostrophes.

    | hers | its | ours |
    | his | theirs | whose |
    | yours | | |

Remember to write *its* (with no apostrophe) to show possession.

> The briefcase lost its handle.

To Show Omission

Use the apostrophe to replace the missing letters or numbers in a contraction. Put the apostrophe where the letters or numbers would otherwise be.

she'll	=	she will	I'm	=	I am
he's	=	he is	'99	=	1999
o'clock	=	of the clock	it's	=	it is
can't	=	cannot	it's	=	it has

Remember to write *it's* (with apostrophe) for the contraction.

> It's too late for a meeting.
>
> It's been three years since the merger.

To Prevent Misreading

Use the apostrophe for clarity when you are using numbers, letters, or symbols as words.

> The chart omits all *0's* to the left of the decimal point.
>
> Technical people often use a crossbar on *7's* to avoid confusing them with *1's*.
>
> In my handwritten memo, he misread my *a's* for *e's*.

If there is no chance of confusion, the apostrophe can be eliminated.

EXERCISE 2

Apostrophes

Use editing marks to correct any apostrophe errors in the following sentences.

Insert apostrophe ⋎   Delete apostrophe ⌒

1. Cartographers greatest challenge is to map the round Earth on a flat surface.

2. More than 200 projections of Earth have been produced; four of it's well-known projections are Robinsons, Van der Grintens, Mallweides, and Mercators.

3. The four mapmaker's projections were produced over a period of 400 years' of cartography.

4. Arthur Robinsons' projection was produced in 1963, and in 88, the National Geographic Society began using the Robinson view for it's world map.

5. In Robinsons projection, Canada's, Greenlands, and the former Soviet Unions' shapes are fairly true to relative size.

6. Chicagos Alphons Van der Grinten patented his projection in 1904. Its biggest fault is that Greenland looks the same size as South America, although its about the size of Mexico.

7. All regions of the Earth are depicted in correct relative size in Mallweides 1805 map, but the higher latitudes' show compression, elongation, or warping.

8. The school childrens' map is still the Mercator projection, which was introduced in 1569.

9. Gerardus Mercator's map exaggerates the size of landmasse's in the high latitudes. For example, Alaska and Brazil look about the same size, while theyre vastly different: Brazil is six times larger.

10. Robinsons' work was done at the University of Wisconsin in Madison, where Robinson was a professor of cartography and geography. In World War II, Robinson directed the Office of Strategic Service's map division.

11. The projections based on an ellipse, with lines of longitude curving toward the pole's. However, the latitude lines are straight, evenly divided by the longitude lines, as on the globes' surface.

11.3 Colons

Professional writers often use colons because they can convey subtle shades of meaning to the reader. You can use a colon in four situations:

To Introduce a Formal List or Series, Especially One Including the Word following

The words preceding the colon must be an independent clause.

> A surge suppressor can perform all three of the following functions: suppress surges, suppress spikes, and suppress line noise.

However, you shouldn't use a colon to separate a verb from its object or complement or to separate a preposition from its object.

> NOT   The three criteria for evaluation are: technical quality, maintenance ease, and cost.
>
> BUT   The three criteria for evaluation are technical quality, maintenance ease, and cost.
>
> NOT   Surges and spikes can cause damage to: microwave ovens, video equipment, and computers.
>
> BUT   Surges and spikes can cause damage to microwave ovens, video equipment, and computers.

You *can* use a colon if the objects or complements are presented in a list, although you do not need to.

> Four skills needed by editors are:
> - organizational
> - analytical
> - interpersonal
> - communicative

To Introduce a Word Group or Clause That Summarizes or Explains the First Word Group

In this case, the colon replaces the words *that is* or *namely*.

> The intended readers are managers: people who are concerned with planning, budgeting, and allocation of resources and personnel.

If the clause following the colon is a complete sentence, the first word is usually capitalized.

> Milk is a good attractor of foreign flavors: It provides both a watery and fatty environment and therefore can dissolve a variety of substances.

To Introduce a Long or Formal Quotation

> The National Environmental Policy Act says it clearly: "An Environmental Impact Statement (EIS) is more than a public disclosure document. It should be used by federal officials in conjunction with other relevant material to plan actions and make decisions."

To Show Typographical Distinctions or Divisions

> Dear Ms. Wong:
>
> 7:30 A.M.
>
> "Ethics Case Study: The Boundaries of Marketing Integrity"

EXERCISE 3

Editing for Colons

> Using standard editing marks, insert colons in the following sentences if they are needed. Below the sentence, write the reason for inserting the colon. If no colons are needed, delete the colon and insert the correct mark, giving the reason.
>
> Insert colon

1. As the nation's central bank, the Federal Reserve System's current goal is the following to balance expanding economic growth against increasing pressures of inflation.

2. The Fed has: taken a leadership role and implemented new methods to control the economy.

3. Congress established the Federal Reserve System after the Panic of 1907 to create a pool of money a reserve from which banks could borrow in a crisis.

4. By 1913, the central bank was in place: all member banks were required to make deposits to the system.

5. Soon the directors of the Fed learned that the reserves could be used: to control banking activity and to influence the business cycle.

6. This is how it works: When the Fed wants to give the economy a boost, it purchases U.S. government securities in the open market.

7. The sellers of those securities begin the following cycle they deposit their proceeds in banks, increasing bank reserves. The banks can then lend more money, which stimulates demand and raises prices.

8. In contrast, the Fed can restrain the economy by selling government securities on the open market. Then the following events occur buyers withdraw funds, and reserves fall; interest rates go up; consumer demand for loans decreases.

9. Alan Greenspan became the chairman of the Fed in 1987: all seven members of the Board of Governors serve 14-year terms.

10. The seven members are: a majority of the 12-member Federal Open Market Committee (FOMC), which directs purchases and sales of U.S. government securities in the open market.

11.4 Commas with Introductory and Interrupter Words and Phrases

Commas are used to set off introductory or interrupter words and phrases from the rest of the sentence. A phrase is a group of related words.

Introductory Words

- Nouns of address.

    Ms. Andreas, do you have information on that product?

- Transitional words that break the continuity of the writing.

    The technical review and copyedit review will be concurrent. Therefore, the writer will need only one additional rewrite.

- Conversational openers like *yes*, *no*, *now*, *well*, or *why*.

    Yes, we revised that document last week.

Introductory Phrases

- Prepositional phrases (especially those of three words or more).

    For the final presentation, he designed 20 new overheads.

- Short prepositional phrases that could cause misreading if unpunctuated.

    For writers, the system has been reconfigured.

- Verbal phrases (made from verbs, but not acting as verbs).

    To open the window, click on the programs icon.

    Blocking the text, she moved it into the new document.

- Long modifiers out of usual order.

    Less hazardous than acetone, the new solvent is now widely used.

Interrupter Words and Phrases

- Direct address.

    Attend the meeting, Carolyn, if you can.

- Appositives (a second noun that amplifies the meaning of the first noun).

    Phuong Le, marketing manager, is attending a conference in Atlanta.

- Parenthetical elements.

    The RESET command, like the ABORT command, cancels the operation in process.

- Contrasting statements.

    The temperature inversion, not the humidity, caused the smog.

- Transitional expressions like *for example, especially,* and *however*.

    The disk controller can miss a seek, however, during peak CPU loads.

EXERCISE 4

Introductory and Interrupter Words and Phrases

Using standard editing marks, edit the following sentences for comma errors.

Insert comma ⋀   Delete comma ⌒

1. Arbitration a method of settling labor disputes is a substitute for costly and time-consuming litigation.

2. Long used in commercial disputes arbitration became common in government employment in the 1960s.

3. In fact until Executive Order 10988 union recognition and collective bargaining were not found in the federal service.

4. Four Supreme Court cases according to Donald Woolf effectively made arbitration awards in the private sector nonappealable.

5. On rare occasions courts have returned cases to an arbitrator for reconsideration.

6. Most labor disputes in government and in the private sector are brought by employees against management.

7. To bring a dispute to an arbitrator the union must "make a case" in other words show probable cause that the contract has been violated.

8. For the union charge to be valid the case must be documented.

9. The documentation must include names, dates, times, events, and witnesses. In addition the union must show specific violation of the agreement and the specific relief sought.

10. Management's defense Woolf says is to attack the credibility, logic, or consistency of the union case and to offer rebuttal.

11.5 Commas and Semicolons in a Series

In a series of items, commas and semicolons serve both to separate the items and to show the reader how the items are related to one another.

Commas

Commas are used to separate three or more individual items in a series. Series items include the following:

- Dates.

    The convention is scheduled for October 14, 2002, in New York City.

- Places.

    The distribution facility is at 3011 N. First Street, Minneapolis, Minnesota.

- Sentence elements.

    The XX command reads, writes, and edits the file.

Always use the comma before the conjunction that precedes the last item. This comma is never wrong, and it may prevent confusion.

The diagnostic identifies the failing drive, file name and type, and address.

Semicolons

Semicolons can be used to separate items in a series if the sentence also contains commas for other purposes. The semicolons are stronger marks of punctuation than commas.

The county-wide commission had 10 members: Brockton City Council, 3; Summerville City Council, 3; County Board of Supervisors, 4.

EXERCISE 5

Commas and Semicolons in Series

Use editing marks to correct any errors in the following sentences.

Insert comma ⋏   Delete comma ⌒
Insert semicolon ⋏̇   Delete semicolon ⌒̇

1. The Internet can connect a user to electronic mail, news or libraries of information.

2. A PC can connect to the Internet by telephone, or by direct wire connections to nearby computers.

3. A user needs a personal computer, (PC) and a modem or connection to a local area network or LAN, and a communications program.

4. A user gets an account through a system administrator a local computer expert or a dial-up service.

5. The account identifies the user to the computer system in two parts: the username, userid or loginname and the password.

6. A password usually consists of eight characters made up of upper, and lower case letters symbols and numbers.

7. The terms *loginname userid* and *username* are all common names for the user's identity.

8. For dial-up services, the procedure is as follows: start up the communications program; turn on the modem's speaker, have the program dial the access number of the Internet host; listen to the dial tone, program and modem tones; listen to the whistles and hisses that indicate "handshaking."

9. With communication established, you can log in by typing your loginname, and pressing the ENTER key.

10. One book that explains the Internet is by David Sachs and Henry Stair. It was published in 1994, by Prentice Hall Englewood Cliffs New Jersey.

11.6 Dashes, Commas, Parentheses, and Brackets

Dashes, commas, parentheses, and brackets are all used to set off inserted words from the rest of the sentence. Dashes give the strongest emphasis to the words they surround; commas give equal emphasis; and parentheses de-emphasize their enclosed words. Brackets are used in a slightly different way; they are conventionally used within sentences to indicate words that are marked off because they are not part of the sentence itself.

A writer can subtly influence the reader's interpretation of a sentence by choosing dashes, commas, or parentheses to surround the inserted words. However, a writer can also use dashes when other marks of punctuation would be more accurate. For example, a dash is sometimes used to connect a sentence or clause that expands on the first sentence. If you have studied the use of colons earlier in this chapter, you know that a colon is a better choice. In editing, you need to carefully evaluate the effect of the sentence before suggesting revision, and you should be able to justify any suggested changes.

What is usually called the dash is more specifically known as the *em dash*. The name comes from the width of the em dash—about the same width as the capital M. If the em dash is not available, it can be shown by two hyphens. No spaces are used before or after the dash.

Use these marks for the following reasons:

To Indicate Parenthetical Information

Use dashes, commas, or parentheses to set off parenthetical information depending on the desired effect.

- For the *strongest* emphasis on the parenthetical information, use dashes.

    Peripheral devices—monitors, printers, and keyboards—are controlled by the central processor.

- For *equal emphasis* between the parenthetical information and the sentence, use commas.

    The parallel port, located on the text monitor adapter card, enables the computer to be connected to the largest number of text and graphics printers.

- To *reduce* the importance of the parenthetical information, use parentheses.

    The controller card (see Figure 3) interfaces with the processor over the B6 bus.

To Show a Break in Thought, Emphatic Idea, or Summary

Use a dash to show a break in thought, to emphasize an idea, or to summarize.

There is always a variety—sometimes a bewildering variety—of ways a thing may be said.

Science experiments can be simulated on a computer—artificially tested inside the computer using a mathematical model of actual phenomena.

For Clarity When Commas Are Already Used in the Sentence

Use dashes or parentheses to help the reader understand which words are "set off" and which words are part of a series.

> A unit of measurement—ft, rpm, cc—may be abbreviated when it follows a numeral.

To Set Off Material Not Integral to the Sentence or Not Part of a Quotation

Use brackets to show that words have been inserted that are not part of the regular text.

> This chapter explains how microcomputer [desktop] graphics fit into the larger family of computer graphics on minicomputers and mainframes.

> We might amend George Summey's statement to include editors: "Skillful writers [and editors] have learned that they must make alert and successful choices between periods and semicolons, semicolons and commas, and commas and dashes, dashes and parentheses, according to meaning and intended emphasis."

EXERCISE 6

Editing for Dashes, Commas, Parentheses, and Brackets

If necessary, insert or delete dashes in the following sentences or change inappropriate dashes to another punctuation mark. Below each sentence explain why you took the action you did.

Insert dash ___ Insert comma ___ Insert semicolon ___
Insert colon ___

1. The candidate company for the merger is the Ocean Containers Group—consisting of Ocean Containers, Inc. and Oceanco.

2. The Group is in the business of leasing containers, chassis, and cranes—it also charters container ships.

3. Ocean Containers owns 200,000 cargo containers (it specializes in refrigerated containers) and 25 container ships.

4. The company is also a manufacturing firm—owning its own facilities for refrigerated containers, dry cargo containers, and specialized containers.

5. These facilities—in England, Chicago, and Singapore—enable the company to market its products easily to more than 70 countries.

6. Leases are made to a variety of transportation entities—ocean carriers, exporters, forwarders, truck lines, railroads, and terminal operators.

7. In addition, Ocean Containers has operations not related to the transportation industry—hotels, a train service, a large condominium complex, and undeveloped real estate.

8. The undeveloped real estate—300 acres in Colorado—may be difficult to sell.

11.7 Hyphens

Hyphenation use varies widely among scientific, business, and technical publications. For this reason, the best source of information for hyphen use is the style guide of the company or profession for which you work. Until you have access to such a style guide, however, you can use the guidelines below, which have been adapted from the United States Government Printing Office *Style Guide*.

Use hyphens in the following situations:

To Indicate Syllable Breaks

The most common use of hyphens is to break a word at the end of a line. Many word processing programs either do this automatically or wrap the whole word to the next line. When you edit, however, remember to do the following:

- Hyphenate between syllables only. If you don't know where the break is, look in the dictionary.
- Do not divide one-syllable words.
- Do not hyphenate names, contractions, abbreviations, acronyms, or numerals.
- Do not hyphenate the last word on a page or the last word in a paragraph.

In Compound Words

Hyphenation of compound words varies widely. Often, a compound word begins as two separate words, then, over time, is hyphenated, and finally becomes a single word. You need to:

- Be consistent. Choose a single dictionary, handbook, or company style guide and follow it faithfully. Set up your own style sheet for hyphens, if necessary, so that you always hyphenate the same words in the same way.
- Keep your reader in mind. Hyphens are supposed to make the reader's task easier. Sometimes you must use hyphens simply to avoid confusion. For example:

    an un-ionized compound (not unionized)

    re-cover it (make a new cover, not recover)

    de-emphasize (not deemphasize)

    re-solve (not resolve. Even better, say solve again.)

In Compound Words Used as Adjectives

Use hyphens in the following compound words when they are used as adjectives *preceding* nouns:

- Numbers from twenty-one to ninety-nine when they are written as words.

    Twenty-seven trucks were at the loading terminal.

- A number plus a unit of measure used as an adjective.

    12-minute interval

    8-hour delay

- Spelled-out fractions used as adjectives.

    one-half circumference

    seven-eighths yard

- An adjective or noun plus a past or present participle used as an adjective.

    heat-activated sensor

    text-editing software

- Do not hyphenate compound words that follow a linking verb.

    The sensor was heat activated.

    The amoeba were free swimming.

- An adjective plus a noun used as an adjective.

    short-term memory

    high-performance workstations

- Two colors used as an adjective preceding the noun.

    yellow-brown residue

- Words with prefixes, if the root is a proper name or a number.

    pre-1980

    non-English

In Compound Words Used as Nouns

Use hyphens in the following compound words used as nouns:

- All compound nouns beginning with *self*.

    a self-evaluation

- A compound made of two nouns of equal value.

    writer-editor

    motor-generator

- Numbers from twenty-one to ninety-nine when written out.

    eighty-six

    Twenty-five was the enrollment limit for the seminar.

Do Not Use Hyphens in the Following Cases

- With most prefixes and suffixes.

    semiautomatic

    transdermal

- With names of chemicals, even when used as adjectives.

    hydrogen peroxide solution

- With an adverb ending in *ly* plus an adjective or participle.

    a highly complex program

    rapidly changing demographics

- If the individual words modify the noun separately.

    a new digital analyzer

EXERCISE 7

Editing for Hyphens

Insert or delete hyphens in the following sentences if necessary. In the space below each sentence, write the reason for your action.

Insert hyphen ⚏ Delete hyphen ⌒

1. As U.S. companies do increasing business with international clients, writers must consider second language readers of documentation.

2. One aid to internationalization is machine-translated text, but such text must still be edited for accuracy by expert native speakers.

3. Management is concerned that an overly-optimistic forecast is contained in the program planning guide.

4. The testing unit is self contained, but it needs a well designed outer shell.

5. Thirty five contributors are listed in the program notes.

6. So called "lite" salt contains sodium-chloride crystals as well as potassium chloride.

7. No Salt®, on the other hand, contains potassium-chloride crystals and potassium bitartrate.

8. Because the draft was double spaced, the final copy (on single spaced pages) had to be reformatted.

9. Use of the blood-pressure gauge was explained in a five page brochure.

10. The style guide is currently in a three ring binder, but the department plans to move it to an online softcopy format.

EXERCISE 8

Editing for Hyphens

Insert or delete hyphens in the following sentences if necessary. In the space below each sentence, write the reason for your action.

Insert hyphen ⍿   Delete hyphen ⌒

1. Cross matched units of blood were made available in case of hemorrhage.

2. The patient was monitored for seventy two hours after the surgery.

3. A star's red giant phase lasts until the hydrogen in the layer around the core is exhausted.

4. The battery operated models have both improved case geometry and sophisticated electronics.

5. It is a little known fact that the Earth's geographical north pole must be, in reality, a south magnetic pole.

6. The company owns four carts that are battery operated and four carts that are gasoline powered.

7. The waves divide right and left around the pilings and then reunite.

8. The blue green water was very clear to a forty foot depth.

9. Air purifying devices use a mechanical filter to remove particulate matter such as dust or fumes.

10. Closely-monitored results involve high technician costs.

11. Carbon tetrachloride fumes can be tested by organic vapor cartridges.

EXERCISE 9

Putting It All Together: Editing For Colons, Hyphens, Dashes, and Parentheses

Punctuate the following sentences with colons, hyphens, dashes, or parentheses as required. Write below each sentence why you made the choices you did.

1. To at least some degree whether a company's base be in digital circuits software medical research petrochemical processing or aerospace development management expects the technical communicator to function also in part as a product engineer.

2. A product engineer's primary responsibility is to analyze and verify the design data in the following areas mechanical drawings schematics specifications and calibration procedures.

3. As a writer editor you can assume varying degrees of the product engineering function within your company.

4. When you can go to the lab and verify a test and inspection requirement and change it as necessary you are functioning as more than an editor.

Adapted from Richard McClain

5. In the first years of the microcomputer industry software was produced with minimal instructions called documentation for a technically oriented audience.

6. For those in policy making positions graphic aids can draw attention to important information statistics comparisons and recommendations for example that might otherwise be overlooked.

7. However some documents such as a weekly field report from a swimming pool salesperson seldom if ever require graphic aids.

8. To make an intelligent decision what to exclude and what to include in a graphic aid you must understand your reader's knowledge and limitations.

9. In the design of a page using graphics the reference containing the page should immediately precede the graphic itself.

10. A reduction related problem in graphics use is the following if an existing graphic is reduced to fit on one page the letters and numbers may be too small to read easily.

11.8 Quotation Marks and Ellipses

Quotation marks and ellipses must be used accurately in text because readers depend on them for clarity. Quotation marks differentiate direct statements, technical terms, or parts of a larger work from the surrounding text. Ellipses indicate where cuts have been made in quoted text. Use quotation marks for the following purposes:

Setting Off Someone's Exact Words

- Use quotation marks to set off someone's exact words.

    Dan Shafer, an expert-system consultant, says, "Most companies are using expert systems, but the use has been pretty invisible."

- Use a comma or (in formal writing) a colon after the identifier.

    Shafer's report concluded: "Expert systems ensure consistency in support, allowing you to move customer support from engineers to people with less product knowledge."

- Use commas before and after the identifier if it divides a complete sentence.

    "The systems have their limits," Shafer noted, "because they aren't suited for reasoning or analysis that relies on subjectivity, emotion, or common sense."

- Place commas and periods inside end quotation marks; place semicolons and colons outside. Question marks go inside or outside depending on whether the quotation is a question or is part of a question.

An "expert system," also called a knowledge-based system, is a computer program containing the knowledge of an expert in some field.

Expert systems are the most common way to dispense "artificial intelligence."

At one company, an expert system "reduced the percentage of time spent on customer calls to the senior engineer . . . to 2 percent"; before the system was installed, the engineer spent 30 percent of his time on such calls.

Consultant Avron Barr asks, "What is the potential of such systems?"

In other words, how can companies take better advantage of the computer as an "expert system"?

Identifying New Technical Terms or Words Discussed as Words

- Use quotation marks to call attention to new technical terms or words that are discussed as words.

    The source information is obtained by an expert often called a "knowledge engineer."

- Sometimes such words are printed in italics instead.

    The source information is obtained by an expert often called a *knowledge engineer*.

Identifying Parts of a Larger Published Work

- Use quotation marks to set off parts of a larger work, such as a chapter or section of a book.

    One section in the style guide is titled "Preferred Terminology."

- Use quotation marks to distinguish article titles in reference lists in some citation systems. See Chapter 17 for details and examples.

Using Single Quotation Marks or No Quotation Marks

- Enclose a quotation within a quotation with single quotation marks.

According to Smith, "Continuous-wave CO_2 lasers are excellent for cutting operations. Because there is no physical contact between the 'cutting tool,' the beam, and the workpiece, the contamination of the workpiece is reduced or eliminated."

- Do not use quotation marks around indirect quotations.

  He stated that the cause was unknown.

Using Ellipses and Quotation Marks

- Surround quoted text with quotation marks if the quote is incorporated into the paragraph. Do not use quotation marks if the quote is long and set off separately from the paragraph.
- Use ellipses to show where words have been omitted in quoted text. An ellipsis is a series of three spaced periods inserted in quoted text to show that words have been omitted from the quotation. If the ellipsis appears in the middle of a sentence, it consists of three spaced periods; if it appears at the end of a sentence, it consists of four spaced periods—the fourth period for the end of the sentence. For example:

  According to Carol Gerich of the Lawrence Livermore Laboratory, "The author and supervisor in this case study . . . asserted that editors perform a valuable function by increasing readability. Editor B says editing can do even more. . . . 'Surface editing gets rid of the distractions, while substantive changes can enhance the author's reputation.'"

See a general style guide like *The Chicago Manual of Style* for more detail.

EXERCISE 10

Cutting and Punctuating Text for a Quotation

Study the following portion of an unpublished paper by Thripura S. Naidu-P titled "Localization." This document has about 1,200 words. Assume that you need to cut the information to about 300 words to be quoted in a document you are composing on "Writing for the International Marketplace." Make the cuts, using ellipses as needed and quotation marks as required. Introduce the quote by crediting the author in an appropriate way. Write out a clean copy of the new version and be prepared to discuss why and how you made the cuts you did.

A recent trend in technical writing, whether it is a published manual or online documentation, is localization. Many American "high-tech" companies find that marketing their products in foreign markets is the best way to counter the decreased profits from the U.S. market. To market their products effectively, American companies have found that they cannot assume the rest of the world is going to learn American English and culture: American companies have to localize their products to each foreign market. . . .

There are several assumptions which have been made about localization. One of these is that localization is the translation of documentation from the source language to the target language, with some thought given to the changed cultural contexts. An example of such a situation is the localization of an American document into French. According to the *Guide to Macintosh Localization* (1992, 150) by Apple Computer, Inc. (which will be referred to as Apple's *Localization Guide* in the rest of this paper), the following countries use French:

> Algeria, Andorra, Belgium, Benin, Burkina Faso, Burundi, Cambodia, Cameroon, Canada (Quebec), Central African Republic, Chad, Comoro Islands, Congo, Djibouti, France, French Guiana, French Polynesia, Gabon, Guadeloupe, Guinea, Haiti, Ivory Coast, Laos, Lebanon, Luxembourg, Madagascar, Mali, Martinique, Mauritania, Mauritius, Mayotte, Monaco, Morocco, New Caledonia, Niger, Reunion, Rwanda, Senegal, St. Pierre and Miquelon, Switzerland, Syria, Togo, Tunisia, Vanuatu, Vietnam, Wallis and Futuna Islands, Zaire.

For each of these countries, there is a different social and cultural context, which will alter the use of the language. Information about these contexts can be found in texts about these countries. However, this information, more often than not, specifies legal requirements. The common practices of the country may not be specified.

The localization specialists at ASK Computers take into consideration the matter of common practices when they localize business and management system software and documentation for their various European customers. Since the European Union (or EU) is trying to standardize financial and legal procedures, the assumption American companies might make is that there is only one standard to follow. This is not always the case.

France, for example, has a financial practice which is not common to all members of the EU, or to all countries which speak French. This practice is the issue of promissory notes. In France, a promissory note can be issued by Company A to Company B. If Company B has to make a payment to Company C, then Company A's promissory note can be handed over to Company C. Company B does not have to cash the note to pay creditors; the promissory note can simply be passed along to them. Should the original issuer of the note default on the payment, then whoever holds the note last will not get paid.

A technical writer cannot assume that the published social and cultural information about a country is complete. Localization requires knowledge of common practices in a country, along with social and cultural information about the country, and the language of the country. To do an accurate localization of a document, a writer has to be prepared to make extensive inquiries, and do extensive research about the common practices in a country.

One assumption which the introduction of this paper implies, and which should be clarified here, is that American localization specialists work to localize American products for the foreign market. This is not always the case, as trade is a two-way process. American companies have bought software, for example, from other countries, and localization specialists had to localize these products for American use.

The localization of foreign software and documentation can be simple, at times. ASK Computers, for example, bought a business and manufacturing systems software package from Great Britain. Most of the localization in the documentation was in terms of vocabulary and contextual and cultural references. The financial module of the package, however, was localized by the software specialists because the tax structure for Great Britain is different from that of the U.S.

At other times, localization of foreign software can be extensive, especially when non-Roman characters are involved. Languages like Arabic, Armenian, Burmese, Chinese, Ethiopian, Hindi, Japanese, Tamil, and Urdu are examples of languages which have non-Roman characters. Chinese, Japanese, and Korean are non-Roman languages which require a 2-byte character set.

Another consideration is the syntactic structure of a language. English, for example, has the **S-V-O/C** structure, while Japanese has the **S-O/C-V** structure (where **S** = subject, **V** = verb, **O** = object, **/** = or, **C** = complement). To allow for the translation of software, the system software has three tools to generate the syntactic structure of any particular language. These tools are grammar, a huge dictionary, and a semantic knowledge base.

Related to the preceding assumption is yet another assumption which a writer might make about localization: that the product to be localized has an international base component. The international base component is that "... part of a product which is sold worldwide without modification" (*Digital Guide to Developing International Software* by Digital Equipment Corporation, which will be referred to as DEC'S *Digital Guide* in the rest of this paper, 1991, 6), and which has been internationalized so that localization can be easily implemented. DEC's *Digital Guide* defines "internationalization" as the "process that includes both the development of an international product and the localization of the international product for delivery into worldwide markets" (1991, 360). A product does not have to have an international base component, because it is not always the case that a company planned to market a product internationally. The business and management systems software that ASK Computers purchased from Great Britain did not have an international base component—it only had a base component.

A technical writer has to be prepared to localize products from foreign countries into American Standard English. The localization might be a simple process, or it might be an extensive process.

Another assumption which a technical writer might make is that the target language uses the same version of the wordprocessing software as the source language. The American language version of a wordprocessing software may be two or three versions ahead (as in WordPerfect® 6.0 to WordPerfect® 5.0) of the Japanese language version. The fact that there is a market for software upgrades in the U.S. is an indication that Americans like to use the most current version of a software. Any document which is written in the latest version of any software cannot be accessed by an older version of that software. When an American writer sends a document to a foreign country to be localized, the localization specialists of the target country may have difficulty accessing the document, because they have the older version of the software. This scenario was provided by Norbert Lindenberg, an internationalization consultant and software development specialist at Apple Computer, Inc.

A technical writer in the U.S. should not only have a library of the different versions of any software, but should also make inquiries as to which version the target company in the foreign country is using before writing an internationalized document. Such inquiries will demonstrate courtesy to the target company—courtesy that will be appreciated and not forgotten. . . .

EXERCISE 11

Quotation Marks

Using standard editing marks, edit the following sentences for the punctuation associated with quotation marks.

Insert comma ⋀ Delete comma ⊘ Transpose ∽
Insert period ⊙ Delete period ⊘ Insert quotation mark ∨

1. The following is an excerpt from a larger report titled Laser Cutting of Sheet Metals."

2. Hudson reported The Q-switched Nd:YAG laser offers high-peak powers for cutting thin pieces of metal.

3. "Because the output wavelength of the Nd:YAG laser is 1,064nm, Hudson continued, it can produce narrower widths than the CO_2 laser, which operates at 10,600nm."

4. He stated, however, that the CO_2 laser can cut through thicker metals than the Nd:YAG laser.

5. According to Hudson The cutting efficiency of a laser can be improved by applying a jet of oxygen through a nozzle located coaxial with the beam."

6. It is the oxygen said Hudson that actually makes the cut, while the laser sustains the cutting reaction.

7. "When used with a reactive gas such as oxygen, both CO_2 and Nd:YAG lasers can produce burr-free cuts with a minimum heat-affected zone"

8. The next section of the report is titled How the Process Works;" in this section, details are given about placement, controls, fastening systems, and cutting activities.

9. "The Computer Numerical Control (CNC) unit controls the table's movement, the type of cut, and the depth, width, and angle of the cut Hudson explained.

10. He went on, Lasers offer many advantages over conventional sheet-metal cutting techniques: the cutting edge never dulls or breaks; "heat-sensitive areas are not affected; cutting beams are easily manipulated and can be computer controlled.

EXERCISE 12

Using Punctuation Skills: A Comprehensive Exercise

Punctuate the following two paragraphs correctly. Then number each punctuation mark in order. On a separate page, list the corresponding number and explain why you used the punctuation mark you did.

The major step in moving a complex electronic product from the design stage to hardware involves getting the boards built. Computer aided design CAD can speed this task by producing accurate circuit board designs from your engineering groups specifications. To benefit from CADs potential however you must acquire a thorough understanding of the CAD process and what it can and cant do. And if you choose to use a CAD service shop rather than acquire a CAD facility you must be able to establish a successful relationship with that shop.

    CAD offers numerous advantages many shops provide a full line of computer based aids such as numerical control NC drill tapes as well as circuit test and logic simulation data. In addition to faster turnaround of board designs CAD provides the following improved line quality and tighter trace spacings improved board space utilization improved artwork durability and better documentation quality.

EXERCISE 13

Using Punctuation Skills: A Comprehensive Exercise

Punctuate the following two paragraphs correctly. Then number each punctuation mark in order. On a separate page, list the corresponding number and explain why you used the punctuation mark you did.

Discrete rotary switches might be a mature technology but they still provide the best way of controlling multicircuit functions. Although they are mature they've adapted to changing design needs in addition to the familiar bulky open frame designs a new generation of rotary switches has developed in response to the microprocessors needs. These new units are much smaller but they sacrifice nothing in performance or control capability. Switches with as many as 12 decks in packages measuring less than 1 in in diameter can replace older single or double deck switches in 2 in diameter units.

The newest rotary switches also provide simpler design and construction thus they're more compatible with automated manufacturing techniques and are less labor intensive than their predecessors. Many come in enclosed or sealed versions that can be mounted on pc boards and wave soldered along with other components which eliminates costly hand soldering operations. Furthermore coded output rotaries are now available for use in logic level applications where they convert decimal inputs into computer codes.

Reference

Thripura S. Naidu-P. 1993. Localization. San Jose State University. Unpublished paper. Reprinted by permission.

12 Choosing the Right Words

Contents of this chapter:

- Ensuring Accurate Terminology
- Choosing Concrete and Denotative Language
- Avoiding Euphemistic and Sexist Terms
- Replacing Pompous Words
- Choosing Understandable and Usable Terms

When editing, you need to pay special attention to individual words and the effect those words have on readers. Words are, after all, the basic building blocks of text. If the words are inaccurate or unclear, the text is likely to be confusing and unusable. These days *usability* is a key word, especially in instruction and manual writing. Martha Cover is manager of a technical writing department at a company that produces automation-design software. Cover defines editing as "reviewing, correcting, and critiquing for usability. We're not making a document pretty," she says. "We're making it usable."

At the word level, editors are concerned with both word choice and word usage. Another term for word choice is *diction,* defined as the choice of words and the force and accuracy with which they are used. *Usage,* on the other hand, is defined as the customary manner in which words are used. Thus, usage can change over the course of time or from one locale to another in which the same language is spoken.

Experienced writers and editors—and even top administrators—are well aware of the importance of word choice and usage. Sherri Sotnick, who edits the monthly journal of a networking company, says, "When I was a student, I wish I'd had more exercises on word choice, such as when to use *enable, allow,* or *permit.*" Jo Levy, freelance editor, notes that when writing for an international audience you need "simplicity of writing with no idioms or Americanisms like 'stepping up to bat.'" And Andrew Grove, chief executive officer of Intel Corp., the major manufacturer of silicon chips, said:

> Using complex terms, in my view, is a symptom of one of two things. Either the person using them intends to make himself [sic] sound more knowledgeable and sophisticated than he is, or he is too lazy to work on removing terms which may have become shorthand expressions among members of the profession, but have no meaning to outsiders.
>
> Either reason is wrong. The more knowledgeable any one of us is in a given field, the more we should be able to explain our thoughts in the language of our audience. In fact, I tend to view the simplicity by which an expert expresses himself as an indicator of both the depth of his knowledge and his self-confidence (1985, 9E).

When you write, you will compose cleaner sentences if you remember that you are the expert—writing, almost always, to a reader who knows less about the subject than you do. Your responsibility as the expert is to convey your information as clearly as possible, always keeping in mind the following:

1. **Your specific purpose.** Ask yourself: "What am I trying to accomplish with this? What action do I want?"

2. **Your particular readers.** Ask yourself: "How much does she already know? Why is he reading this? What will she do as a result of what I say here?"

When you edit, you can check for diction and usage in the several ways this chapter explains. Take the advice of senior technical writer Christopher Salander. Salander works at a company producing software for computer-aided engineering. "Editors," says Salander, "can't just skim. They need the patience and diligence to look at *every* word on *every* page."

12.1 Ensuring Accurate Terminology

One way to achieve clarity in diction is to ensure that the chosen word exactly conveys the writer's intended meaning. Many English words in general use have close counterparts in spelling, sound, or meaning: words that are spelled almost the same, sound almost the same, or mean almost (but not quite) the same thing. An editor's job is to help the writer by securing exactly the right word for each occasion.

The more you read, both in your specific field and in general literature, the more aware you will be of the fine distinctions in word choice. But you can also take specific steps to sharpen your diction skills:

1. Look up every word of which you're unsure. If you use an online thesaurus, check the exact meaning of every word offered as a synonym.
2. Choose one word for one action or object and use it consistently. Aim for clarity, not variety.
3. Question everything. Is it *different than* or *different from*? Should the sentence use *normally* or *typically*? Should it be *used for* or *used with*? Is the correct term *insure, ensure,* or *assure*?
4. Read in your technical field and learn the language used by writers to describe products, services, and actions. Learn what technical dictionaries and acronym-abbreviation source books apply to your field. Use them for reference.
5. Rely on subject-matter experts (SMEs) in your field to clarify meanings of technical terms. Remember that the same word can mean something slightly different in two different fields. For example, consider the variety of meanings of these words:

    case   register   default   bus

EXERCISE 1

Discriminating between Similar Words

To sharpen your understanding of the following similar words, look up each word in the dictionary. Do not simply copy the definition. Instead, write a sentence that explains the difference *between the two words. Then use each word correctly in a sentence.*

Note: *In an editing class, this exercise is best done in collaboration. Each student should be responsible for two or three word pairs, researching the differences in meaning to explain to the class. Keep this list for reference on the job.*

1. adjacent/contiguous
2. advise/inform
3. affect/effect
4. alternative/choice
5. among/between
6. amount/number
7. anxious/eager
8. anticipate/expect
9. appreciate/understand
10. appraise/apprise
11. assure/insure/ensure
12. bimonthly/semimonthly
13. capital/capitol
14. censor/censure
15. complement/compliment
16. continual/continuous
17. credible/creditable/credulous
18. depreciate/deprecate
19. deteriorate/degenerate
20. discreet/discrete
21. disinterested/uninterested
22. dispense/disburse
23. disprove/disapprove
24. economical/economic
25. fewer/less
26. filtrate/filter
27. flair/flare
28. hoard/horde
29. homogenous/homogeneous
30. indite/indict
31. inequity/iniquity
32. infer/imply
33. its/it's
34. liable/likely
35. lose/loose
36. mantle/mantel
37. maximum/optimum
38. oppose/appose
39. oral/verbal
40. parameter/perimeter
41. peddle/pedal
42. perfect/unique
43. phase/faze
44. predominate/predominant
45. principal/principle
46. reaction/opinion
47. respectfully/respectively
48. site/cite/sight
49. subsequent/consequent
50. think/feel
51. verbiage/wording
52. universally/generally

EXERCISE 2

Checking for Accurate Diction

Edit the following sentences to change any inaccurate words or terms. Check your dictionary to be sure of the meanings. Justify your changes below each sentence. Selected answers appear in the Appendix.

insert words 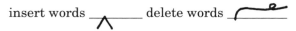 delete words

1. Work assignments outside the noted shift hours will be coordinated on an individual basis with a supervisor.

2. Harrison County is growing substantially less in job creation than the country overall; the number of jobs has increased only fractionally in the last year.

3. Are freedom today is lessening as the price of gas becomes more expensive.

4. There are millions of microcomputers in use today in both the home and the workplace. Of these, perhaps as little as 10 percent are protected by electrical stabilizers known as surge protectors.

5. Monitor the record levels and tone adjustments to make sure they stay within exceptable bounds.

6. In recent years, a vocal band of patients has focused public attention on the plight of blepharospasm and peaked the curiosity of researchers.

7. Less than 100 students were eating lunch in the dining room at the time of the explosion.

8. If you can disapprove the allegation of misuse of funds and site evidence of need, the ruling could be reversed.

9. She assumed the mantel of leadership at XYZ Corporation, and her consequent actions proved she was a good choice.

10. Because the building has degenerated, the consultant will need to apprise it's value and assure management of its worth.

12.2 Choosing Concrete and Denotative Language

Editors can help writers focus on specifics by substituting concrete and denotative words for those that are abstract, general, or connotative.

Concrete and Abstract Words

To choose a concrete term, you need to visualize a word's place on a "ladder of abstraction." The bottom rung on a ladder of abstraction will hold the most specific term, which might be a model name or a part number. Thus, the bottom rung has the most *concrete* word: a specific place, object, action, or person. Concrete words help the reader understand by creating pictures in the reader's mind. For example, in writing about trees, the concrete term *Pinus palustris* is the scientific name for the longleaf pine tree. A specialist reader could picture this tree exactly. A nonspecialist, however, would need the common term *longleaf pine tree.*

With each level that you move up the ladder, the term becomes less concrete and more general, until you reach a rung where the term becomes too inclusive to be visualized at all. Now you have a word labeled *abstract:* a general condition, quality, concept, or act. Let us put *Pinus palustris* on a simple ladder of abstraction as shown in Figure 12.1.

As you move up the ladder, the words become more general and less specific. When you read the word *vegetation,* can you picture anything? What do you picture? Does it bear any resemblance to the longleaf pine tree? At what level on this ladder do you stop visualizing an object?

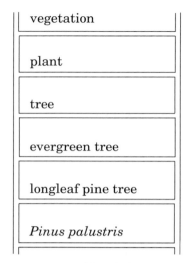

FIGURE 12.1 Ladder of Abstraction

Objects (named by nouns) are not the only words that can be placed on a ladder of abstraction. Verbs (action words) can also be concrete or abstract.

Consider the verb *end*. What visual image do you have of *end?* Do you see a manager concluding a meeting? A writer completing a document? An exterminator killing a rat? Or do you see someone finishing a session at the computer? *End* could be the chosen verb for each action, but you'd help your reader by choosing a verb specific to that situation. Furthermore, *end* is also a noun, and it has as many as 17 more meanings as a noun. When you use an abstract term like *end,* you force readers to supply a more specific term in their minds. For translators and second-language speakers, abstract or general terms cause particular difficulties.

However, when editing, you can't always insist on words at the bottom of the abstraction ladder. As Professor Carolyn Matalene notes, "Good writing moves up and down the abstraction ladder from the abstract to the concrete and from the specific to the general. The most readable writing, such as detective fiction, stays low on the ladder, concentrating on the concrete details of corpses and weapons, exemplifying rather than explaining. Philosophical discourse stays high, offering abstract propositions and explanations. Writing that explains or convinces or informs, yet is easy on readers, moves up and down, connecting data with conclusions, propositions with support, generalizations with specifics, explanation with exemplification" (Matalene 1995, 61). Thus, when you edit for diction, you can help the reader if you:

1. Choose the most concrete words (at the lowest level on the abstraction ladder) whenever possible.

DOS, not operating system

beaker, not glass

2. Clarify abstract words by giving concrete examples—several of them if you wish to indicate a range.

Summaries [abstract noun] can include abstracts, overviews, and executive summaries [concrete nouns].

As a writer if you choose general or abstract words when specific or concrete terms are available, you will be accused of *vague* or *ambiguous* usage. In other words, your reader is not sure of your meaning. One editor says, "We have to put ourselves in the shoes of the reader. The true meaning is not what the writer thinks it means, but what the reader thinks it means. As writers and editors, we have to keep our brains turned on and figure out what makes the most sense to the reader" (Bush 1974).

EXERCISE 3

Concrete and Abstract Words—Abstraction Ladders

Construct an abstraction ladder of four to six levels for each of the following terms, putting the most abstract or general term at the top and the most concrete or specific term at the bottom. You may have to use a dictionary and a thesaurus to help you choose accurate words. The given words can appear any place on the abstraction ladder. Write your abstraction ladders on a separate sheet of paper.

EXAMPLE

    thing
    instrument
    weapon
    gun
    SHOTGUN
    Winchester Model 12

ANALYSIS   VEHICLE   CALIPERS

RAN   SPEAK   SUSTENANCE

EXERCISE 4

Concrete and Abstract Words—Adding Specifics

Each of the sentences below is vague because it uses too many general and abstract words. First, circle the general words and phrases. Then, replace each with words and phrases that are more concrete and specific. Make up your own details and specify who the reader is.

1. The fact of the matter is that a manned trip to the moon did cost a lot; it definitely would have been much cheaper to send an unmanned craft, but it just wouldn't have been the same.
2. The committee met for several hours and got a lot done.
3. Because the load factor increased by a significant amount, we will be unable to complete the testing for several weeks.
4. People hold different views on technology and the consequences that have happened.
5. The tax rate for business is higher this year.
6. The electronic tests represent how truly remarkable these units sound.
7. We're still sticking to the early summer because of our own election. We've explained this to them, that this would be kind of complicated and heavy duty for us to try and combine the two things.

Denotative and Connotative Meanings

The *denotative* meaning of a word is its direct, objective, and neutral meaning. The word *father,* for example, denotes simply "male parent."

The *connotative* meaning of a word includes associated meanings that are indirect, subjective, and often emotionally loaded. When you hear the word *father,* what are your immediate associations? Probably you first think of your own father, and then your responses are triggered by your association with your father—maybe leadership, caring, and sharing, or perhaps rigidity and meanness. Other terms for *father* also indicate subjectivity; *father* tends to be formal, *dad* less formal, and *daddy* informal or childish.

In theory, workplace writing is objective and denotative, but in fact, much of it must be persuasive in forms like proposals, memos, letters, and reports. Thus the *associated* meanings of words are important because they affect the reader's attitudes. Often when the tone of a letter or memo is criticized, the offensive tone is caused by connotations of particular words.

One way to check for accurate use of words is to determine the positive, neutral, or negative connotations. How would you classify these three related words: *argument, fight, discussion?* It's easy to see that *discussion* has positive connotations, *argument* is less positive, and *fight* is definitely negative. You can sharpen your awareness of connotative meanings by such classifications.

EXERCISE 5

Classifying Words by Connotation

Decide whether each of the following words has positive, neutral, or negative connotations. On a separate page, list two other words or phrases that have similar denotative meanings but contrasting connotations. Use your dictionary and thesaurus and be prepared to discuss in class.

For example, the given word is *tardy*. I label tardy as negative. Then I add to the list a positive word, *delayed,* and a neutral word, *late.*

positive: delayed
neutral: late
negative: tardy

1. to bargain
2. speculator
3. inexpensive
4. stubborn
5. profession
6. liberal
7. to talk
8. compensation
9. undecided
10. generous

12.3 Avoiding Euphemistic and Sexist Terms

Because words affect attitudes, editors must also watch for euphemisms and sexist language.

Euphemisms

Writers sometimes try to soften unpleasant facts by using words that muffle the meaning. Such "avoidance" words are called *euphemisms*.

Most euphemisms are used in reference to subjects that cause Americans discomfort, such as war, death, body functions or parts, and social problems. As a writer or editor, you must use common sense. You might choose to use the term *passed away* in writing a letter of condolence to a colleague, knowing that you are using a euphemism for *died* out of courtesy to your friend.

However, when you write or edit a technical document that deals with facts, your obligation is to be as straightforward and plain as possible. According to editor Don Bush, "The technical content of an article is directly influenced by the language that's used to describe it. You can't separate the two. When there is something wrong with the language, the technical content is affected" (1974). And as George Orwell said, "But if thought corrupts language, language can also corrupt thought" (1946). By muffling unpleasant facts with pleasant words, you can distort meaning. For example, here is a company's recent announcement: "We have completed a company-wide restructuring to reflect an increased focus on operating profitability." What did the announcement really mean? "We have just fired part of the work force."

Euphemisms that can distort meaning include the following terms: pre-owned for used; revenue enhancements for taxes; furlough for layoff, dismiss, or fire; and performance-based assessment for essay test.

EXERCISE 6

Removing Euphemisms and Abstract Language

Rewrite the following sentences as needed to eliminate euphemisms and abstract language.

1. "We must turn our attention to the logistics trail of spare-parts acquisition as we fine tune our intricacies and push them down to the operating level."

   Wm. Howard Taft IV

2. The president is cautiously optimistic that the scenario described by his aides will be viable.
3. The two heads of state drafted a communique saying that the problems of nuclear arms and space-based defense systems will be "resolved in their interrelationship."
4. Negative factors in the economy will impact the ability of the government to employ new revenue enhancers.
5. While processing your request, your order was disassociated from your mailing label.
6. The military spokesman said that the enemy's assets had been suppressed by our weapons systems with only minimal collateral damage.
7. "The impending time of parting for you and a beloved pet brings deep emotion. . . . It's also a time to consider the means by which your treasured companion will be put to rest. The common methods of burial and cremation are not the only alternatives. Pet preservation is both the oldest and newest of alternatives."

   From an ad.

Sexist Language

Editors must also be sensitive to the subtle ways that language can influence attitudes toward persons of the opposite sex. Within the American workplace, the following terms are accepted usage:

1. job titles

police officer	instead of	policeman
flight attendant	instead of	stewardess
sales representative	instead of	salesman
mail carrier	instead of	mailman
camera operator	instead of	cameraman
chairperson or chair	instead of	chairman
cleaner	instead of	cleaning woman

2. letter salutations and attention lines

Dear Supervisor:	instead of	Dear Sir:
Attn: Usability Dept.	instead of	Gentlemen:

3. references to people in general

humanity	instead of	mankind
user-system interface	instead of	man-machine interface

However, no commonly accepted substitutes exist for the use of *he* and *his* to refer to both men and women. Most editors choose one of the following solutions, depending on the context.

1. Make the sentence plural.

    NOT Each writer is responsible for creating his own graphics.

    BUT Writers are responsible for creating their own graphics.

2. Eliminate the pronoun or replace it with an article.

    Each writer is responsible for creating the graphics.

3. Change third-person pronouns to first-person or second-person pronouns.

    We are responsible for creating our own graphics.

    You are responsible for creating your own graphics.

4. Repeat the noun (that person or that child).

    A writer is responsible for creating that writer's graphics.

5. Use both *he* and *she* or *his* and *her*.

    Each writer is responsible for creating his or her graphics.

6. Make the sentence passive.

    Graphics must be created by each writer.

7. Alternate pronouns from chapter to chapter or paragraph to paragraph, using male pronouns in one and female in the next.

The first three solutions in this list are the most commonly used. However, each solution changes the meaning slightly, so you need to think of the effect the writer wants to create. Most important, you need to keep the purpose of the document and the intended readers clearly in mind.

EXERCISE 7

Finding Examples of Euphemisms and Sexist Language

To sharpen your awareness of word choice, find three examples of connotative or sexist language (or, conversely, three examples of denotative or nonsexist language). Look in newspapers, magazines, and books. Clip or copy the examples and bring them to class. Be prepared to discuss them.

EXERCISE 8

Choosing Appropriate Nonsexist Revisions

For each of the following texts, choose one of the nonsexist editing solutions and revise accordingly. Be prepared to discuss why you made the changes you did. If you decide not to change the passage, be prepared to explain why.

1. Alcohol intoxication, whether due to an acute overdose or to prolonged abuse, is treated as follows:

   a. If the person is sleeping quietly, his face is of normal color, his breathing is normal, and his pulse is regular, no immediate first aid is necessary.

   b. If the person shows such signs of shock as cold and clammy skin, rapid and thready pulse, and abnormal breathing, or if he does not respond at all, obtain medical aid immediately.

   c. Maintain an open airway, give artificial respiration if indicated, and maintain body heat.

   d. If the victim is unconscious, place him in the coma position (Fig. 58) so that secretions may drool from his mouth. This position will usually allow for good respiration.

   e. Remember that an intoxicated person may be violent and obstreperous and will need careful handling to prevent him from harming himself and others.

   The alcoholic should be encouraged to seek help from Alcoholics Anonymous or from a drug abuse treatment center.

   American Red Cross 1979, 129–130

2. What we know about children and how they learn would dictate that we accept the premise that all children in a given classroom do not need identical amounts of phonic instruction. Most phonics instruction manuals do not make provision for pupil differences. Differentiation of instruction in this area is primarily the task of the teacher, just as it is in all areas of the curriculum.

To provide less instruction than a child *needs* would deny him the opportunity to master a skill that he *must* have in order to progress in independent reading. To subject other children to drill they do not need runs the risk of destroying interest in the act of reading. It is a simple matter to turn off a potential learner by requiring that he sit through group drill on sounding letters; or complete a series of workbook pages which force him to deal with minute details of word attack when he is already capable of applying these skills in sustained reading.

The key to providing children with what they need in the way of instruction is knowledge of their weaknesses. This knowledge is acquired through diagnosis. The best diagnosis is observation and analysis of *reading behavior*. It would seem that discovering what a child needs in the area of code cracking ability should be relatively easy, since he cannot help but disclose his needs. Every technique he might use to cover his weakness is an added clue. Some of the more obvious behaviors are omitting, miscalling, or substituting words.

Listening to a child read a sentence or two should provide the teacher with clues to his word attack ability. A hypothesis that a particular skill is lacking can be tested by having the child read words or sentences containing words which call for him to make the letter-sound relationships which fit the hypothesis. If this simple informal test discloses a problem, the teacher selects or develops appropriate materials, and works simultaneously with all pupils who can profit from the instruction she has decided upon.

Heilman 1968, 21

3. Competition fails to provide ideal motivation for learning, according to our four principles; worse, it has harmful effects. By focusing the spotlight on the very best performer, it damages the self-respect of the remainder. The boy whose craftsmanship is mediocre needs to handle tools around home and to have a belief in his adequacy in this respect, just as much as the boy who can win prizes for his skill. The student who learns

that he is hopelessly outclassed in public-speaking competitions will avoid the speech activities, when the school should be developing his self-expression and his confidence that others are interested in his statements. Class discussions, panels, and non-competitive talks can help him develop. Competition to see who is best draws emphasis away from bringing each person to his full potentiality. Moreover, by emphasizing the false standard that one should take pride only in fields where he excels, it discourages the pupil from developing his lesser talents. When competition becomes a principal form of motivation in school, tension mounts. If a person is emotionally aroused by the threat of failure, his performance deteriorates; he simply cannot do his best because he is so tense.

Cronbach 1954, 479

4. Man's enjoyment of wildlife is evidenced by the large daily attendance of visitors to the many zoological parks, botanical gardens, and museums in this and other countries. There is little doubt that some of man's intangible needs are thus satisfied through his opportunities to watch and study plants and animals. Even hunters, trappers, and fishermen derive more satisfaction from their relationships with wildlife than the mere bringing home of a group of dead bodies. They admire the graceful flights of fowl and the flashing of iridescent colors, the prodigious leaps of a startled deer and the symmetry of its antlers, the song of birds, and many other natural phenomena connected with wildlife. Such values as these cannot be measured in dollars and cents; they can be estimated only in their effects upon the moral fiber of man. The mass effect of such values becomes of greater and greater importance with increases in density of population. National parks and refuges are visited annually by thousands of people who return to their homes refreshed in body, mind, and spirit.

Not only does man derive much esthetic value from wildlife, but he also derives considerable factual information. By studying plants and animals developing under natural conditions he discovers ecologic principles which have sociologic and biologic application to man himself.

Harbaugh 1953, 562–563

12.4 Replacing Pompous Words

The word *pompous* is the adjective form of *pomp,* which can mean (positively) splendor or magnificence or (negatively) ostentatious or vain display. The Latin source word *pompa* means display, parade, show. All these definitions provide clues to what is behind pompous language: The language is calling attention to itself and the cleverness of the writer. "Look at this," it says. "Doesn't this show intelligence, mastery, and erudition?"

What pompous language really does, of course, is distract the reader from the message behind the words. In workplace communication, the content is the key factor, and if the reader cannot get past the showy language to the meaning, the document has failed.

Many writers think "technical" and "professional" writing means they must use long, fancy words, but, in fact, the opposite is true.

For example, here is a pompous sentence:

> As the engineer became the focal point of hiring and success for the production of a product, the technical writer/communicator is on the cusp of becoming the new focal point.

What does this sentence say? Something like this:

> As the hiring of engineers once determined the successful production of a product, so the hiring of technical communicators now is becoming critical to successful [documentation?] of that product.

Notice that the editor's job is to cut and to simplify: to replace pompous words with everyday language and to eliminate extra, weak, or overused words and phrases.

EXERCISE 9

Eliminating Pompous Words

For each of the pompous words below, write a simpler equivalent. As you edit, replace these pompous words when appropriate.

1. circumvent
2. utilize
3. attempt
4. unequivocal
5. initiate
6. facilitate
7. subsequently
8. sufficient

CHAPTER 12   CHOOSING THE RIGHT WORDS   **189**

9. ascertain
10. peruse
11. initial
12. implement
13. minimal
14. effectuate
15. endeavor
16. increment
17. rescind
18. prioritize
19. terminate
20. preponderant

EXERCISE 10

Rewriting for Clarity

On a separate page, rewrite each of the sentences below in simple, clear language. Be prepared to discuss problems of interpretation in class.

1. From a technical definition of a tape recorder.
   A properly designed three-head machine can have its record and play heads optimized for their individual duties. Playback frequency response is improved by the narrower gap possible in a play-only head, but a wider gap in a record head.

2. From the California Labor Code.
   When payment of compensation has been unreasonably delayed or refused, either prior to or subsequent to the issuance of an award, the full amount of the order, decision or award shall be increased by 10 percent. The question of delay and the reasonableness of the cause therefor shall be determined by the appeals board in accordance with the facts. Such delay or refusal shall constitute good cause under Section 5803 to rescind, alter or amend the order, decision or award for the purpose of making the increase provided for herein.

3. From a federal government memo.
   This memorandum is a consolidation of all previous memoranda identifying Transition Office contacts, transition staff authorized to contact federal departments and agencies, and Transition Team members (see Section II). Note that agencies newly designated for transition activity and newly certified individuals are marked with an asterisk. As contacts are now being initiated with the departments and agencies, we would appreciate your ensuring that the proper authorities have been notified of these individuals' authorizations.

4. From a proposal.
   I will also be in contact with several airlines, embassies, rental services, and travel agencies throughout the next two months by the utilization of telephone and written communications.

5. From a government news release.
    Because the heavy mistletoe infestation in the Hoffman Flats area has rendered the residual timber useless for timber production, the ultimate goal is to establish a healthy new stand of Douglas Fir.
6. From a director of telecommunications.
    Yesterday the telecommunications technician had the system restored by 5 P.M. We then used the occasion to reallocate disk space so that we could further improve the response time on the system. Upon the recommendation of our vendor, we will be shutting the system down one afternoon a month for this purpose. We will be giving all users advance notice of such shutdowns. Given the high volume of traffic through our system, this is an extremely critical maintenance function to perform if we are to maximize memory utilization as well as maintain synchronicity in message delivery and notification.

12.5 Choosing Understandable and Usable Terms

The introduction of this chapter defined *usage* as the customary manner in which words are used. *Usage* is an important concept in editing because usage varies from occupation to occupation. When editing, you must understand the usage in a particular field. Over time, the people who work in various technical and scientific industries develop their own ways of using certain words that are special to those industries and may be incomprehensible to outsiders. Such specialized use of words is called *jargon*. Jargon can include a range of terms from abbreviations and acronyms to symbols, words, and phrases that have specific meanings to people who work in that field.

In the segment of the computer industry devoted to multimedia, for example, common terms include *access time, transfer rate, cache buffer, seek time, audio compression* and *decompression,* and *CD-ROM interface adapter.* In the specific area of science dealing with electophoresis, people use terms like *stock solutions, Metler scale, degassing, vortex mixer, dialyzing tubing,* and *supernatant.* Jargon like this is very useful for communicating with coworkers because it saves time and limits meaning. However, such specialized language can be used effectively only for readers who already understand the terms.

In editing, then, you need always to ask who the intended readers are. Unless you are positive that all readers are specialists in this field and will understand the terms, you need to replace the jargon with common terms. If a specialized term is the only word that is accurate, make sure it is defined the first time it is used. A good general procedure is always to assume your readers have slightly less knowledge of jargon than you think they *actually* have. Few readers are insulted by having a jargon term defined if they already know it; they simply skim over the definition. Readers *are* insulted by language they cannot understand.

For user documentation, editor Nancy Allison suggests: "If the powers that be don't want you to explain terms in the text—making it too basic for their taste—insist on providing a glossary. You may even be able to include an appendix explaining basics" (1993, 523).

Editors, as well as writers, need to learn all the jargon of the field in which they work. You lose credibility with a writer if you change the meaning of a sentence by substituting an inaccurate word in an attempt to simplify.

Unfortunately, the word *jargon* also has a second meaning: meaningless, unnecessarily obscure, or pretentious terminology. This definition of jargon is related to pompous language, and many people think of jargon in this sense only. For example, the semanticist S. I. Hayakawa said:

> A learned vocabulary has two functions: first, it has the *communicative function* of giving expression to ideas—including important, difficult, or recondite ideas; secondly, it has a *social function* of conferring prestige upon its users and arousing respect and awe among those who do not understand it. ("Gosh, he must be smart. I can't understand a word he says!") It can be stated as a general rule that *whenever the social function of a learned vocabulary becomes more important to its users than its communicative function, communication suffers and jargon proliferates.* (1972, 237)

One way to understand the difference between this meaning of *jargon* and the meaning of *usage* is given by writer-editor William Zinsser. In IBM's *Think* magazine, Zinsser said:

> ... What is good usage? Perhaps one helpful new way of looking at the question is to try to separate usage from jargon.
>
> I would say, for instance, that "prioritize" is jargon—a pompous new verb that sounds more important than "rank"—and that "bottom line" is usage, a metaphor borrowed from the world of bookkeeping which conveys an image that we can picture. As every businessman knows, the bottom line is the one that ultimately matters. It tells how things stand after all the gains and losses have been totted up. If someone says, "The bottom line is that we just can't work together," you know what he means. I don't particularly like the phrase, but the bottom line is that it's probably here to stay....
>
> I would suggest a similar guideline for separating good English from technical English. It's the difference between, say, "printout" and "input." A printout is a specific object that a computer emits when it is asked for information. Before the advent of computers, it wasn't needed. Now it is. But it has stayed where it belongs—in the world where computers are used. Not so with "input," which was

invented to describe the information fed to a computer. The word has broken out of the machine and run wild. Our input is sought on every subject, from diets and pets to philosophical discourse ("I'd like your input on whether God really exists.") I don't want to give somebody my input and get his feedback, though I'd be glad to offer my ideas. (1981, 23)

A problem related to "bad" jargon is the use of trendy or vogue language. This is often language that begins within one industry and then moves out into general circulation. Zinsser talks about *input* as such a word. Other words and phrases in recent vogue include the following:

virtual	managed competition
proactive	reinventing (government, management)
interruptive	political correctness
real time	ramp up
deconstruction	facilitate
counter productive	orient
nonconcur	target (the verb)

EXERCISE 11

Compiling a List of the Jargon in Your Field

Write down as many jargon terms as you can from your field or the field in which you are editing. After each term, write a definition in simple language that is appropriate for a generalist reader. Add to the list when you encounter more jargon terms.

EXERCISE 12

Rewriting Jargon for the General Reader

The following sentences are taken from a variety of specialized fields and are written for experts in those fields. Using a general or specialized dictionary to find common equivalents, rewrite each sentence for a general reader. Be ready to discuss what you can and cannot translate.

1. Psychology (from a textbook).
   The counselor and client should engage in direct, mutual communication whenever it is useful, especially in times of stress and when the

relationship seems aimless. Both underuse and overuse of immediacy in interpersonal relationships, including counseling, are stultifying.

2. Personal computers (from a salesperson).
   Word processing on this system is really a snap. You just boot up your DOS, put your diskette in your drive, wait for the prompt, then check your defaults and your formatting, and you're ready to go.

3. Medicine (from a journal for paramedics).
   Jugular venous distension (JVD) is absent, as is hepatojugular reflux (HJR). Auscultation of the lungs reveals good air exchange with normal vesicular breath sounds. Rhonchi and rales are absent.

4. Law (from a promissory note).
   In the event the herein described business property or any part thereof, or any interest therein is sold, agreed to be sold, conveyed or alienated by any of the promisors, or by operation of law or otherwise, all obligations secured by this instrument, irrespective of the maturity date expressed herein, at the option of the holder thereof and without demand or notice shall immediately become due and payable.

5. Ski manufacture (from a technical report).
   Aluminum and fiberglass laminates are commonly used in the load carrying layers of Alpine skis due to their great elastic properties relative to their weight. Despite the high elasticity of aluminum, permanent deformation will occur in an aluminum structure prior to occurrence of permanent deformation in an equivalent fiberglass structure.

6. Administration of justice (from a journal article).
   First responders to violent client behavior are often in a vulnerable position, because they may not have prior knowledge of the client's behavioral characteristics, history of aggressiveness, or amenability to intervention.

EXERCISE 13

Editing to Eliminate Jargon and Vogue Words

Rewrite each sentence to eliminate jargon or vogue words. Make up details.

1. To make a long story short, his acceptance is contingent upon our optimizing the work situation.
2. We should arrive at a determination as to the best program we can implement based on technological and financial considerations.

3. In our vision, we see this high school as a competency based community learning center organized around interdisciplinary thematic units. Students, as members of their assessment panels and curriculum committees, will be pro-active participants in the entire education program. The program will produce critical thinkers and lifelong learners who make effective use of interpersonal skills, and use technology to show a capacity to formulate and articulate original thoughts.

4. We are endeavoring to secure proposals for the development and leasing of the facilities described in the attached documents and would be pleased to receive a proposal from you at your earliest convenience, provided you are interested in the matter.

5. We are in concurrence with your plan for the purpose of reviewing local purchasing activities.

6. Dicta from several authors working in the fields of instructional design, human behavior, and hypermedia are coalesced into useful summary lists.

7. If the budget restrictions continue, the purveyance of university graduates will drop by 60,000 in the course of the next decade.

8. We have an applicant-tracking system which allows us to source for required skills. We retain applications in our active files for a period of one year; therefore, it is not necessary to submit more than one application per year. During that time frame we will not respond to any additional applications, résumés, or updates submitted.

9. Exclusions: Any expenses for services rendered by employees or physicians or other persons employed or retained by the Policyholder or for use of the Policyholder's facilities except those benefits specifically listed in the Policy Schedule for Benefits as payable at the Policyholder's Health Service, Infirmary or Hospital; or for any expenses for services rendered elsewhere which are available at the Policyholder's Health Service, Infirmary or Hospital except in cases of emergency nature.

10. The socially dynamic environment of nonacademic settings has a rhetorical impact on nonacademic writers because their company processes take into account, to some degree, transaction with people and texts which help them create documents.

Peter Smudde

References

Allison, Nancy. 1993. Eliminate jargon? Dream on! *Technical Communication* 40.3 (August).

American Red Cross. 1979. *Standard first aid and personal safety*. 2nd ed. Garden City, NY: Doubleday.

Bush, Don. 1974. Semantics (Words are chameleons). In Lola Zook, Ed. *Technical editing: Principles and practices.* Anthology Series 4. Washington, D.C.: Society for Technical Communication.

Cronbach, Lee J. 1954. *Educational psychology.* New York: Harcourt.

Grove, Andrew. 1985. Managing at work. *San Jose Mercury News,* October 10.

Harbaugh, M. J. and A. L. Goodrich, Eds. 1953. *Fundamentals of biology.* New York: Blakiston.

Hayakawa, S. I. 1972. *Language in thought and action.* 3rd ed. New York: Harcourt.

Heilman, Arthur W. 1968. *Phonics in proper perspective.* 2nd ed. Columbus, OH: Merrill.

Matalene, Carolyn. 1995. Of the people, by the people, for the people: Texts in public contexts. In John Fredrick Reynolds et al. *Professional writing in context: Lessons from teaching and consulting in worlds of work.* Hillsdale, NJ: Erlbaum.

Orwell, George. 1946. Politics and the English language. In *Shooting an elephant and other essays.* New York: Harcourt.

Zinsser, William. 1981. What's the point? *Think* 47:3 (May/June).

13 Building Effective Sentences

Contents of this chapter:

- Syntax, Style, and Reader Expectations
- Finding and Fixing Syntax Problems
- Improving Sentence Structure and Style

13.1 Syntax, Style, and Reader Expectations

"Get the basic architectures of the English sentence straight . . . and everything else will follow." Professor Richard A. Lanham (1992, VII) is talking about sentence building (or rebuilding), a task every writer and editor faces at the editing stage. Writing, after all, is about getting ideas down on paper or on screen; editing is about restructuring the sentences that carry those ideas so that the reader clearly understands.

Syntax

Another term for the architecture of sentences is *syntax:* the structure or pattern of the words in a sentence, clause, or phrase. Editors need to be concerned about the position and arrangement of words in sentences because meaning can change as a word shifts position. For example, look at the following sentence:

> To activate the program, press the control button for *only* three seconds.

Suppose you move the word *only*.

> To activate the program, press *only* the control button for three seconds.

Do you see how important the *position* of a word can be in an English sentence? By the simple act of moving one word, you completely changed the meaning of the sentence.

Style

In editing, you can change word order to fix problems and avoid ambiguity. You can also change word order to improve the style. *Style,* as used in this chapter, is the total effect created by the pattern of words in the sentence. Style in workplace writing should be characterized by simplicity and lucidity. Or, as Matthew Arnold said, "Have something to say, and say it as clearly as you can. That is the only secret of style" (Williams 1997, 3). See Chapter 5 for more discussion of style.

How important is sentence structure to working writers and editors? Joy Thomas, a technical writer at a networking company, calls "knowing how sentence structure affects readability" a critical skill. "For example," she says, "which is more readable—*chassis front panel* or *the front panel of the chassis?*" Paul Briscoe, who writes for a company involved in circuit design, says that in editing you need "an ability to recognize awkward, colloquial, or otherwise bad phrasing and be able to suggest something more appropriate." And Connie Lomansky, writer at a company that produces design-automation software, says that students preparing to be technical writers and editors need "more practice in spotting common sentence construction problems."

This chapter will cover sentence patterns and word order for three reasons:

1. to help you understand what readers are used to reading and what they therefore expect
2. to show you how to avoid common syntax problems
3. to help you learn the specific sentence patterns that will make your writing or editing more effective

Reader Expectations

Readers of English—even those for whom English is a second language—come to any piece of writing expecting sentences to be built in certain ways. Those expectations may be subconscious, but they are real and based on much experience in reading English. In effect, readers have constructed a *schema*—or framework—for sentences, and they will read more easily and comprehend more readily if the sentences are put together in the ways they expect. In English the most common patterns are these:

subject-verb-object (SVO)

subject-verb (SV)

subject-verb-complement (SVC)

For example:

Uploading [S] puts [V] information [O] into the system's main memory.

The meeting [S] is scheduled [V] for Monday.

A laser printer [S] is [V] a non-impact printer [C].

How do we know these patterns are common? In 1967, Professor Francis Christensen studied samples of 200 sentences from each of 20 well-known American writers—10 fiction writers and 10 nonfiction writers. He analyzed the kinds of sentences they wrote and counted the different sentence patterns. He found that 75.5 percent of the time these writers constructed subject-verb-object, subject-verb, or subject-verb-complement sentences.

Another 23 percent of the time, the sentences opened with some kind of adverbial modifier—a single word, a phrase, or a clause—and the rest of the sentence was in S-V-O or S-V-C order. Most of the adverbial openers served as transitions from a previous idea to a new idea. For example:

> Thus, it's possible to create a "binary wire" made up of a row of cells.
>
> As a result, they can now see how convection currents can influence the growth rate and curvature of dendrites.

Adding these two percentages together, you can see that respected professional writers use nearly the same sentence pattern 98 percent of the time. The very few exceptions are sentences in inverted order (with a delayed subject) and sentences that open with verbals. For example:

> Central to the project is the idea of cluster computing. (delayed subject)
>
> Lacking any physical explanation, many scientists were prepared to write the correlation off as coincidence, not cause and effect. (verbal opening—present participle)

What can you learn from this as a writer or editor?

1. You communicate best when you meet your reader's expectations.
2. Those expectations are for sentences that follow the standard S-V-O, S-V, or S-V-C order most of the time.

EXERCISE 1

Analyzing Your Own Sentence Patterns

Choose a 200-word sample of your writing (about a page of typed, double-spaced material) from a major paper that is nonfiction prose. Count the total number of sentences. Analyze each sentence by marking subjects, verbs, direct objects, and complements to determine the type. Count the number of each type and record them on the following chart:

> Total number of sentences in sample _____
> Number of sentences of S-V-O, S-V, or S-V-C type _____
> Number of sentences beginning with adverbial modifiers _____

Number of sentences beginning with a verbal _____

Number of sentences with a delayed subject _____

How does your own writing compare with Christensen's results?

13.2 Finding and Fixing Syntax Problems

Verbs

Readers expect writers to give them sentences that are constructed with:

1. standard word order (subject-verb-object, subject-verb, or subject-verb-complement)
2. standard punctuation (to avoid fragments and run-ons)

Readers also expect the writer to be clear about how many items are under discussion. They expect that a plural verb will have a plural subject, and a singular verb will have a singular subject. This is called *subject-verb agreement*.

Sometimes, though, writers separate a subject from its verb with several intervening words. In the process, they forget about agreement and write a sentence like this one:

> Other advantages of a rotary hook is less thread jamming due to the 360 degree rotation, and it tends to be more durable than an oscillating bobbin.

You can analyze and repair sentences with subject-verb problems by taking the following steps:

1. Enclose prepositional phrases in parentheses. Remember that the object of a preposition *cannot* also be the subject or the direct object. Remove prepositional phrases first.
2. Eliminate expletives (*there* and *it* as introductory words). Expletives cannot be subjects.
3. Underline subjects once and verbs twice.
4. Check to see if the subject agrees with the verb in number and change either the subject or the verb if necessary.

If you apply this procedure to the example sentence, you first remove *of a rotary hook* as a prepositional phrase. Then you can easily see that *advantages* (the subject) is a plural word, but *is* (the verb) is singular. Obviously, you need to say "Other advantages . . . are . . ." You can do more to fix this sentence if you make the two advantages parallel. With added parallelism, the sentence might read:

> Other advantages of a rotary hook are greater durability and less thread jamming due to the 360 degree rotation.

A related problem is caused by sending readers conflicting clues about how many items are being discussed. For example:

> The only additional accessory the Juno 106 comes with without additional cost is a connection cord and a detailed owner's manual.

First the reader is told there's *one* accessory ("the only")—and then is given two.

EXERCISE 2

Correcting Subject-Verb Agreement and Other Number Problems

Edit the following sentences for clarity, if necessary correcting subject-verb errors or other number problems. Selected answers appear in the Appendix.

1. The cost of the ovens were rated as follows: under $300, 5 points; $300–$400, 3 points; over $400, 1 point.
2. An important safety feature for night riding are reflectors.
3. The link between the hardware (the physical components of a computer system) and the software (programs that are composed of high-level language instructions designed to solve a problem or perform a certain task) is the operating system.
4. The prices of both Model A and Model B generally depends on the number of features a model offers.
5. On all VCRs there exists certain standard features, which vary little from model to model.

6. It is this tactic and the methods used which has produced untenable labor relations.

7. The findings of this report indicates the advantages and disadvantages of the personal computer.

8. It is the interaction of these microwaves about the oven and food which cook the food.

9. In the previous section of this report, each of the four instruments in question were compared relative to each other in a number of different areas of interest or criteria.

10. Given that the velocity at which a shock wave travels in the ground is dependent upon the density of the ground itself, discrepancies in arrival times indicates changes in composition of the subsurface.

11. The possibility of TSC machining its own tools in the future exist; hence, a need for N/C tools and CAM is established.

12. The specific area of concern for the natural recharge loss are the sand hills surrounding the basin.

Pronouns

Pronouns are words that "stand for" nouns. Writers put them into sentences as generic substitutes for the real thing: the noun itself. You might call a pronoun the pinch hitter or stand-in of the sentence. The noun makes its grand appearance at the beginning (and probably at strategic intervals), but the colorless, hard-working pronoun appears more often and in more roles. It also can cause problems in sentence clarity if the pronoun and the noun that it replaces do not agree.

Definitions of Terms

- *Pronoun:* a word that can take the place of a noun.

    Shoko organized the presentation.

    She organized the presentation.

- *Antecedent:* a word that comes before a pronoun and to which the pronoun refers.

    The writers [antecedent] submitted the proposal on time, but they [pronoun] neglected to include a budget.

- *Case:* the change that pronouns undergo based on their function in the sentence. The three cases in English are subjective, possessive, and objective.

    He designed a Web page. [subject]

    His Web-page design won a prize. [possession]

    The manager sent *him* to the meeting. [object]

One problem in pronoun use is that, whereas a noun rarely changes form no matter how it is used in a sentence, a pronoun shifts form depending on use and position. Such changes are called *inflections.* For example, look at these sentences—both with nouns and with pronouns.

noun as subject	*Charlie* called the meeting.
pronoun as subject	*He* called the meeting.
noun as direct object	The president called *Charlie.*
pronoun as direct object	The president called *him.*
noun showing possession	Give me *Charlie's* address.
pronoun showing possession	Give me *his* address.

The noun—Charlie—changed only to show possession, while the pronoun changed case (and thus form) with each use and position in the sentence.

When editing, you need to examine those areas where pronoun problems confuse readers. Problems with pronouns fall into three areas:

1. pronoun-antecedent agreement
2. vague pronoun reference
3. the who-whom-whose dilemma

Pronoun-Antecedent Agreement

In the section on subject-verb agreement problems, you learned that readers need to know exactly how many items are being discussed. Thus, if a singular subject appears with a plural verb, the reader is confused.

A pronoun problem called *pronoun-antecedent agreement* is similar. Remember that a pronoun is a stand-in or pinch hitter. Because it is a generic word, it lacks the sharp outline and individual identity of a noun. But because it replaces a noun in a sentence, it must have the same characteristics as that noun: It must agree in both number and gender with its antecedent (the noun that comes before it and which it replaces).

Read the following explanation of tire warranties.

> Almost every brand of tire is backed by a warranty based on the normal wear of the tire. That means that once the tire is purchased, the customer can return the tire for an adjustment if unusual treadwear takes place. Then, depending on how long the customer had the tire, they would receive a percentage off the next tire.

How many customers is the writer talking about? The noun *customer* is singular, but the pronoun that replaces it in the last sentence is plural. When a writer does this to readers, confusion results. "All right," you might say, "since the noun *customer* is singular, I'll make the pronoun singular too." But what pronoun will you choose? *He?* What if the customer is a woman? *She?* What if you don't know? *He or she?* All those pronouns are possible choices; alternatively, you could make the word *customers* plural and use *they* for the pronoun. This solution eliminates problems with sexist language as well as pronoun-antecedent problems.

EXERCISE 3

Solving Pronoun-Antecedent Agreement Problems

Underline subjects once and verbs twice. Enclose prepositional phrases in parentheses. Analyze each sentence to determine what noun is the antecedent for each pronoun. Circle the antecedent. Make any necessary changes.

1. The products will be compared in each category, and each of the units will be rated according to their performance and/or availability.

2. Each oven has specific features, and if the consumer wants an oven with a carousel, they are going to buy an oven with a carousel without considering any others.

3. There is a need to study different brands of microwave ovens because, with so many brands available to the consumer, one has to be careful about what they purchase.

4. I suggest that before a family unit purchases their first microwave oven, they should analyze their cooking needs.

5. The swing shift, which begins at 3:00 P.M., has no opportunity to buy food during their breaks because the food truck's last stop is at 5:00.

6. The colors in RGB (Red, Green, Blue) monitors help distinguish different parts of a design, it accents detail, and it adds a dimensional look.

7. The oscilloscopes were given one point for each feature that it is equipped with.

8. We don't send everyone a personalized sample. We only mail to people like you who take an extra measure of pride in themself and their company.

9. The audience for our documentation are adult learners who have a problem-centered approach to learning.

10. Miller's experiments on transmitting information and recording information is essential for understanding how this process works.

Pronoun Reference

Pronouns are useful substitute words for nouns; sometimes, they can even substitute for phrases or clauses. When pronouns appear in sentences, though, editors must ensure that the reader understands the reference. In editing, you need to ask what *real* word, phrase, or clause this pronoun is replacing. Problems in pronoun reference typically occur in three ways:

1. The pronoun could refer to more than one noun in the preceding portion of the sentence.

> The manager told Bill Chitten the news that he was being transferred to the Houston plant.

Who's being transferred? The manager? Bill Chitten?

> When the electric cart rammed the workstation, it was demolished.

What was demolished? The cart? The workstation?

2. The pronoun is so far away from its antecedent that the reader forgets what the antecedent is.

> In fully automatic turntables, the tone arm automatically moves to the start of the record and lowers itself. When the record is finished, it raises itself and returns to a resting position.

What's raising itself? The record?

3. The pronoun reference is general and refers to a whole idea, or the pronoun has no specific reference.

> The delay/start function can be used to begin cooking even when you are not at home. Only the Panasonic is equipped with this, which can be set up to 12 hours in advance.

What *this* does the Panasonic have? What can be set? The Panasonic?

You can fix many pronoun reference problems by making sure that each pronoun is placed immediately *after* the word or word group to which it refers. Because such placement is typical in an English sentence, readers expect to be able to backtrack from a pronoun to the immediately preceding noun, phrase, or clause to find the specific reference. So in the first example, you can clarify who is moving to Houston.

> The manager told Bill Chitten the news. "You are being transferred to Houston."

or

> From his manager, Bill Chitten learned that he was being transferred to Houston.

Sometimes, however, it is difficult to place the pronoun and its antecedent side by side because so many other words intervene. In this case, you should repeat the noun.

> In fully automatic turntables, the tone arm automatically moves to the start of the record and lowers itself. When the record is finished, the *tone arm* raises itself...

In the third example, you can help your reader by repeating the noun.

> The delay/start function can be used to begin cooking even when you are not at home. Only the Panasonic is equipped with a *delay/start function, which* can be set as far in advance as 12 hours.

EXERCISE 4

Rewriting to Clarify Pronoun Reference

Underline simple subjects once and verbs twice. Analyze each sentence to determine what noun is the antecedent for each pronoun. Circle the antecedents. Make any necessary changes.

Effective immediately on all magazine jobs will be a new mode of operation concerning the weighing of the finished product. Instead of weighing every pallet that comes off the conveyor, shipping and receiving personnel will weigh every fifth pallet. This will significantly reduce our handling time for these jobs. In my talks with Mr. Smith, it was of significant concern that we take precautions to back this up with alternate methods. With this in mind, shipping and receiving will also visually check a completed line of products, looking for consistency in height. Also we will take sample lifts periodically through the job, weighing these, multiplying times the number of lifts per skid, and coupling those numbers with the tare. This should allow us numbers with which to compare irregularities.

The Who-Whom-Whose Dilemma

The pronouns *who—whom—whose* cause trouble for both writers and editors. Part of the problem stems from the difference between informal use of language in speech and formal use in writing. When we speak informally, we seldom worry about placement; most of us use *who* in nearly all constructions, except immediately after a preposition. So, for example, you will often hear sentences like these:

> Who do you wish to speak to?
>
> Who is the letter addressed to?

In writing, however, many writers feel they must be more formal. They are uneasy with sentences like those above; they think they should use *whom* more often, but they are unsure about when and how to use it.

These same writers may remember some English teacher from their past telling them "Use *whose* only to talk about people; never use *whose* to talk about animals or things." Remembering that rule, they will worry about a sentence like this:

> The CRTs whose cables needed repair were stacked in the warehouse.

You need to remember that your primary goal as a writer or editor is *clarity*; clear sentences are more important than adherence to a form. Thus, most readers will readily accept

> Who are you looking for?

instead of

> For whom are you looking?

In either sentence, the meaning is clear, and the first sentence sounds less stuffy than the second. In the same way, readers have no problem with

> The proposals, whose details were published yesterday, must be revised by November.

instead of

> The proposals, the details of which were published yesterday, must be revised by November.

Guidelines for Informal Use. Generally, the word *who* is replacing *whom* in most cases of informal speech and writing. *Whose* is also widely accepted to show the possessive of things as well as people.

Guidelines for Formal or Semiformal Use. However, if your company or personal style requires more formal use, here are some general guidelines for *who* and *whom:*

- *Who* is the subject form whether in a question, a sentence, or a clause.

  > Who will attend that meeting?

  > The manager asked who was on vacation.

- *Whom* is the object form, used as direct object or object of a preposition or verbal.

  > T. Smith is the engineer whom you should contact.

  > The letter was written "To whom it might concern."

- To determine whether you need a subject or an object form, do these things:

1. Underline subjects once and verbs twice.
2. Rearrange the sentence and clauses in normal subject-verb-object order if necessary.
3. Remove intervening words.
   Who ~~did the memo say~~ was assigned that job?
4. Test by substituting *he* or *him* and choose the corresponding *who* or *whom*.

EXERCISE 5

Finding Examples of Who—Whom—Whose

To sharpen your awareness of the ways writers actually use the pronouns who, whom, *and* whose, *gather three examples of each from magazines, newspapers, or other printed sources. Bring the examples to class and submit them to your instructor.*

Modifiers

Modifiers are words, phrases, or clauses that limit or describe another word. Some modifiers act as adjectives; they modify nouns or pronouns. Other modifiers act as adverbs; they modify verbs, adjectives, or other adverbs. *Prepositional phrases* can function as modifiers, as can *verbal phrases*.

He graduated in 1997. (prepositional phrase acting as adverb)

The pendulum, swinging in an arc of 20 degrees, activated the clocking mechanism. (verbal phrase acting as adverb)

Just as pronouns should be placed next to the noun or phrase they replace, so modifiers should be placed next to the word they modify whenever possible. Modifiers can either precede or follow the word they modify, but if they simply appear somewhere in the sentence, they can confuse the reader. Many unclear or ambiguous sentences are made that way by modifiers in the wrong position.

Definitions of Terms

Grammarians have coined three terms for wrongly placed modifiers. Even though the terms are not as important as the problems they describe, you may find it helpful to know what each term means:

- *Dangling modifier:* a modifier at the beginning of a sentence that has a different implied subject from the real subject of the sentence.

    > Being programmable, you can set the VCR to record your show even if you're not home.

- *Misplaced modifier:* a modifier that appears by its position to modify the wrong thing.

    > Pour each of the solutions (developer, fixer, and water) into a separate tank at 68° F.

- *Squinting modifier:* a modifier that could modify either the preceding word or the following word; it's unclear which.

    > The calendar which is published regularly lists meeting times.

How to Repair Sentences with Wrongly Placed Modifiers

Follow these general rules for good placement:

1. Single-word adjectives usually *precede* the words they modify.
2. Adjective phrases or clauses usually *follow* the words they modify.
3. Single-word adverbs can either precede or follow the word they modify.
4. Adverb phrases or clauses may be placed in many positions. Study each sentence from the reader's perspective and rephrase for clarity.

How to Repair Specific Modifier Problems

- Dangling modifiers. To fix a dangling modifier, you can 1) rewrite the introductory phrase, giving it a subject of its own, or 2) make the subject of the main clause the same as the implied subject of the modifier.

NOT	Being programmable, you can set the VCR to record your show even if you're not home.
BUT	Because a VCR is programmable, you can set it to record your show even if you're not home.
OR	Being programmable, a VCR can be set to record your show even if you're not home.

- Misplaced modifiers. To fix a misplaced modifier, you can 1) move the modifier closer to the word you intend it to modify, or 2) rewrite the sentence so that the modifier is clear.

CHAPTER 13 BUILDING EFFECTIVE SENTENCES **211**

NOT   Pour each of the solutions (developer, fixer, and water) into a separate tank at 68° F.

BUT   Pour each 68° F solution (developer, fixer, and water) into a separate tank.

OR   Bring each solution (developer, fixer, and water) to 68° F, then pour each into a separate tank.

- Squinting modifiers. To fix a squinting modifier, you can 1) move the modifier closer to the word you intend it to modify, or 2) repeat the modified word so the intention is clear.

    NOT   The calendar which is published regularly lists meeting times.
    BUT   The regularly published calendar lists meeting times.
    OR   The calendar, which is regularly published, lists meeting times.
    OR   The calendar, which is published regularly, is the kind of calendar that lists meeting times.

EXERCISE 6

Moving Modifiers for Clarity

Underline subjects once and verbs twice. Analyze the sentence to determine the modifier problem; then rewrite or reorganize as needed.

1. Entering through the front door of the house the living room is to your immediate left.

2. Her references were interesting and would no doubt be good reading for anyone considering freelancing as well as various small businesses.

3. Walking down the long carpeted hallway, it eventually comes to an end at a rather large white door.

4. Looking down, there is a hardwood floor.

5. Curious as to how he manages such highly technical concepts without being an engineer himself, he explains that "It's easy to learn the jargon in any field."

6. Leaving the heater alone the repairmen thought they had caused the damage.

7. He hitched up his livestock and my grandmother on the wagon train also.

8. Standing in the entrance the 8-foot desk is situated against the middle of the south wall.

9. Going across to the other side of the room there is a small table.

10. After walking along the left side of the house the two-car garage appears at the bottom.

11. A California native, Nancy's home has been the San Francisco Bay Area.

12. Now entering her fourth year, the experience has provided her with an excellent portfolio.

EXERCISE 7

Moving Modifiers for Clarity

Underline subjects once and verbs twice. Analyze the sentence to determine the modifier problem; then rewrite or reorganize as necessary.

1. P. T. Smith, library access director, told Forsythe he had until April 29 to settle the case in an April 5 memo.

2. After parking your automobile on level ground and after locating the crankcase which the drainage bolt is located on the bottom of, you can now proceed.

3. A keyboard split point is the actual point programmed into the memory, where one half of the keyboard will produce different sounds than the other half simultaneously.

4. Based upon those goals, each school will establish a concise set of expectations which will describe student behaviors clearly that are to be achieved.

5. Since dirt from the air may become embedded in the film or the film may become scratched, extreme care is to be taken to protect the film.

6. As a student of Technical University, a formal report comparing alpine skis is required.

7. Putting the old Dodge into gear, I saw Smith drive down the back side of the summit.

8. Capital input is divided into several stages, each stage being related to specific milestones, and is detailed in subsequent sections of this plan.

9. If people are near a shelter, Black said that they should go to it, wearing a coat to protect them from the falling debris.

10. Although written in 1996, the prospective technical writer will get a very clear and realistic preview of the expectations of a writer in any size company.

Wordiness

A very common problem in workplace communication is wordiness; sentences, clauses, or paragraphs are stuffed with too many words. Whenever readers must work their way through excess words to get at meaning, editing is required.

To control wordiness, some companies even recommend limiting sentence length. The 1993 Cadence Technical Publications *Style Guide,* for example, says: "Write simple, short sentences. You are less likely to confuse a user if your sentences have 15 or fewer words. Even complex ideas are understood one piece at a time. Take a complex idea apart in logical units and rebuild it for the user" (G-1).

But writing a short sentence or a series of short sentences does not guarantee improved communication. The most important criterion is that each sentence matches reader expectations for sentence construction and makes the relationship between ideas clear. Thus, rather than counting the number of words in each sentence when editing, you should look for and correct specific constructions that cause wordiness.

Professor Richard A. Lanham maintains that most sentences have a "lard factor" of one-third to one-half; these are words that can be removed to improve the sentences. Extra words that confuse readers occur for several reasons:

1. redundancy (unnecessary repetition)
2. long modifier strings
3. delayed subjects or verbs

Redundancy

Redundancy is a general term that could be applied to all the kinds of wordiness, because the word *redundant* comes from a Latin word meaning *overflowing* or *excess*. Sometimes excess words have simply been repeated unnecessarily, as in the following examples from an IBM contest to encourage "Straight Talk" (1987, 9).

> Literals are considered relocatable because the address of the literal, rather than the literal itself, will be assembled in the statement that employs the literal.
>
> There is an existing problem which I would like to bring to your attention. It is a recurring one and one I feel is a hindrance to the project.

To fix this problem, analyze the sentence to determine where the key term is needed. Then restructure the sentence, perhaps using pronouns to replace the redundant term.

Some redundancy is caused by time, quantity, type, and dimension words. For example:

> The program carries out all tests in *a time* of 3 seconds.
>
> The temperature ranged from *a minimum* of 20° C to 85° C.
>
> The monitor screen measures 12 by 15 inches *in size*.
>
> She has been working in *the area of* usability testing.

In each case the italicized words are unnecessary; the information has been adequately conveyed by the specifics.

Finally, there is the specific kind of redundancy sometimes called *deadwood* or *repetition:* The writer uses two words that mean the same or nearly the same thing. Three kinds of redundancy show up in workplace writing:

1. grandiose pairs
2. general plus specific pairs
3. implied plus stated pairs

Grandiose Pairs. Both legal and insurance documents overflow with words presented in grandiose pairs. Some other workplace writers also think a pair of words somehow has a more professional sound than a single word. These writers give us:

> each and every
>
> basic and fundamental
>
> made and accepted
>
> provisions and stipulations
>
> ordinance and law
>
> reconstruction and repair

The truth is that even in legal writing such pairs are seldom necessary. Law schools these days stress plain writing, which generally means using only one word instead of a pair.

The solution with grandiose pairs is simple: Retain the more effective of the two words and cross out the other.

General Plus Specific Pairs. Many writers habitually use general words and then modify them with more specific adjectives or adverbs. In workplace writing, it is easy to reach into the word bag and pull out general-purpose nouns like:

area	concept	projection
subject	degree	option
condition	matter	capability
factor	policy	order

But because these are general-purpose words, they must be made more specific with modifying adjectives. Add an adjective or two and you get a resounding, professional-sounding phrase, like:

> initial policy projection
>
> initial work projection
>
> initial data projection

It might be possible in all three cases to drop the word *projection*.

The same problems can occur with verbs. We all have a cache of colorless general verbs that we then modify with adverbs in order to specify. We say:

walk quickly	instead of	trot or jog
studied intently	instead of	examined
talked loudly	instead of	shouted

In reducing general plus specific pairs, the solution may be as simple as choosing the more specific word of the pair and crossing out the other.

serious crisis	can become	crisis
perfectly accurate	can become	accurate
various samples	can become	samples
span of time	can become	time

Sometimes, though, you must choose a new word—for example, a specific verb or a concrete noun.

jumped ahead	might become	leapfrogged
at this point in time	might become	now

Implied Plus Stated Pairs. Sometimes writers don't give their readers enough credit for thinking. They strongly imply a fact and then state it too—just to make sure. That kind of redundancy appeals to some students, who say the writing "flows better." The truth is that the flow of words is just that—a flow, much like the auctioneer's patter or the huckster's spiel. For example:

> Proposals that are submitted to the committee on procedures will need to be updated and revised at regular intervals. Such updates should occur after Phase 1, Phase 2, and Phase 3 review procedures. Review procedure committee chairpersons should submit their changes according to standard updating format, and see that those revisions are in the hands of the members of the committee on procedures within three working days after each phase review procedure is completed.

Redundancies from implied plus stated pairs may be tricky because these repetitions often occur from sentence to sentence. You should read an entire paragraph or section before you begin to revise at the sentence level. For practice, we can boil the excess juice out of the paragraph on proposals.

What's the main idea here? That the chairperson of each phase review committee must submit a revised proposal within three working days after the review is finished.

You may need to add a few more details, but the idea is simple. Now go back and look at the individual sentences. In sentence #1, *update* implies *revision,* so you don't need both words. In sentence #2, you have the specifics (Phase 1, 2, and 3) for the previous general statement *at regular intervals.* In sentence #3, *submit their changes* implies *in the hands of the members of the committee.* Also in sentence #3, *after each phase review procedure* repeats the earlier *Phase 1, Phase 2,* and *Phase 3.*

EXERCISE 8

Rewriting to Eliminate Redundancies

Read through each sentence or paragraph below to determine where the redundancies occur. Then rewrite to eliminate each redundancy.

1. It does have a built-in paper feed that is tractor fed.

2. Another irritating sore spot is the way the layoffs were handled.

3. Hopefully, in the future, with these improvements, another situation like the boiler room fire will never happen again.

4. This device is highly useful in speech analysis, and the results it produces will be very acceptable to the professional community.

5. This model measures 17 × 4 × 13 inches and weighs 12.8 pounds. It is of a smooth metallic substance and comes in a silver color.

6. The author examines three types of mistakes that engineers make. First, there is human error; then, there is lack of imagination; and, finally, there is blind ignorance.

7. Certain elements of automobile performance are critical in respect to the performance characteristics of a particular car.

8. Interchangeable items: Upon approval, changes may be made to interchangeable parts. Non-interchangeable items: A new part number is required in order to make a non-interchangeable change.

9. I have personally reviewed the new initiatives and am glad to report that they will have an influence on our future plans.

10. School is only one part of the far-reaching and all-inclusive process we call education. Schooling requires commitment and caring on the part of the student, parent, teacher, and the community. It is truly a cooperative effort, with each member doing his/her part.

Long Modifier Strings

Another kind of wordy sentence is created when the writer tries to pack too many technical words in a phrase, particularly in a long modifier string. Such packing delays the subject or verb and yields sentences like this:

> Among the most difficult components in the safety systems of a nuclear reactor to quantify are the activated charcoal beds used to remove radioactive noble gases.

This sentence is in reverse order: *activated charcoal beds* is the subject. Because of the reversal, the sentence opens with a 15-word series of prepositional phrases that makes the reader search for the subject and verb.

Here is another example of a sentence that confuses because of a long modifier string:

> The increasing use of the icons, flowcharts, graphs, and other graphic devices in documents is one of the more obvious changes in how we design information.

In this sentence, the subject *use* does appear early in the sentence, but the 10-word prepositional phrase that modifies the subject keeps the reader from the verb for too long.

Delay of Subject or Verb

Readers have the most trouble with piled-up words when those words come at the beginning of the sentence. Readers expect subject-verb-object order in most sentences. When they read a sentence, they search subconsciously for the subject—to lead them to the verb—to lead them to a completer of some kind. If they encounter a barrier of words before they even get to the subject, they may abandon the whole search and quit reading.

Not only are readers expecting subject-verb-object order, but they are also influenced by the limits of short-term memory. Short-term memory contains the number of items a person can remember long enough to carry out a task like entering a phone number or reading a sentence. As psychologist George Miller demonstrated (1967), that number is seven plus or minus two items. Each bit of information, and sometimes each word, counts toward that magic number. Thus, for ease of reading, readers should have to deal with no more than nine items of information in a sentence, and current research says that fewer is better. If a reader's short-term memory is at capacity before reaching the subject, the sentence needs editing. For example:

> *To ensure operational readiness, annual system familiarization* inspections of all sprinkler and standpipe systems are conducted by the fire companies.

The subject of that sentence is *inspections,* but the reader must wade through seven words before finding the subject.

Sometimes the simple subject is complicated by a following modifier phrase, which keeps readers from the verb with too many words.

> The decline *of natural recharge* and elevation *of runoff rates due to pollution and residential use of watersheds* can be controlled by integrated water management.

Fourteen words intervene between subjects and the verb, and these words cause problems for readers.

Most readers can handle extra words better at the end of the sentence than at the beginning. Therefore, you need to look for ways of editing that give readers a reasonably simple subject and verb, followed if necessary by modifiers.

You might rewrite the first sentence this way:

> Each year, fire companies conduct system familiarization inspections of all sprinkler and standpipe systems to ensure operational readiness.

In the second example, the best fix is to move the whole first part of the sentence to the end, making the subject shorter and the verb active.

Integrated water management can control both the decline of natural recharge and the elevation of run-off rates due to pollution and residential use of watersheds.

(Note: A sensitive editor would question the phrase "due to pollution." Are the run-off rates due to pollution?)

Sometimes you can simply eliminate excess words within the first part of the sentence. You need to evaluate each sentence both as an individual sentence and in the context of the paragraph.

In summary, you can improve sentences with long modifier strings or delayed subjects and verbs by taking these actions:

1. Move key nouns (especially subjects) to the beginning of the sentence.
2. Keep subjects and verbs close together.
3. Use possessives and hyphens rather than a series of prepositional phrases.
4. Turn nominals into verbals (*documentation* into *documenting*).
5. Put prepositional phrases *after* important nouns.

EXERCISE 9

Rewriting to Clarify Long Sentences

Analyze each of the following sentences to determine where the problems occur. Start by looking for the main subject and the main verb. Rewrite to clarify, choosing the method most appropriate to that sentence.

1. Apparent transit problems with the company handling the equipment repair work have put the schedule back more than two weeks.

2. The reason this report was made was to find methods of improving the quality of groundwater in the Clinton Basin.

3. The OSHA language concerning the pressroom and the combustion engine needs to be cleared up before any major changes can be made.

4. A crucial part of polishing the sound that is produced by a cassette tuner/radio deck deals with the type of noise-reduction circuitry that is used in the deck.

5. Perimeter sensors are and interior sensors sometimes are connected to a separate control unit, either with wires or with small radio transmitters.

6. The complexity of the technology required to solve the performance problems was enormous.

7. Only effort expended in preparing and presenting reviews to our customer is allowable in the JN 378 contract.

8. One of the criteria indicated as determinate of who should be considered was regular participation at meetings.

9. Basic service rates, timing of local measured service usage, calls placed from a coin telephone, calls to Directory Assistance, and labor charges for initial work at the customer home are among the services affected by the new rates and charges.

10. The amount of time that each group requires in order to meaningfully go through the process of identifying ideas ranges from a low of approximately 4 hours to an entire 8-hour day.

Inappropriate Expletives

Subjects of sentences can also be delayed if the writer uses expletives extensively or inappropriately. In English grammar, expletives are the introductory words *it* and *there* used in the usual position of a subject. An expletive can be useful if the writer wants to move the real subject to the end of the sentence. However, using expletives also adds extra words to the sentence for these reasons:

1. The introductory expletives themselves provide no content in the sentence; they're simply extra words serving as "place holders" for the delayed subject.
2. When the usual subject-verb-object order is reversed, the subject often appears in a clause or phrase, a process that adds extra words.
3. Expletives often combine with passive voice, which also adds extra words.

For example, consider the following sentences:

1. Besides engineering, there are many departments that significantly contribute to the company's success.
2. It is asked that supervisors divide their groups into two sessions for each shift to minimize the impact on production operations.
3. It is in the processor and the program store (instruction and translation store) that the key elements of the stored program really lie.

In sentence 1, the expletive *there* encourages use of a linking verb, which then requires a relative clause to give us the action. Extra words include *there, are,* and *that*. Why not say:

Besides engineering, many departments significantly contribute to the company's success.

In sentence 2, the expletive *it,* combined with the passive voice, reduces the strength of the sentence. Again the action is lodged in a clause. Why not say:

Supervisors should [must?] divide their groups into two sessions for each shift to minimize the impact on production operations.

In sentence 3, the expletive *it* and the weak verb *is* keep the reader from the important information as well as from the real subject of the sentence: *key elements*. Why not say:

The key elements of the stored program control lie in the processor and the program store (instruction and translation store). [or . . . instruction and translation program store?]

Often you can eliminate expletives, but sometimes you will need an expletive for emphasis, to avoid an awkward construction, to move a long subject to the end of the sentence, or to lead the reader to a new topic. You need to assess each sentence individually and in the context of the paragraph to see if the expletive functions effectively or is superfluous. In the following sentences, the expletives work well:

- To emphasize:

    There are 14 subassemblies included in the working drawings.

    There is often a good reason for using an expletive.

- To avoid an awkward construction:

    There was no attempt to interrupt the speaker with questions.

    There was no time to lose.

- To move a long subject to the end of the sentence:

    It is clear that the project will be delayed six months.

    It was contrary to company policy for the interoffice memo to be issued to the public.

- To lead the reader to a new topic:

    There are four possible reasons for the equipment breakdown: poor installation, inadequate maintenance, power surges, and defective parts.

Notice that when *it* is used as an expletive, it acts as the grammatical subject of the sentence. Since *it* is singular, it is followed by a singular verb.

It was the district managers who favored the accounting change.

The pronoun *it* can also begin a sentence. Then the sentence follows standard word order because the *it* refers to a specific noun in the previous sentence.

It is designed to help you collect the information you need to configure your local node.

There, on the other hand, requires verb agreement with the delayed subject.

There are 14 students attending the seminar.

EXERCISE 10

Evaluating Expletives

Analyze the following sentences to see if the expletive can be eliminated. Rewrite if appropriate, and underline subjects and verbs in the new sentence.

1. There are three key paragraphs in the proposal's introduction.

2. It is preferable for editors to meet with writers before they begin editing.

3. There are many different ways to evaluate multimedia components.

4. There are so many punctuation and syntax errors in the document that its technical credibility is endangered.

5. There are three reasons for changing to the new word processing package: ease of use, compatibility with the graphics software, and the ability to mount on many platforms.

6. It is important to understand that technical communicators add value to products by presenting information clearly enough so that users need not use hot-line or help services.

13.3 Improving Sentence Structure and Style

Good editing involves more than correcting errors; it means improving sentences and paragraphs by revising in the following areas as appropriate:

1. choosing active or passive voice
2. locating action in verbs
3. preferring positive over negative constructions
4. emphasizing by position

Choosing Active or Passive Voice

When editing, you can often improve the readability of a document by changing passive voice constructions to the active voice. Active voice is so named because the emphasis is on someone or something *acting*. In passive voice, the emphasis is on someone or something *being acted upon*.

Mike Belef, a technical writer at a medical-equipment manufacturer, stresses the importance of understanding the rhetorical effects of voice. "Identifying passive voice is vital," he says. "It took three years of editing before I mastered the skills in identifying inappropriate passive voice construction and found the best formulas for converting passive voice. One thing I discovered is that passive voice is nearly always the result of 'formative thought,' the stage when people are trying to understand what they actually want to convey. Explaining this fact to engineers often helps them to identify and eliminate passive constructions in their second drafts. The more planning and review an author puts into a document, the less passive voice you will find."

Both active and passive voice constructions are useful, depending on the effect the writer wants to create. Thus, as an editor, you need to understand how voice works and how to change voice to meet readers' needs.

Definitions of Terms

- *Voice*: the verb form that shows whether the subject is acting or acted upon.
- *Active voice*: the sentence construction in which the subject of the sentence *performs* the action of the verb.

   V. Nguyen attended the conference.

   The software engineer debugged the program.

○ *Passive voice:* the sentence construction in which the subject of the sentence *receives* the action of the verb.

> The conference was attended by V. Nguyen.
>
> The program was debugged by the software engineer.

Reader Expectations and Active vs. Passive Voice

Readers expect **active voice:**

1. When they must know who is doing what.

    > The system delivers electronic mail quickly.
    >
    > Acme Refrigeration will pay all costs if the unit fails.
    >
    > I designed the overload circuit.

2. When they must follow a series of sentences that present new information. The beginning of each sentence should set the context or orient the reader to what follows.

    > This product is a member of a family of high-performance super microcomputers. An entirely new and innovative multiprocessor architecture sets this family of super-microcomputers apart from other systems. This new architecture runs an enhanced version of the standard UNIX operating system. The system provides for incremental growth in performance and connectivity while offering both hardware and software standards for ease of integration into the commercial marketplace. (Wyse Technology n.d., 6-1)

Readers will accept **passive voice:**

1. When what is important is the receiver, events, or results of an action.

    > The first versions of UNIX were developed around 1970 at Bell Laboratories. In recent years, UNIX has become the de facto industry standard for use in software and hardware development, document preparation, research, and instruction. Several versions of UNIX have been developed since 1970, both at Bell Labs and at the University of California at Berkeley. Many other companies and institutions have developed systems that are based on UNIX or UNIX look-alikes, and an increasing number of college students are studying UNIX as part of their computer science curricula. (Wyse Technology n.d., 6-1)

2. When who or what carried out the action is not important or is unknown.

    > Error messages are displayed on the screen.
    >
    > Programs are executed according to their priority level.

3. When the writer does not want to assign responsibility for an action.

   The graphics in the manual were badly designed.

   The discount has been reduced to 8 percent.

4. When the agent (doer) is modified by a long phrase that would separate the subject from the verb by too many words.

   NOT Staff members from the Inertial Confinement Fusion (ICF) Program and the Science Education Center (SEC) have developed a special program about lasers and fusion energy.

   BUT A special program about lasers and fusion energy has been developed by staff members from the Inertial Confinement Fusion (ICF) Program and the Science Education Center (SEC).

EXERCISE 11

Evaluating Passive Sentences

Analyze each sentence to determine if the passive is used to advantage or disadvantage. Rewrite if necessary. Explain why you made the changes you did.

1. The common point is connected to ground through the speed regulator circuits.

2. Registers are loaded from memory addresses, but the memory locations from which the data are taken are not altered in the process.

3. An offline CPU card switcher test can be performed at any time in the cycle.

4. The commands are summarized in Appendix A.

5. The need to upgrade the image of our documentation was realized at the end of 1997. With the new system, a major setback was that the artwork was not networked into the desktop publishing system, so it had to be scanned in.

6. Several writers have discussed the limitations that are imposed on power by back-biased diodes.

7. Before a thorough study is to be initiated, a survey of all persons that might be interested or affected by the opening of a flight instruction program must be conducted so that information obtained through the survey can be used to guide further research.

8. The possibility of becoming a work-study project will also be examined.

Locating the Action in the Verb

A main cause of wordiness and stuffiness in workplace writing is overuse of noun forms combined with weak verbs. When writers are groping for meaning, they often forget to use active verbs; instead, they hide the action in a noun somewhere in the sentence and then use a weak verb like *is, made, take*. You can improve such sentences by relocating the action in the verb itself, a process I call *verbalizing the action*. This editing change removes excess words and energizes the sentence.

I usually avoid words ending in *-ize* because such words can easily lead to pompous diction. In this case, though, *verbalize the action* says very directly what you want to accomplish. First you want to find the action in a sentence and then make sure that the action is lodged in the action word: the verb. If you don't put the action in the verb, you may have a sentence problem called *nominalization*: the excess use of nouns made from verbs or adjectives.

How to Spot the Hidden Action

Sometimes you must hunt for the action; it can be well hidden by writers who think they are sounding professional by using long words that end with *-ment,*

-tion, -ance, and *-ence.* Here are the four most common places writers can hide the action:

1. In the *subject.*

    The *contention* of the reviewers is that Section C should be revised.

    Notice that placing the action in the subject *contention* weakens the verb. If you move the action into the verb, the sentence becomes shorter and more readable.

    The reviewers contend that Section C should be revised.

2. In the *direct object* or *complement.*

    The senior writer does the *review* for technical accuracy and logical presentation.

    The following paragraphs are an *explanation* of the access commands.

    In the first sentence, notice the weak verb *does.* Move the action into the verb, and you can also eliminate excess words.

    The senior writer *reviews* for technical accuracy and logical presentation.

    In the second sentence, remove the weak verb *are* and move the action from the complement into a verb for a shorter, clearer sentence.

    The following paragraphs *explain* the access commands.

3. In a *verbal* or *prepositional phrase.*

    Users of the system must take *into consideration* the power requirements and heat-generating capacity.

    If you move the action to the verb *consider,* you can cut two dull words from the sentence.

    Users of the system must *consider* the power requirements and heat-generating capacity.

4. In a *noun that follows an expletive.*

    It is the *conclusion* of the report that the method of weighing the finished product should be changed.

    Turn the noun into a verb, and you can erase the ineffective expletive as well.

    The report *concludes* that the method of weighing the finished product should be changed.

    For an even better sentence, you can make the noun clause active.

The report concludes that we should change the method of weighing the finished product.

Notice in each case that by verbalizing the action, you have also tightened the sentence by eliminating extra words.

How to Verbalize the Action

You can follow this three-step sequence to put the action in the verb:

1. Search through the sentence, asking "Where is the action?" Be especially suspicious if the sentence includes nominals that end in *-ment, -sion, -tion, -ance,* or *-ence.*
2. Turn the word that contains the action into a verb, or find a verb to express that action.
3. Rewrite the sentence, eliminating the excess words. If the sentence has no agent or "doer," see if you can determine the agent from context. If you can, you can also write the sentence in active voice. If you can't, you may have to retain the passive voice.

EXERCISE 12

Verbalizing the Action in a Sentence

Analyze each sentence to find the hidden action. Mark the word that contains the action. Then rewrite, putting the action into the verb and eliminating the excess words.

1. The preceding line graphs are an illustration of the fluctuations in power consumption.

2. The conclusion of the review board was that the proposal was clearly written.

3. Make corrections on the draft with standard copyeditors' marks.

4. In the following discussion, the theory of operation is given.

5. Figure 5–4 provides a clarification of the maintenance procedure.

6. The alignment of the guide was accurate.

7. The leads on the wafer do not make contact with the pins of the second layer.

8. Selection of a division manager will be made in June.

9. The adult units are housed in Bldg. D.

10. In each figure, the totals are given.

11. In the translation, I was forced to choose just one of those words and convey the other implication by context.

Preferring Positive over Negative Constructions

Readers can be needlessly confused by sentence constructions that are negatively stated when the writer's ultimate purpose is to make a positive statement. For example:

It is *not uncommon* for the checklist to be located at the end of the training module.

What does the writer really mean? That checklists are commonly located (or can be located) at the end. Why not say:

The checklist is commonly located at the end of the training module.

Locate [place?] a checklist at the end of the training module.

Not only is a positive statement clearer, but it also helps readers remember the information better. Consulting industrial journalist Frederick Harbaugh says: "Clinical psychology confirms that the human mind prefers to receive information in positive form and remembers it best that way. Therefore, when we want to communicate an idea with the least possible risk of misunderstanding and the greatest probability of the listener's or reader's remembering what we said, we should find a way, which is usually possible, to express the idea in positive rather than in negative language" (1991, 73–74).

Other examples of negative constructions:

NOT   You cannot afford not to identify the problem areas in the program.
BUT   You must [should] identify the problem areas in the program.
NOT   Do not hesitate to call us if you have problems.
BUT   Call us if we can help you.
NOT   You cannot print without reconfiguring the file.
BUT   To print, reconfigure the file.

A more complicated negative appears in the following sentence:

> Spacing the time between composing and proofing *impairs* the writer's ability to predict what is on the page.

Nothing is wrong with the sense of this sentence: Writers should allow time to elapse between writing and proofing because they will then be more likely to see the page from a fresh point of view. But readers do not expect a sentence that is meant positively to use a verb like *impairs,* and they will be forced to reconstruct the sentence positively in their minds. A better sentence would read:

> Spacing the time between composing and proofing *allows* the writer to edit the page without interference from intended meaning.
>
> Allow time to elapse between writing and proofing. The added time will help you edit what is actually written on the page instead of what you intended to write.

Emphasizing by Position

Editors can help readers understand what they are reading by applying two concepts we have learned from research on how adults read.

1. Short-term memory affects the relationship of old and new information. Reading forces us to keep a number of ideas in our short-term memory while we process the words that supply new ideas. Those new ideas must then be related to the old ones for understanding to occur. Because short-term memory can efficiently hold only up to nine items, those items need to constantly be rearranged and re-emphasized as new information is introduced. A sensitive editor will revise to ensure that most of the time the beginning of a sentence refers to previous or "old" information, while the end introduces new information.

This balance of old and new information in sentences clarifies meaning for readers by meeting their subconscious expectations. Readers expect to build on a foundation of what they already know; therefore, they expect a sentence opening to point back to previous explanations, while the sentence ending moves them to an understanding of the new information. Often, old information appears in the subject, and new information in the predicate. As Professor Carolyn Matalene explains, "Readers cannot process too many items of new information in a single sentence, nor do they like new before old very often; that kind of sentence has to be turned inside out to be processed. Complex material needs to be presented to readers slowly; keeping the subject the same and adding new information in successive predicates gives readers a break, as in A + B, A + C, A + D, and so on" (Matalene 1995, 62).

2. A word's position in a sentence affects the emphasis it receives. Editors should know that:

- The most important position in a sentence is the end.
- The next most important position is the beginning.
- Words in the middle of sentences tend to be skipped over.

This concept is related to the old-to-new information flow, but it is also influenced by the way a reader's eye moves over the page. Readers read word groups, not individual words, and they use the capital letters and periods of sentences as orientation points. The natural tendency is to give more attention to the beginnings and endings of sentences.

In editing, therefore, you should try to make sure that the key words of a sentence appear at the end and the beginning. This may mean:

1. Moving a dependent clause.

The revised manual is now ready for release, even though it needed extensive rewriting.

Where does the emphasis fall in that sentence? On the release date? No, it seems to fall on the need for rewriting. But if you move the dependent clause to the beginning, the emphasis shifts.

Even though it needed extensive rewriting, the revised manual is now ready for release.

2. Moving modifier phrases or words.

Like most people, knowledge engineers are rarely trained in practical interviewing techniques.

How important is the phrase *like most people*? The key words in this sentence appear to be *knowledge engineers* and *practical interviewing techniques.* If you shift the prepositional phrase to the middle of the sentence, the key words appear at the beginning and end.

> Knowledge engineers, like most people, are rarely trained in practical interviewing techniques.

You should also consider individual adverbial connectors like *however, consequently,* and *nevertheless.* For example, which of these sentences stresses the word *nevertheless* the most?

> Nevertheless, the parts will be shipped on time.
>
> The parts, nevertheless, will be shipped on time.
>
> The parts will be shipped, nevertheless, on time.
>
> The parts will be shipped on time, nevertheless.

Following the rule, you might well say that the last sentence gives *nevertheless* the most stress. But is that the most effective sentence? What other words do you want to stress? If your customer is waiting, the words *on time* are key words. If you put *nevertheless* at the end, you move those words to the middle, which is the least emphatic position. Would you agree, then, that the first sentence—with *nevertheless* at the beginning and *on time* at the end—is most effective?

3. Making the sentence passive, using an expletive, or both.

> In Japan, the private sector has carried the burden of research and development, even in basic research areas.

What are the key words in this sentence? Probably not *even in basic research areas,* although these words appear at the end. More likely, the key words are *burden of research and development* and *private sector.* Those key words should not be buried in the middle of the sentence. If you rewrite the sentence in passive voice, you might achieve the needed emphasis.

> The burden of research and development, even in basic research areas, has been carried in Japan by the private sector.

Note how the two emphatic positions are filled by the key words *burden of research and development* and *private sector.* However, you have to ask if you want the sentence to be passive.

4. Breaking a sentence into a numbered or bulleted list.

Then after inserting the card into the slot, press one of the one-touch keys located on the operator control panel to immediately transmit your document.

This series of instructions would be easier to follow if the items were stacked in a numbered list.

To transmit your document:
1. Insert the card in the slot.
2. Press one of the one-touch keys located on the operator control panel. [Better to specify a key.]

In lists, you need to think about the order of your items. The items should be chronological, the most important should come first, or the most important should be last—with the items building in importance to the end.

In general, sentence revising for emphasis requires thought and analysis because you must balance the principle of emphasis with other demands of the sentence. You need to think about:

- the advantages of subject-verb-object order
- the choice of a passive or active construction
- parallelism (stating similar items the same way)
- the location of the action in the sentence
- clarity

And you must balance all these enhancements at the same time. Thus, in order to emphasize certain words, you may even disregard other specific principles. Whenever possible, you should play with sentence arrangement at the editing stage, shifting words around for various effects until you find an order that is pleasing and also fulfills the purpose of the document.

EXERCISE 13

Revising Sentences to Emphasize by Position

Determine the key words in each of the following sentences. Then rewrite if necessary, moving those key words into positions of emphasis.

1. Where the stays join on a ten-speed bicycle is where the wheel and the rear gear mechanism are hooked up.

2. One of the most common methods to reproduce the image on the fax machine is thermal.

3. Using definitions selectively and presenting them in an objective style is how convincing documents are written.

4. Because of the unavailability of any comparative analysis between brands of surge suppressors in the microcomputer-user marketplace, and because of the need for those microcomputer users to have the proper pieces of equipment, powerwise and costwise, to protect their computers, I chose to research and report on the differences, similarities, effectiveness, and cost of at least one dozen samples available to the general public.

5. On the agenda was whether to recommend approval of the development of a 35-acre mobile home park 3 miles north of River City.

6. If a student is attending school, and not living with the aided parent, an evaluation should be made to determine the parent's and child's intent: Should the child be considered emancipated, is a home being maintained, is the parent contributing to the needs of the child? Statements from both the parent and the child will be needed to determine present status and future intent, and then determination of FBU composition and exempt status of any loans or grants can be made.

EXERCISE 14

Putting It All Together: Editing for Clarity and Economy

1. Bring to class a sample of your own writing, preferably something technical, of at least three pages.
2. Exchange writing samples with a classmate.

3. Choose at least five sentences from the sample and edit them to improve their effectiveness.
4. Return the sample and rewritten sentences to the writer.
5. When you receive your own edited sentences, read them carefully and consider their effectiveness as rewritten.
6. Write a response to the editor, commenting on each rewritten sentence and then on the editor's work overall.
7. Turn in all the papers to the instructor.

EXERCISE 15

Putting It All Together: Editing for Diction, Syntax, and Punctuation

Read the following section of a procedures document to get a sense of the whole. Write a paragraph assessing the effectiveness of the whole document and suggesting how you would fix it. Then, choose 15 sentences that you can improve by editing for diction, syntax, and/or punctuation. On a separate sheet, copy the original sentences, writing your edited version below each one. Do not simply choose the sentences that are easiest to edit; choose those with a variety of problems. The document's purpose is to set standards for reviewing reports of complaint investigators. The audience is federal investigators for Equal Employment Opportunity complaints. The type style is that of the original.

GUIDANCE FOR INVESTIGATORS #3 — SUGGESTED TECHNIQUES FOR INVESTIGATING COMPLAINTS INVOLVING PERFORMANCE OR RELATED ISSUES/ALLEGATIONS

THERE HAS BEEN SOME SLIPPAGE RECENTLY IN THE INVESTIGATION OF COMPLAINTS INVOLVING PERFORMANCE OR RELATED ISSUES. THE PERFORMANCE OR RELATED ISSUE COMPLAINT IS ONE IN WHICH THE LEVEL OR MANNER OF COMPLAINANT'S PERFORMANCE IS IN DISPUTE. THESE COMPLAINTS MAY INCLUDE SUCH MATTERS AS PERFORMANCE APPRAISALS, PROGRESS REVIEWS, ASSIGNMENT OF DUTIES, WORK CONDITIONS, NON-SELECTIONS, DENIAL OF PROMOTION, DENIAL OF AWARDS/WITHIN-GRACE INCREASES, CERTAIN CLASSIFICATION ACTIONS, AND DISCIPLINARY ACTIONS THAT RELY ON PERFORMANCE. [THIS LISTING IS NOT INTENDED TO BE EXHAUSTIVE.]

THE FOLLOWING GUIDELINES ARE PROVIDED FOR PURPOSES OF SETTING FORTH STANDARDS BY WHICH REPORTS OF INVESTIGATION OF COMPLAINTS RELATING TO SOME ASPECT OF PERFORMANCE WILL BE REVIEWED AND APPROVED.

CHAPTER 13   BUILDING EFFECTIVE SENTENCES   **239**

A. COMPLAINANT'S AFFIDAVIT

ESSENTIAL TO THE SUCCESSFUL INVESTIGATION OF A PERFORMANCE OR RELATED ISSUE COMPLAINT IS COMPLAINANT'S AFFIDAVIT. CARE MUST BE TAKEN TO ASSURE THAT COMPLAINANT FULLY ADDRESSES ALL OF THE ACCEPTED PERFORMANCE OR RELATED ISSUES/ALLEGATIONS AND BASES OF HIS/HER COMPLAINT IN HIS/HER AFFIDAVIT. MOREOVER, SOME ACCEPTED PERFORMANCE OR RELATED ISSUES/ALLEGATIONS IMPLY PERFORMANCE OR RELATED SUB-ISSUES/ALLEGATIONS THAT ALSO NEED TO BE FULLY EXPLORED AND ADDRESSED BY COMPLAINANT. IT IS SELDOM LIKELY THAT COMPLAINANT WILL PRODUCE AN ADEQUATE AFFIDAVIT IF LEFT TO HIS/HER OWN DEVICES. INSTEAD, NEARLY EVERY COMPLAINANT WILL HAVE TO BE ACTIVELY LED, GUIDED, QUESTIONED, EXAMINED AND INTERROGATED ABOUT EACH PERFORMANCE OR RELATED ISSUE/ALLEGATION, PERFORMANCE OR RELATED SUB-ISSUE/ALLEGATION IN ORDER TO ARRIVE AT AN AFFIDAVIT THAT IS FULLY RESPONSIVE TO THE MATTERS UNDER INVESTIGATION.

UNLESS A COMPLAINANT HAS THOROUGHLY ARTICULATED IN THE AFFIDAVIT SUCH BELIEFS REGARDING ALL OF THE ACCEPTED ISSUES/ALLEGATIONS AS THEY RELATE TO PERFORMANCE OR RELATED ISSUES, THIS INSTRUMENT WILL BE FLAWED. IT IS LIKELY TO LEAD TO INADEQUATE RESPONSES FROM ALLEGED DISCRIMINATORY OFFICIALS (ADO'S), MANAGEMENT OFFICIALS, AND FRIENDLY/HOSTILE WITNESSES.

ONCE COMPLAINANT HAS ARTICULATED HIS/HER BELIEFS ABOUT HIS/HER PERFORMANCE, IT IS REQUIRED THAT COMPLAINANT SUBMIT OR INDICATE WHERE CAN BE OBTAINED DOCUMENTARY EVIDENCE THAT S/HE BELIEVES WILL SUPPORT HIS/HER ASSERTIONS THAT S/HE PERFORMED AT THE LEVEL OR IN THE MANNER S/HE MAINTAINS. THESE DOCUMENTS ARE TO BE APPENDED TO COMPLAINANT'S AFFIDAVIT AS ATTACHMENTS. IF COMPLAINANT SETS FORTH THAT S/HE HAS NO IDEA OF OR IS UNAWARE OF THE EXISTENCE OF SUCH DOCUMENTARY EVIDENCE TO SUPPORT HIS/HER CONTENTIONS, THIS MUST BE RECORDED IN COMPLAINANT'S AFFIDAVIT.

B. DESIGNATION OF ADO'S

IN VIRTUALLY EVERY PERFORMANCE OR RELATED ISSUE COMPLAINT, THERE ARE AT LEAST TWO ADO'S; IN SOME, THREE; AND, OCCASIONALLY, FOUR (PARTICULARLY SOME

NON-SELECTION AND MANY DISCIPLINARY ACTIONS). AN ADO IN A PERFORMANCE OR RELATED ISSUE COMPLAINT MAY BE ANY PERSON WHO, IN AN OFFICIAL CAPACITY, WAS RESPONSIBLE FOR AND/OR CONTRIBUTED TO THE ACT, DECISION, EVENT OR OCCASION COMPLAINED OF. IT WILL ALSO BE ANY PERSON WHO RECOMMENDED, ENDORSED, APPROVED OR SUSTAINED THE ACT, DECISION, EVENT OR OCCASION COMPLAINED OF. A COMPLAINANT MAY NOT HAVE DESIGNATED ANY ADO OR ALL OF THE ADOs. THIS DOES NOT DEVIATE THE NECESSITY FOR TAKING AN AFFIDAVIT FROM PERSONS THAT THE INVESTIGATOR IDENTIFIES AS ADO'S. THUS, THE INVESTIGATOR MUST CLOSELY ANALYZE THE FINAL EEO COUNSELING REPORT, THE FORMAL COMPLAINT AND OTHER APPROPRIATE DOCUMENTS IN THE COMPLAINT FILE IN ORDER TO DETERMINE EXACTLY WHICH PERSONS ARE TO BE TREATED AS ADO'S.

C. ADO'S AFFIDAVITS

AS INDICATED IN THE FOREGOING, AN AFFIDAVIT MUST BE TAKEN FROM THOSE PERSONS IDENTIFIED AS AN ADO. HAVING OBTAINED A FULLY ARTICULATED AFFIDAVIT FROM COMPLAINANT, IT IS REQUIRED THAT A FULLY RESPONSIVE AFFIDAVIT BE TAKEN FROM EACH IDENTIFIED ADO WHETHER PREVIOUSLY IDENTIFIED IN THE ACCEPTANCE LETTER OR NOT. IN ORDER TO FACILITATE THE ADO'S RESPONSE, THE INVESTIGATOR SHOULD SHARE WITH THE ADO: THE LETTER OF ACCEPTANCE, COMPLAINANT'S AFFIDAVIT TOGETHER WITH APPROPRIATE ATTACHMENTS, OTHER AFFIDAVITS IN WHICH THE ADO IS NAMED (THESE MUST BE SANITIZED OF NAMES OF AND IDENTIFYING INFORMATION ON PERSONS OTHER THAN THE COMPLAINANT AND THE ADO TO AVOID INVASION OF PRIVACY) AND THE FINAL EEO COUNSELOR'S REPORT. (WHILE AN ADO IS ENTITLED TO SEE THE FINAL EEO COUNSELOR'S REPORT, THE INVESTIGATOR MUST TAKE CARE TO CONFINE THE INVESTIGATION TO THE ISSUES/ALLEGATIONS SET FORTH IN THE LETTER OF ACCEPTANCE.)

AS WITH COMPLAINANT, CARE MUST BE TAKEN TO ASSURE THAT EACH ADO FULLY RESPONDS TO EVERY PERFORMANCE OR RELATED ISSUE/ALLEGATION AND PERFORMANCE OR RELATED SUB-ISSUE/ALLEGATION. SIMILAR TO COMPLAINANT, AN ADO WILL ALL TOO FREQUENTLY NOT BE FULLY RESPONSIVE TO THE PERFORMANCE OR RELATED ISSUES/ALLEGATIONS UNLESS EACH IS ACTIVELY PROMPTED INTO THOROUGHLY ARTICULATING A RESPONSE. ALSO, AS WITH COMPLAINANT, REQUIRE EACH

ADO TO SUPPLY OR INDICATE WHERE CAN BE OBTAINED DOCUMENTARY EVIDENCE THAT SUPPORTS THAT ADO'S CONTENTIONS. IT IS CRUCIAL THAT THE ADO SUBMIT DOCUMENTATION THAT SUPPORTS HIS/HER POSITION. THESE DOCUMENTS ARE TO BE APPENDED TO THE ADO'S AFFIDAVIT AS ATTACHMENTS. IF THERE IS NO DOCUMENTARY EVIDENCE, THEN THIS FACT IS TO BE RECORDED IN THE BODY OF THE ADO'S AFFIDAVIT.

ADO'S ARE TO BE GIVEN AN OPPORTUNITY TO SUGGEST WITNESSES WHO CAN BE CONTACTED TO CORROBORATE THE TESTIMONY.

AGAIN, AS WITH COMPLAINANT, THE ADO, LEFT TO HIS/HER OWN DEVICES WILL NOT LIKELY PREPARE A FULLY RESPONSIVE AFFIDAVIT. THUS, IF A GIVEN ADO HAS NO KNOWLEDGE ABOUT A GIVEN PERFORMANCE OR RELATED ISSUE/ALLEGATION, THEN A CLEAR STATEMENT OF THAT FACT IS TO BE SET FORTH IN THE ADO'S AFFIDAVIT. A MERE STATEMENT OF GENERAL OR ALL INCLUSIVE DETAIL IS UNACCEPTABLE.

References

Arnold, Matthew. Quoted in Joseph Williams. 1997. *Style: Ten lessons in clarity and grace.* 5th ed. New York: Longman.
Cadence Design Systems. 1993. *Style guide.* 6th ed. San Jose, CA: Cadence Design Systems, Inc.
Christensen, Francis. 1967. *Notes toward a new rhetoric.* New York: Harper, 41–51.
Harbaugh, Frederick W. 1991. Accentuate the positive. *Technical Communication* 36.1 (February).
IBM. 1987. *Think* 53:5. White Plains, NY.
Lanham, Richard A. 1992. *Revising business prose.* 3rd ed. New York: Macmillan.
Matalene, Carolyn B. 1989. *Worlds of writing: Teaching and learning in discourse communities of work.* New York: Random.
———. 1995. Of the people, by the people, for the people: Texts in public contexts. In John Frederick Reynolds et al. *Professional writing in context: Lessons from teaching and consulting in worlds of work.* Hillsdale, NJ: Erlbaum.
Miller, George. 1967. *The psychology of communication,* New York: Basic Books.
Wyse Technology, Inc. n.d. *Series 9000.* Technical Overview.

14 Developmental and Organizational Editing

Contents of this chapter:

- Developmental Editing
- Organizational Editing

Editing at the sentence level—copyediting and proofreading—is necessary to ensure a readable and correct document. Such editing, as Chapter 1 explains, occurs *after* the document has been drafted. However, some of the most important editing work you can do is in the early stages of a writing project, often even before a series of screens or a document is written. Whether you are a full-time editor, or a writer who acts as a some-time editor for another writer, you can be most effective in this kind of whole-document editing if you understand who the readers are, what they expect from a particular kind of document, and how much information they can manage as they seek to comprehend and apply what they are reading.

When she talks about levels of edit, freelance editor Jo Levy differentiates between two kinds of whole-document editing: developmental and substantive (also called rhetorical). Developmental editing, she says, is "a view from 30,000 feet, which can assist [the writer] in deciding what information to convey and how to convey it." Once that overall structure is in place—for example, manuals in a library, chapters in a manual, screens in a help system, sections within a chapter—"a substantive edit focuses on presentation and organization within that overall structure." Thus, *developmental editing* is the process of planning: working with the writer to determine the overall content, define the readers, and plan the general organization and method of presentation.

Substantive or rhetorical editing is the next phase, and it usually includes two parts: review of technical content and review of emphasis, logic, and organization. Review of the accuracy and completeness of technical content is primarily the responsibility of subject matter experts (SMEs) like engineers, programmers, scientists, or managers. Review of emphasis, logic, and organization, however, is the responsibility of the editor and is therefore the focus of the second part of this chapter. Whenever possible, whether through hands-on testing or fact-checking, editors should perform content editing. But even before the document goes to SMEs for their evaluation, editors should perform organizational editing. Both developmental and organizational editing require that you know some of the basic concepts that researchers have learned in the past 20 years about how readers actually read and use workplace documents.

14.1 Developmental Editing

In developmental editing, a writer and an editor work together to determine how information can best be packaged to meet the reader's needs. Two important factors in choosing that packaging are what the reader already knows about this kind of information, if anything, and how that information is stored in the reader's long-term memory.

Understanding Long-term Memory

Cognitive psychologists tell us that information and experiences are stored in our long-term memory in structures or frameworks called *schemata* (singular *schema*). This "prior knowledge serves as a framework which makes new information more meaningful and easier to absorb" (Huckin 1983, 92). A schema can be of "action" or of "form," and both use the reader's previous knowledge or experience to aid in understanding new information.

Action schemata "capture general information about routine and recurring kinds of events" (Williams 1994, 88). One frequently cited example of an action schema (also called a "script") is the sequence of events Americans think of as "going to a restaurant." In this sequence, adult experiences and knowledge lead us to expect that the following events will occur when we go to a restaurant:

1. A host or hostess leads us to a seat and gives us menus.
2. A server comes to take the food order.
3. Water, bread, and perhaps an appetizer and beverages are delivered.
4. The server brings our food, and we eat.
5. The server brings our check.
6. We give the server payment, which he or she takes to the cashier, or we take the check to the cashier as we exit.
7. We leave a tip for the server.

A schema is also considered *active*, which means that it is frequently modified or updated. Thus, Americans modify the restaurant schema for a fast-food restaurant, knowing that we usually order food standing at a counter and pay for it before we sit down; we often clear away our own dishes; we do not tip. Likewise, travelers will modify a restaurant schema to account for local customs: In Germany, for example, the server carries a large wallet from which he or she makes change at the table; in Japan, there is no tipping.

Form schemata help readers process and remember written documents. "Readers, in fact, appear to be quite sensitive to the overall organizational plan of an expository text, and typically use the perceived text structure as a guide for processing it." In fact, ". . . subjects presented with unstructured text recall structured text. Readers, in other words, impose a conventional structure in their recall of text content that was presented in an unstructured way" (Williams 1994, 90–91). What this means to editors is that the organization of a document should be conventional and explicit.

The general kinds of organizational plans that readers are sensitive to include:

1. *collection* (ideas or events related by what they have in common)
2. *causal* (relationship by cause and effect)
3. *response* (relationship by problem and solution)
4. *comparison* (relationship by similarities and differences)
5. *description* (attributes of the thing being described)

"A single passage, of course, may employ several of these organizational schemes. However, . . . the top-level structure of a passage is likely to employ only one" (Williams 1994, 92).

In technical communication, the organizational schemes may be more highly structured. For example, each of the three "genres" of presentation explained later in this section has a distinctive form. A proposal has standard parts arranged in a standard order, as do a set of instructions and a report. See "Packaging Information to Meet Reader Needs" later in this section.

Readers of technical information also have varying amounts of prior knowledge about what they are reading. Novices, or first-time users of a personal computer system, for example, already know how information is stored (in a filing cabinet) or managed (on a desktop). These concepts are therefore often used in describing storage and management of information on a computer. Experienced workers at a textile plant or a paper mill already know a typical sequence of events in their work; the schemata in their long-term memory can be used, with new information simply added to the framework or the framework modified as needed.

In regard to readers, the "power of schemata . . . resides largely in their ability to induce inferences from the reader . . . [The] writer does not have to refer explicitly to all the details of the schema for those details to be conveyed to the reader . . . [The] reader can simply fill in the slots" (Huckin 1983, 93).

Cognitive psychologists also remind us that people learn best when they are actively involved. This idea is especially important for writers of manuals or instructions because "a majority of readers will mentally revise instructions into scenarios: hypothesized situations in which a human agent performs a particular action" (Mirel 1988, 115). At the core of a scenario is the reader's thinking of "who did (or might do) what to whom."

Using Short-term or Working Memory

If our long-term memory is like the computer's hard disk (or perhaps a CD-ROM), our short-term memory is like the computer's buffer or temporary storage. Short-term memory holds the amount of information we remember only long enough to carry out a task, such as looking up a telephone number and then making the call. It also gives us the ability to remember one idea we have read long enough to associate it with new information. Short-term memory is working memory.

However, short-term memory is limited. Psychologist George Miller demonstrated in a 1967 study that the limit of short-term memory was seven plus or minus two items. That meant, for ease of use, readers should have to deal with no more than nine items of information at one time. Since then, researchers have determined that the "chunk capacity of short-term memory has been shown to be in the range of five to seven" (Simon 1974, 487). This new "magic" number means that readers process information best when the number of items is fewer than seven.

Short-term memory affects organization at both the developmental and organizational stages. When more than seven items are involved, whether they are chapters in a book or items in a list, the writer needs to group (chunk) them into a smaller number of related items. For example, in a set of instructions, 15 or 20 steps will overwhelm readers, whereas 3 or 4 "main steps" with 5 to 7 "substeps" will assure readers that they can carry out the task.

Analyzing Readers

Editors, as well as writers, need to understand exactly who their readers are because developmental decisions depend on two primary factors: the purpose of the document and the needs of the reader.

Reader Type

One way to understand readers is to examine their previous experience, training, and reasons for reading. In general terms, as Figure 14.1 shows, we can break readers into five groups that form a continuum from least technical to most technical (Rew 1993).

Generalists	Managers	Operators	Technicians	Specialists
←Least Technical				Most Technical→

FIGURE 14.1 Types of Readers

Generalists are lay persons who have no special experience or training in the field. They are reading for interest or information.

Managers may be engineers or scientists by training, but their current concerns are products, people, schedules, and budget. They read primarily for help in making decisions.

Operators or users may be nontechnical or highly technical, but their motivation for reading is to accomplish a task. That task may be as simple as entering data in a spreadsheet or as complex as operating a desalinization plant.

Technicians work in laboratories building, testing, maintaining, and repairing equipment. Like operators, they read in order to accomplish tasks, but the tasks may be varied and individual rather than repetitive. Technicians also have both technical knowledge and experience, thus possessing more elaborate schemata on which to build.

Specialists are experts in their field of work, with advanced degrees and years of experience. Specialists read to expand their knowledge base and to examine the work of colleagues. They have highly developed schemata and make inferences from what they read.

Reader Tasks

Another consideration in analyzing readers is to ask what their primary task is. In other words, how will they use the information the writer is providing? One way to determine reader tasks is to think about the area in which they work.

In *research laboratories,* readers are looking for information that can be synthesized and applied to new situations. Researchers want to find new materials and methods in a given technology.

In *preliminary design,* readers are setting standards for new products, establishing goals, and spelling out requirements. They may be evaluating proposals and planning for the long term.

In *development*, readers are converting design theory into performance. Developers translate the criteria of designers into specifications and working models.

In *production*, readers must take the processes developed in the laboratory and refine them to work efficiently in mass production.

In *marketing*, readers must sell the product, focusing on customer needs and the ways to meet those needs.

In *installation*, *operation*, and *maintenance*, readers' tasks are clear: They want to be able to do the job efficiently, easily, and correctly.

In *administration*, readers must coordinate activities; manage people, products, and budgets; and meet time constraints. Administrators often must make decisions, and they look to documents for help in that decision-making.

Reading Style

Finally, editors and writers need to ask: How will the reader probably read this information?

According to reading expert A. K. Pugh (1978), readers use one of the following five reading strategies, depending on their purpose:

1. skimming—reading for the general drift of the passage
2. scanning—reading quickly to find specific items of information
3. search reading—scanning with attention to the meaning of specific items
4. receptive reading—reading for thorough comprehension
5. critical reading—reading for evaluation

In reading technical documents, most readers will use one of the first three strategies, and editors can help improve organization and presentation to assist readers as they proceed.

Packaging Information to Meet Readers' Needs

At the developmental stage of editing, editors can help writers package information by choosing the *genre* of presentation. Writing theorists Killingsworth and Gilbertson define genre as "a *kind* of writing recognized as distinctive by writers and readers" (1992, 78). In literary studies, the commonly accepted genres are fiction, poetry, nonfiction, and drama. In workplace communication, these authors suggest three genres:

1. *proposals* (which deal with the future and are persuasive)
2. *manuals* or *instructions* (which deal with the present and are operational)
3. *reports* (which deal with the past and are narrative)

Each genre, or kind of writing, also has a distinctive form: the traditional way in which that information is presented to the readers. While the forms are not always rigid and prescribed, editors can help writers by following the general guidelines for each form. By fulfilling what readers expect, the proper schemata are triggered.

Proposals

A *proposal* is a written offer to solve a problem or provide a service. Thus, a proposal is a persuasive document that asks its readers to commit to action (accept the proposal) so that the writer or agency is given the authority to proceed. Proposals are evaluated on the basis of their specified procedure for solving the problem or providing the service; their identification of people with needed expertise; their timetable and budget.

The typical form of a proposal includes the following parts in this order:

- executive summary
- introduction, including statement of problem or need
- procedure or methods (scope of work)
- evaluation method
- schedule
- personnel
- budget
- appendixes

For example, a "Proposal to Design, Fabricate, Test, and Demonstrate a Coal Mine Hopper-Feeder-Bolter" is directed to the requesting agency, the U.S. Interior Department's Bureau of Mines. Such a proposal describes the procedure to be followed in the design, manufacture, testing, and demonstration in a section called *Scope of Work*. The proposal details how the work will be evaluated: in this case, how successfully it loads coal while concurrently bolting the mine roof. The proposal defines a schedule for the work. Available personnel are described, and their experience is validated with résumés. Finally, the proposal includes the projected costs and terms.

Manuals, Procedures, and Instructions

A *manual* usually gives both information and instructions or procedures. Manuals can be written for a variety of purposes, including training, installation, operation, maintenance, trouble-shooting, and repair. Some manuals are written for novice users, while others target experienced users; sometimes one manual must serve both kinds of users. Finally, manuals often contain reference information, or there may be a separate reference manual. With differences in purpose and readers, manual forms can vary widely.

The typical form of a set of instructions includes the following parts in this order:

- overview
    - materials
    - tools and equipment
    - summary of steps
    - time required
- individual steps in sequential organization

For example, a simple one-page set of instructions tells "How to Dress for a Cleanroom." The instructions begin with a list of materials—such items as eye or facial hood, cleanroom smock, white nylon gloves, clear vinyl gloves. The overview announces the number of major steps (four) and says the procedure takes 10 minutes. The body of the instructions breaks down the four steps in chronological order, with substeps presenting details.

An example of a complex manual is an Administrator's Manual for *Lotus cc: Mail Router*. This manual of almost 400 pages contains a preface with an overview, definition of types of readers, definitions of terms, and requirements. It contains nine major sections of instructions and seven appendixes detailing such information as command options, connections, networks, and error messages.

Reports

Purpose, not form, determines the two major classifications of reports. *Recommendation* reports either examine alternative products or courses of action, or they examine the feasibility of action. Their purpose is to recommend some action to a decision maker. Subcategories of recommendation reports are comparative reports and feasibility reports.

Completion reports present information gathered by research in the laboratory or field or at the close of a project. Completion reports do not usually evaluate or recommend; instead, they provide information or record accomplishments. Subcategories of completion reports are field or laboratory research reports and project reports.

Recommendation reports generally follow this order:

- executive summary
- introduction
- body, including:
    - background
    - theory
    - description
    - definition
    - methods
    - criteria

evaluations and comparisons
results
analysis of results
- conclusions and recommendations
- references and appendixes

Research reports and completion reports follow this order:

Research reports

- abstract
- introduction, including literature review
- materials and methods
- results
- discussion
- references

Completion reports

- abstract or executive summary
- introduction, including background information
- methods
- factual and technical information
- conclusions
- appendixes, including references

Online Documentation

When writers move from paper (hardcopy) to online, the "document" disappears and is replaced by "documentation." In other words, a document with a discernible size, shape, and organization is replaced by a body of information with no visible physical structure and usually no predictable reading order or entry and exit points. While instructions, context-sensitive help, and reports can all appear online, the changes of form pose challenges to both writers and editors.

At the developmental editing stage, editors can look at several attributes that contribute to the logical organization of online documentation from the reader's perspective.

1. **Chunks.** Is the information presented in short units or chunks that are self-contained? Is each chunk confined to a single screen?
2. **Navigational aids.** Do maps or flowcharts provide orientation? In other words, can readers see where they are at any time within the larger context? Can they keep track of where they have been? Can they backtrack or retrace their steps?
3. **Menus and key words.** Can readers choose where they want to move within an information hierarchy or web? Can they search by keywords? Are the number of menus and menu options limited (fewer than seven), so readers can use short-term memory?
4. **Screen format.** Do screens consistently display similar kinds of information in the same place? For example, is cueing information on the left side of the screen? Is guidance and placement information at the top of the screen? Are directions and responses at the bottom?
5. **Readability.** Are the type of font and the size of font easy to read on the screen?
6. **Amount of text.** Is the amount of text on each screen limited to 50 percent of screen space? Are the line lengths short (40 to 60 characters per line)?
7. **Graphics.** Does the document include a variety of graphics that are simple and easy to understand?

For details of online editing, see Chapter 18.

Working with the Writer in a Developmental Edit

Editors and writers need to communicate regularly during a developmental edit. Most often, the decisions about choosing a genre or form are straightforward and dictated by the purpose of the assigned document, the identity of readers, and what the readers want the document to do. But an editor can help a writer clarify genre and form choices and ensure that those choices fulfill the intended purpose and meet reader needs.

In a developmental edit, editor and author often meet to clarify issues. Alternatively, the editor may review the writer's first or alpha draft and write a memo with suggested changes. Figure 14.2 shows the opening paragraph of such a memo, written by editor Anjali Puri (reprinted by permission.)

> To: Rob Berkowitz
> From: Anjali Puri
> Subject: cc:Mail View Alpha Help Editorial Survey
> cc: Ron Kirchem, Gail Haspert, and Sylvia Thompson
> Date: October 19, 19—
>
> The alpha draft of cc:Mail View Help is developing very well. This pass is a developmental edit that focuses on developing patterns (that users expect) and topics that are task oriented. I like the simple organizational approach you've used to structure Help since it makes the information easy to follow; however, I've copyedited some files to develop consistent patterns that anticipate how administrators find information in Help.

FIGURE 14.2 Editor's Comments After a Developmental Edit

14.2 Organizational Editing

While developmental editing concentrates on the broad issues of what information to convey and how to convey it, organizational editing considers emphasis and presentation within the larger structures. Organizational editing is only one part of the process of substantive or rhetorical editing, but it is the part that an editor can most effectively influence. Ensuring accuracy and completeness of content is usually the responsibility of SMEs, although editors who have considerable experience with a product or procedure can also assess its accuracy and completeness. Organization and presentation are closely allied because, in effective documents, the organization is supported and highlighted by the presentation (format, headings, visual cues, and so on).

What Are the Methods of Organization?

Just as the overall purpose of a document influences the writer's choice of genre, so too the purpose of each section influences the writer's choice of a method or pattern of organization. Methods of organization may vary from section to section within a larger document; the editor's task is always to evaluate the writer's chosen method of organization to see if it fulfills the purpose of that section and meets the readers' needs.

The following sections briefly review the various methods of organization commonly used in workplace documents and give examples of each method. In editing, you need to remember that the same material can often be organized in more than one way. As an editor, you need to "re-view" the material to test the writer's chosen method.

Organization by Sequence

Sequential organization methods have items that follow one another in order. In *chronological* organization, the points are arranged on the basis of time, from first to last. Chronological organization is useful in documents like the following:

- instructions (how to install an upgrade card in a computer)
- process descriptions (how hail forms)
- the methods section of a laboratory report (how the experiment was conducted)
- the procedures section of a proposal (how the field research will be carried out)

In *spatial* organization, the points are arranged as they appear physically, either in relationship to one another or the surroundings. Spatial organization can be left to right, top to bottom, outside to inside, or variations of these patterns. Spatial arrangement is typically used in descriptions of the following:

- places (a building, an airport, a parking lot)
- objects (an incinerator, a keyboard, a centrifuge)

A third type of sequential organization is *alphabetical* (indexes, concordances, glossaries). Alphabetical sequence is often used for reference material because readers are accustomed to "looking up" items in an alphabetical sequence such as that used in a dictionary.

Organization by Hierarchy

Setting up a hierarchy means discriminating between important points and less important points and showing their relationship by means of grouping.

Perhaps the most common kind of hierarchy is achieved by *division* or *classification*. In division, a topic is broken into its constituent parts; in classification, items are grouped by their relationships under some umbrella term. Division is often used for the following:

- descriptions (the parts of the apple blossom, the components of a camcorder)
- procedures (managers' duties for annual employee evaluations)

Classification is used in reporting information like the following:

- field research (the number of respondents who favored the new contract)
- description (the types of roses planted in a test plot, the types of readers involved in a study)

In the kind of hierarchical organization called *comparison*, the organization is achieved by showing similarities or differences between two or more objects or courses of action or by explaining an unfamiliar subject by comparing it to a familiar one. Comparison may form the central part of a report where analysis occurs (word processing system A versus system B); comparison may also be part of an extended definition (computer servers are like managers when they allocate resources and determine order of need).

Pro and con organization depends on presenting information by grouping advantages and disadvantages, presenting one group at a time, and following the presentation with conclusions and recommendations. Pro and con organization is used in evaluations (advantages of method A and method B, disadvantages of both) and recommendations (the advantages of A outweigh those of B).

Organizational Combinations

Many methods of organization have elements of both sequence and hierarchy. In most cases, the major elements show hierarchy, while the subpoints are arranged sequentially. Following are typical combined methods of organization:

Method	Used for
General to specific	*Introductions*. This manual will show you how to install, operate, and repair the XXX.
	Evaluations. The efficiency of the solar collectors is based on five major factors.

	Descriptions. Viewed externally, the workstation consists of . . .
	Recommendations. We recommend that the contract be awarded to — for —.
Most-to-least important	*Explanations.* The primary element in laser safety is . . .
	Work activity reports. Principal activity during the period of — to — included . . .
	Conclusions. Of the three options, option C is most important to the operation of . . .
	Recommendations. The first action should be to . . .
Most-used to least-used	*Manuals.* The most commonly used feature is . . .
	Tutorials. Always save the file before exiting the program.
Questions and answers	*Procedures.* What determines eligibility for emergency funding?
	Instructions. How do I apply for a new license?
	Process descriptions. What is fractional distillation?
	Object descriptions. What is a flat panel?
Cause and effect	*Analysis.* The rise in temperature is due to the following four factors . . .
	Accident reports. The crash was caused by . . .
	Proposals or plans. The effect of the proposed actions will be . . .
Problem and solution	*Analysis.* The problem was caused by rotation of the engine due to . . .
	Recommendations. We recommend installation of restraints as the most cost-effective solution to the problem of . . .

As an editor, you need to understand all these methods of organization and their most common applications so that you can assist the writer in choosing the most effective method of organization for each particular document.

What Does an Editor Need to Know about Organization?

As an editor, you need to be concerned with organization and outlines in four different ways:

1. To be able to read a document and outline it (extract the main points) to see how it is organized
2. To be able to look at an outline, consider it in light of the purpose and the reader, and evaluate whether it works
3. To reorganize the material in either a document or an outline so that it most efficiently presents the material
4. To justify any changes you make

How to Determine the Organization of a Written Piece

Here are the steps you can follow to determine how a document is organized:

1. Look at the table of contents and compare it with the document to see if it shows the main points.
2. If there is no table of contents, read the whole document quickly.
3. Ask the writer or determine from the text the purpose, scope, and intended reader or readers. Find the thesis or main point.
4. If no outline exists, outline the organization:
    - Use headings and subheadings to determine the main points.
    - Use paragraphs for the next main divisions. Look for topic sentences.
    - Pick out key words.

   Write either a sentence or topic outline and use any numbering system with which you are comfortable.
5. Compare your outline with any summary or abstract of the document, or any outline or introductory statement that indicates the organization.

EXERCISE 1

Determining the Method of Organization

The purpose of the following paragraph is to describe the products extracted from crude oil. The reader is a generalist. First, determine the best method of organization to fulfill this purpose for this reader. Second, analyze the paragraph to see what method of organization is used. Third, write a paragraph telling how you could improve the organization. Write your analysis and new paragraph on a separate piece of paper. Selected answers appear in the Appendix.

The "boiling range" of crude oil progressively yields lighter petroleum products (such as gasoline and butane), then heavier (naphtha, light solvents such as paint thinner), then kerosene, #1 diesel, home heating oil, and #2 diesel. Further up the boiling range are extracted industrial heating oil and lubricating oil. The *end point*, the temperature at which most of the diesel in a barrel of crude oil suitable for high-speed automotive engines has been vaporized out, is between 690 and 725 degrees F.

How to Evaluate an Outline

Once you have outlined a document or have received the outline from a writer, you need to evaluate it to see if it is doing the job for which it is intended: fulfilling its purpose and meeting reader needs. Here are the steps to follow in evaluating an outline.

1. Consider the purpose, asking:
   - What is this document supposed to do? Should it recommend? Instruct? Inform? Analyze? Report?
   - How does this document fit into the larger series of documents or publication plan?

2. Consider the reader, asking:
   - What is the reader's background? Is the reader a generalist, a manager, an operator, a technician, or a specialist? Is this totally new information, or is the reader already familiar with something similar?
   - What does the reader want to do after reading this document? Make a decision? Install or repair something? Test something? Buy something?
   - How will the reader probably read this information? Skim it? Scan it for specific information? Search-read for the meaning of parts of the document?

3. Consider the scope, asking:
   - How broad or narrow is this document's intended coverage?
   - Should the scope be changed? If so, how?

4. Ask if a thesis is evident. Look at the title, introduction, abstract, and major points for clues about the thesis.

5. Determine the number of major points. Remember that for short-term memory, points should average three to seven, never more than nine.
   - Are all points equally important?
   - Is all the material related?

- Is there a logical and defensible reason for the arrangement of the major points? Review the types of organization in this chapter if necessary.
- Are items of equal importance parallel?
- Has anything been left out? Is information included that is irrelevant?

6. Work through the main points first, asking the above questions; then look at the next level of points, and so on.

How to Reorganize an Outline

In reorganizing, you may have to go back to the steps that the writer (presumably) went through in deciding on the initial organization.

1. Brainstorming or random listing of ideas as they come. As an editor, you are limited by the information in front of you, but you can reproduce the process by jotting main points on separate cards. You can also use the technique of clustering (grouping items and connecting them with lines to show relationships).
2. Translating the random listing of ideas into some kind of sequential or hierarchical organization.
   - Review the purpose and reader.
   - Refine the thesis if necessary, or write one if it is missing.
   - Arrange the items in some kind of meaningful order. Study the kinds of arrangement discussed in this chapter.

EXERCISE 2

Editing an Outline

The following outline is for a report comparing four microwave ovens in order to recommend one for purchase. The reader is a consumer interested in buying a microwave oven. First, determine what method of organization would be most appropriate for this purpose and this reader. Second, analyze the following outline and tell what its method of organization is. Third, reorganize the outline.

1.0 Introduction
    1.1 Definition of a microwave
    1.2 History and development of microwave ovens
    1.3 How it works
    1.4 Why is it safe?

2.0 Types of Ovens
    2.1 Countertop
    2.2 Under the cabinet
    2.3 Over the range
    2.4 Built-in
3.0 Sizes and Their Uses
    3.1 Compact
    3.2 Midsize
    3.3 Fullsize
4.0 Distribution of Microwaves in Oven
    4.1 Dual-wave system
    4.2 Multiwave system
    4.3 Rotowave system
    4.4 Revolving turntable
5.0 Other features
    5.1 Power available
    5.2 Power levels
    5.3 Cooking controls
        5.3.1 Types
            5.3.1.1 Touch dial
            5.3.1.2 Dial timer
        5.3.2 Features
    5.4 Oven internal capacity
    5.5 Combination ovens
        5.5.1 Convection/microwave
        5.5.2 Conduction/microwave
    5.6 Price
6.0 Conclusion
    6.1 Comparison of actual models
    6.2 Choice of a model and why

How to Justify the Changes You Make

Editing for organization is always difficult because you are challenging the very heart of a writer's work. Therefore, you must be both tactful and sensitive, remembering that there may be reasons for the organization as it exists. Your justification must be based on fulfilling the purpose and meeting the readers' needs, and unless you are also a technical expert, you need to realize that you may have changed the meaning by recasting the organization. Proceed very carefully. Ask questions instead of making pronouncements. Try to meet with the writer instead of simply writing your changes.

    If possible, try to edit for organization by examining an outline before the writer has written the first draft. Elizabeth Wilde, who edits for a major computer manufacturer, says, "I often find that writers are grateful for help on restructuring a document, especially if they are stuck and realize that the organization is not what it could be. Editors should always make an effort to be involved in the early stages of information development, so they can have input into the organization before the writer gets too attached to something that might not work. I enjoy doing outline reviews and find that writers often have an easier time filling in the details when they're confident about the basic structure."

Sylvia Thompson, senior editor at a software company, stresses the tact needed in editing. She says that an editor needs to know "how to relate to a writer. You have to respect the writer's right to what he or she has done . . . to be aware of the writer's position."

Yvonne Kucher, senior editor at a networking company, talks about the process. "In editing," she says, "everything is negotiated. The first time I review a document for a writer, we look at the schedule and I ask what I should focus on. Often the schedule is the dominant factor. At other times, it might be the importance of the product to the company.

"If I'm lucky, I have background material like a product announcement. I also look at the document plan, and I compare what I have against the plan. If I don't have a plan or if this is a new product, I call the writer and ask 'What are our assumptions? Has this person [the reader] been trained? Is this a novice?'

"When I begin editing, I read everything carefully. I study the art work to see how it fits in with the document. I start a list of things to watch for. For example, if I see 35 steps, I ask if there are really two procedures instead of one. If the procedure looks complicated, can it be simplified? Are cautions or warnings around items missing?

"Writers know that the only person who reads a manual from cover to cover is the editor, and they respect that. *I am not making work for writers, I am making their work better. My job is to make the writer's work shine.*"

EXERCISE 3

Recognizing the Existing Organization: A Definition

Follow the suggestions in this chapter to determine the existing organization of the following definition. Outline it. Write a paragraph explaining the problems in that organization.

A data base is a particular software program used to store and retrieve data. All data bases perform this common function. How they accomplish this sets them apart from one another. Data bases are comprised of three basic processes: data entry, data accessing, and data reporting.

The term software can be used in many different contexts. It refers to a set of instructions written for a computer to follow. The first process in using a data base is called data entry. Data entry is the process of creating the data base. The data are placed into records. Records are the specific pieces of information. A record is usually just one particular piece of data. This information can be either character or numeric data. Character data are words or sentences. Numeric data are numbers usually used for calculation purposes. The next step in using a data base is data accessing. Data accessing is the process of retrieving data. This is usually done by sorting records for a similar bond. Sorting refers to the process of arranging data in a specific order. The program goes to each record and examines it. If the record does not meet the qualifications, then the program goes on to the next record. Data reporting is the process of comparing the data in a logical way in order to interpret it. This can be done with tables, charts, and even graphs.

These three processes act together in the operation of the data base. Each step is dependent on the other. Every data base performs these specific functions in different ways.

EXERCISE 4

Recognizing the Existing Organization: A Comparison

The following comparison is the conclusion of a technical report analyzing exercise bicycles in order to recommend one for purchase. The audience is generalists. Study the document and write an outline showing how it is organized. Then, in a paragraph, suggest an improved organization using the information in the accompanying table.

V. COMPARISON OF THE MODELS

The criterion scores that are calculated in the previous section are used to compare the four models of exercise bicycles with each other. Table 2 is a list of each model's eight criterion scores and the total number of criteria score points that each model received. Although three of the models, the Schwinn XR8, the Monark 875 Home Cycle, and the AMF Whitney Ergometer, all have total scores that are within one point of each other, their individual criterion scores are dramatically different.

At $249.95, the Schwinn XR8 costs between $50 and $100 less than the other three models. Besides having the lowest price, the XR8 has the best score for Repair and Maintenance Service, and with the AMF Whitney, it is one of the most accessible models in this study. The XR8, however, is only mediocre when it is compared to the construction, pedaling action, and resistance device of either the Monark 875 Home Cycle or the AMF Whitney Ergometer. Although the XR8 offers the widest selection of colors and a good range of extra features, it is one of the noisiest models in this study.

The Monark 875 Home Cycle is the most user-to-user adjustable model, and it has the second best repair and maintenance service accompanying it. It is, however, the most difficult model to purchase, and it offers no extra features. In addition, at $325, the Monark 875 is the second most expensive model in this study.

Like the Monark 875, the AMF Whitney Ergometer's strength lies mainly in its frame design and construction. The Whitney Ergometer, however, lacks both the Monark 875's range and ease of adjustment. Priced at $349.95, the Whitney Ergometer is the most expensive of the exercise bicycles; moreover, it ties with the Tunturi Home Cycle for last place in the repair and maintenance category.

The Tunturi Home Cycle has the lowest total point score of all four exercise bicycles, and the biggest contributing factor to this point deficit is the Home Cycle's construction. However, the Home Cycle is also consistently rated lower because of inadequate user-to-user adjustability, excessive noise, and limited availability.

TABLE 2 Comparisons of Criteria Scores for the Exercise Bicycles

	A	B	C	D	E	F	G	H	I
Schwinn XR8	6	3	4	9	4	3	1	3	33
Monark 875 Home Cycle	2	1	3	14	7	4	3	0	34
Tunturi Home Cycle	3	2	0	2	3	5	1	1	17
AMF Whitney Ergometer	2	3	0	14	3	5	3	3	33
Points Possible	7	3	4	16	8	5	4	5	52

KEY
A= Cost
B= Product Availability
C= Repair and Maintenance Service
D= Construction
E= User-to-User Adjustability
F= Pedaling Action Resistance Device
G= Noise
H= Color Selections and Extra Features
I= Total Points

How Can an Editor Clarify the Chosen Organization of a Document?

After evaluating the document's organization and perhaps making changes, the editor can also clarify the chosen organization for the reader by using any or all of the following three techniques:

1. *Verbal devices* such as headings and subheadings, parallelism, and access words
2. *Visual devices* such as chunking with white space, showing hierarchy with type size and style, listing, and using icons
3. *Page or screen design*, such as placement of text, relationship of text to graphics, consistent placement of similar information

Chapters 15 and 16 provide detailed information about graphics and page design; this chapter explains verbal devices and simple visual devices.

Verbal Devices

Three verbal devices that effectively clarify a document's organization for a reader are headings, parallelism, and access aids.

Headings. *Headings* work well to direct the reader's eye when scanning or search reading, especially when the headings are made visually prominent by boldface type or larger type size. Headings and subheadings can also provide valuable organizational information for the reader if they do two things:

1. tell what information is covered in the following section through key words (content information)
2. tell how that information is handled (organizational indicators)

Content information is conveyed through key words—primarily nouns—that tell what the important information is. For example, here is a partial outline for a technical report describing the design, fabrication, testing, and demonstration of a machine designed to mine coal from underground, while concurrently supporting the roof of the mine shaft.

    4.0   TECHNICAL DISCUSSION
        4.1   Preparation of the Equipment for Mine Demonstration
        4.2   Underground Mine Demonstration
        4.3   Operation of the Hopper-Feeder-Bolter
            4.3.1   Hopper-Feeder-Bolter Operator
            4.3.2   Two Bolter Operators
            4.3.3   Maintenance Personnel
        4.4   Hopper-Feeder-Bolter Components and Performance
            4.4.1   Chassis
            4.4.2   Track Frames
            4.4.3   Conveyor Tail
            4.4.4   Conveyor
            4.4.5   Crawlers

These headings contain primarily content information with general terms like *components* and *performance* and names of parts like *chassis, conveyer,* and *crawlers*. But note too the main organizational indicators: 4.0 is a Technical *Discussion*, and 4.3 details the *Operation* of the Hopper-Feeder-Bolter.

Organizational indicators tell the reader the method of organization and the purpose of the section. Such words include:

How to . . .	Installing
An Analysis of . . .	Analyzing
A Definition of . . .	Defining
Recommendation for . . .	Choosing
Results of . . .	Understanding

Organizational indicators can also use questions—a technique that research has shown to be particularly effective with generalist readers. For example, here are some of the questions used as cross-reference subheadings in the California Driver's Handbook:

How Do I Obtain a Driver's License?
How Do I Renew My License?
How Much Insurance Must I Carry?

Headings or subheadings at the same level of a hierarchy can also reinforce their similarities of content if they are syntactically parallel. In the following partial outline, the writer is setting up the criteria for judging the relative value of cordless telephones. Notice how the points at the same level of hierarchy are made parallel.

    2. Quality of Transmission
        2.1 Clarity must be similar to that of a conventional telephone within a 200-ft range.
        2.2 Signal must be strong enough to cancel reasonable interference of
- Neighbor's cordless telephone
- Automobile traffic
- Electrical appliances

Parallelism. *Parallelism* helps readers understand organization by showing the similarities among items through similar syntactic structure. Parallelism means using patterns—repeating the same structure—for items that are alike, similar, or direct opposites. As you edit the writer's draft, you should use parallelism (1) whenever you want to give the same importance to two or more items, as in a list, an outline, or a sentence, and (2) whenever you want the reader to see the way two or more items are related.

Follow these steps to make items parallel in structure:

1. List the points or words one under the other, even if you will put them in a sentence.
2. Study the list. Determine if the order of the points is effective. Do they go from most to least important? From general to specific? In some other order?
3. Use the same syntactical structure for each element, whether words, phrases, clauses, or sentences. For example:

Words:	Protect, Create, and Restore Habitat
	Protecting, Creating, and Restoring Habitat
Phrases:	As a cover, vetch serves to fix nitrogen, recycle nutrients, reduce soil erosion, and add organic matter to the soil.
Clauses:	The first step in helping migratory songbirds is to learn who they are, what they look like, where they live, and how they are threatened.

4. For long or complex parallel constructions, consider setting each parallel element on a separate line in a list introduced by bullets or numbers. This chapter contains many examples of such lists.

To see how the process works, study the following sentence, which lists the qualities of a good occupational therapist.

> The qualities a good occupational therapist should have are a good understanding of people, flexibility, personality, good observation skills, and to be able to realize the differences between work and play, realizing that both are of equal importance.

Following the suggested procedure, list the items in a list.

> good understanding of people
>
> flexibility
>
> personality
>
> good observation skills
>
> be able to realize the differences between work and play and their equal importance

While not equivalent, most items in the list (except the last) are at least phrases, though some could be more specific. Try sharpening the focus with the help of a dictionary or thesaurus.

good understanding of people	=	empathy?
flexibility	=	flexibility
personality (what kind?)	=	enthusiasm?
good observation skills	=	analytical ability?
realize the difference . . .	=	clarity about the roles of work and play?

Now rewrite the sentence and study the meaning and organization.

A good occupational therapist should have the following qualities: empathy, flexibility, enthusiasm, and analytical ability, and the therapist must understand the roles of work and play and their equal importance.

Alternatively, the sentence could use adjectives instead of nouns.

A good occupational therapist should be empathetic, flexible, enthusiastic, and analytical, and a therapist must be clear about the roles of work and play and their equal importance.

These equivalents may not be those of the author, so after recasting the sentence, a sensitive editor will query the author about debatable revisions.

Access Aids. *Access aids* are also called orienters, indexes, or pointers because they point the way to information and help readers see where they are within a larger context. Access aids that show organization include:

- tabs in binders or books
- headers or footers in chapters
- the table of contents
- overviews of chapters or sections that tell what will follow
- maps, flowcharts, and menus for online documentation

Another important access aid is the index. However, an index with its sequential (alphabetical) organization is valuable to a reader precisely because it provides a different way of getting at information than does the author's organization. Indexes do not represent the organization of the text, but an alternate organization. See Chapter 17 for details on editing access aids.

EXERCISE 5

Editing by Adding Headings and Subheadings

Analyze the following two paragraphs to determine the existing organization. Then recast the information, writing headings and subheadings as needed.

A television tuner is an electronic device that selects and processes incoming broadcast signals. Three types of tuners—rotary, veractor, and quartz electronic—are used in modern sets; though they process the incoming signals the same way, they have different user interfaces. A rotary tuner is a tuner that requires the user to select the channel by rotating a mechanical knob and fine-tune by way of a turnable ring extending around the base of the knob. A veractor tuner is a tuner that the user can pre-tune up to 14 channels by adjusting tiny mechanical switches, each corresponding to one push button that enables instant channel changing. A quartz electronic tuner is a tuner that requires the user either to type in the channel number on a 0-to-9 numbered keypad or to press one of two buttons that automatically changes the channels. The tuner automatically fine-tunes each channel.

The picture adjustment controls are a set of four electronic switches that adjust a picture's color and brightness. Color adjusts the intensity of the colors, and tint fine-tunes each color to produce different shades. (These two are adjusted together to produce a naturally colored picture.) Bright adjusts the overall brightness of the picture, and contrast adjusts the ratio between a picture's light and dark areas. (These two are adjusted according to the user's preference.)

EXERCISE 6

Rewriting for Parallelism

On a separate page, rewrite the following sentences in parallel structure.

1. For ordinary writing, three simple requirements are recommended. These are, to whom and about what one is writing, the number of lines written, and minimal use of long words.
2. More information on the cost (operational cost) of each product is needed to complete the chart as well as evaluating the criterion.
3. Ski bottoms are cleaned, dried, and fairly smooth before treatment.
4. The floppy disk is the most popular and less expensive than a hard disk.
5. Other reasons for choosing these criteria are ease of understanding, applicable to all portable generators, practical for selecting a generator.
6. These variations affect the life span, operation, and using the machine significantly.
7. The thermal printer has the following problems:
   a. poor quality printing
   b. thermal paper is expensive
   c. printed copy will fade over a short period of time

8. To avoid these mistakes, no engineer would quarrel with the objective of being more cautious, establish stricter controls, test as extensively as possible, and strive for perfection.

EXERCISE 7

Editing and Rewriting for Parallelism

On a separate page, rewrite the following sentences in parallel structure and improve diction and syntax as needed.

1. The function of the ring detector is to screen for invalid ring signals and activates the microprocessor, which activates the message and recording modes.
2. They estimate the time a drawing will take, bid on the job, and still the subject must be researched.
3. The largest division is the Audit department which employs 57 percent of their staff, followed by the MICD which involves 26 percent of their employees, and last the tax, which has the remainder of the employees.
4. The graduation handbook tells students where they can obtain their academic regalia, when to expect their diploma, and commencement ceremony procedures.
5. Some further advantages of the cantilever brakes are increased size of brake pads (longer), sturdier (each brake is bolted to the frame), and plenty of tire clearance (ideal for fenders or when tires get caked with mud).
6. Upon tape insertion there are two steps to follow:
   a. Before inserting the tape, the operator must make sure the exposed portion is facing inward.
   b. Make sure the label/logo is facing the operator in the upward position.
7. The combination of the in-house contracted program for flight training has a relatively low startup cost, gives the university a large degree of quality control, and the university always has the option of expanding to a completely in-house school.
8. Additional benefits of working with a tough editor include:
   ○ learn to take orders and carry them out
   ○ gain patience and maturity
   ○ editor's self-confidence will rub off
   ○ your writing standards will rise to match those of the editor
   ○ more assertive

- research ~~will~~ improve your
- learn to take constructive criticism

9. The challenge to an engineer is to design and produce a product in accordance with:
   - market needs
   - corporate objectives
   - reasonable cost
   - produceability
   - incorporation of latest technology
   - ~~meets~~ government legislation
   - is sellable ~~ability~~ marketability

10. The link between engineering and manufacturing activities is provided by manufacturing engineers. They will provide:
    - ~~the~~ lowest cost manufacturing
    - ~~the~~ best application of technologies
    - ~~access~~ producibility/manufacturability
    - ~~assume~~ the technical interface
    - ~~insure~~ product schedule attainment

Visual Devices

In addition to verbal devices to enhance organization, you can use the following visual devices:

1. using icons
2. showing hierarchy
3. chunking
4. listing

Icons. *Icons,* which provide links to a reader's schemata through metaphor, can enhance organization by visual shortcuts in both online documentation and hardcopy. The file cabinet, clipboard, trash can, clock, and camera all have associated meanings to action or locations in software. They are usually easier for users to remember than are the related words, but their meanings may be ambiguous and, therefore, difficult to interpret. In hardcopy, icons can provide interesting visual links between ideas, as Figure 14.5 shows (England, 1994).

Hierarchy. *Hierarchy* can be shown by the white space achieved by successive indentation from the left margin. Without any kind of numbering system, the following example shows the relationship of ideas by indentation.

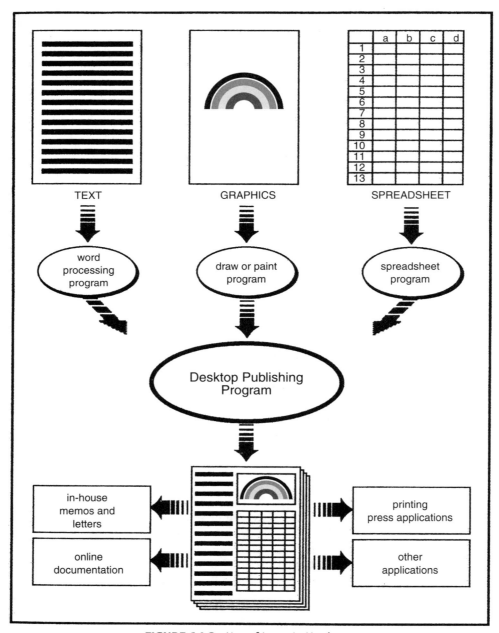

FIGURE 14.3 Use of Icons in Hardcopy

Principles of Capillary Electrophoresis
    Historical background
    Theory
        Electrophoresis
        Electrostatic flow
        Analytical parameters
        Dispersion
        Resolution

Hierarchy can also be displayed by type size (point size) and by type style (boldface, italics, underlining, capitalization). Larger type sizes are used for major headings, and successively smaller sizes are used for subheadings. Most companies have a style guide or templates that show preferred type styles. In desktop publishing, boldface type is often used for headings and italicized type for highlighting individual words. Underlining is seldom used. On screens, white space is probably the most important element because computer screens are harder to read than printed pages. (See Chapter 16 for details on document design.)

Chunks. *Chunks* are segments or logical units of information. Chunks can be individual chapters, single pages or screens, or even paragraphs or lists. By limiting the amount of information in short-term memory at one time, chunking helps readers process that information. Chunks are usually defined by white space—margins, indentation for paragraphs, space around a list.

For example, the following chunks on screen layout for online documentation make the information easy for the reader to understand (Hofmeister 1989, 1):

- Place important information in visually prominent areas to help users sort through information on the screen.
- Limit information to simple, concise text because the area of most screens is smaller than pages in printed manuals.
- Use a single column for text. Research indicates that two-column formats for online text can retard reading speed by as much as 200 words per minute.
- Use unjustified text. Although some research indicates that there is no difference in reading speed between justified and unjustified text in online documentation, studies in print documentation indicate that unjustified text helps novice users read more quickly.

Lists. *Lists* take advantage of the defining quality of white space and also (if the list items are parallel) of the patterning of similar syntactic structures.

Figure 14.4 is an entry in the *Apple Publications Style Guide* (1992, 34) that uses a list to show examples of the use of the hyphen in the variations of the word *double click*.

---

double click (n.), **double-click** (v.), **double-clicking** (n., v.)
Note hyphenation.

> Small children may have trouble with a double click.
> Adults can double-click without difficulty.
> Double-clicking allows you to work faster.
> You do this by double-clicking the icon.

FIGURE 14.4 Use of White Space around a List

EXERCISE 8

A Comprehensive Project in Document Editing

Following is a process description. Edit the document as a whole, using all the editing skills you have learned—copyediting, sentence editing, organizational editing. First, analyze the organization. Second, reorganize. Third, edit at the sentence level. Fourth, write a paragraph in which you explain what the major problems are and why they occurred. The intended reader is a generalist.

HOW BINOCULARS OPERATE

INTRODUCTION

Background

Almost anyone at one time or another has used binoculars to observe distant objects or events. Most that have notice of course these instrument's ability to make far away things look much closer. Another feature which binoculars possess, and one that is not commonly perceived, is their ability to enhance stereoscopic vision. In both cases, the user is employing the features of a sophisticated optical device. It is the purpose of this discussion to explain in general terms how these instruments achieve such useful and interesting effects.

Short Description

Binoculars are optical devices consisting of two parallel telescopic systems that enable the user to comfortably see an enlarged image of a distant object with both eyes open. In addition to giving an enlarged view of a remote subject of observation, the binocular enhances the stereoscopic perception of depth beyond that of normal vision. This enhancement is aided both by the separation of the axes of the objective lenses and by the magnifying power. The term "binocular" can be applied to any device using the parallel telescope principle, but it is commonly restricted to prism binoculars: instruments that use prisms to erect the image and to shorten the distance between the objective (front) lens and the ocular (rear) lens. Each half of a typical binocular contains two prisms and at least two lenses.

Major Process Steps

The major steps involved in the process by which binoculars produce useful images are (1) light collection (light-gathering power), (2) manipulation of collected light into images and processing in order to correct images of inversion or objectional aberrations, and (3) adjustability of the image to the individual focusing requirements and refractive errors of the eyes of the user.

STEPS IN THE PROCESS

Light Collection

The collection of light is the initial stage in the image delivery process of binoculars. This is achieved by the objective lenses. These are the largest lenses of the instrument, and are the pair at the device's functional front which is pointed at the desired subject of observation. The larger the diameter of these lenses, the greater the light-gathering power of the particular instrument. This light collection function of the objective lenses is particularly useful at low levels of illumination, such as at twilight, overcast days or at night, when human eyes are at a disadvantage. Objective lenses are measured in terms of diameter in millimeters. Binoculars are manufactured and sold, classified by a dual-numbered identification system. The first number in the description denoting the magnifying power in terms of increase in apparent diameters of the target image, the second number denoting the diameter of the objective lenses. Thus a 7 × 50 instrument is one that magnifies seven times and has fifty millimeter objective lenses. Objective lenses are convex (bulging outwards from their center) simple lenses which bend parallel rays of light which pass through them behind them to a point of focus called a focal point. The binocular is then ready to further process the light to produce a functional image.

Image Manipulation

The primary problem with light which is gathered by unaided objective lenses is that they are inverted (turned upside down) by the lenses. To correct for this effect, an image erection (uninvertion) system is employed in the binocular imaging process. This usually consists of a set of prism pairs, each consisting of two 45-90-45 degree prisms (cut glass shapes which refract or bend light to certain specifications) oriented at 90 degrees to each other. These prisms not only correct the image so that it will not appear to be upside down with respect to the observer but also free the image from objectional aberrations.

Image Adjustability

After the image is gathered in the form of light waves and the waves are processed to correct them for distracting or distorting effects inherent in the gathering process, the binocular must be adjusted for the eyesight of the individual user. This is achieved in several, clever mechanical ways. The most obvious way is merely a gross adjustment for the actual difference in head shape and eye pupil separation of the observer. This is affected firstly by means of a moveable hinge which is hand pressure activated, found directly in the center of the long axis of the binoculars between the two lens tubes. This is termed interocular control. Secondly, adjustment is done by means of physical rotation of the metal threading of one of the two eyepieces. These eyepieces terminate the ocular or rear lenses and have a moveable control to correct for the difference in refractive distortion found normally in a pair of human eyes. Finally, adjustment can be controlled by focusing both trains of lenses and prisms of the two telescope tubes in tandem. This control is found usually in the form of a large rotating wheel meshed with gears which extends or contracts the focal length of the gained image. This geometric manipulation of the gathered light is what produces the fine or coarse focus necessary to discern images.

Summary of Process

Binoculars may more properly be termed image processors as they are sophisticated devices which gather and process light to deliver the end product of a useful image. The process is three-fold in nature: light-gathering through the use of convex, objective lenses, image reinversion through the use of prisms, and image focus and adjustment through the use of eyepieces and a gear system.

Major Binocular Uses

Binoculars are used for military and naval purposes, field sports, hunting, bird and animal watching, amateur astronomy and observation of theatrical performances (operas, plays, etc.). Any activity where an optical instrument is required which can present an easily discernible image with good resolution is one suitable for binoculars usage. Light-gathering power and magnifying ability, combined with portability and convenience of set-up and adjustment, are the main advantages of binocular devices.

EXERCISE 9

A Comprehensive Project: Editing for Organization and Document Design

As a comprehensive project, you will review organizational principles and the strategies of document design for four types of documents: a conventional printed report, a poster, a pamphlet or brochure, and an online document.

1. Working with two or three others, select the type of document you are most interested in and study the particular constraints and potentials of that type. See Chapters 15 and 16 for more information on graphics and document design. After reading the rest of these instructions, divide duties among group members.

2. Read the entire *Poison Ivy, Poison Oak, Poison Sumac* document to determine the purpose, intended reader, and topic. Figures in the original are not included.
3. Outline the existing organization to the third level of detail.
4. Analyze that outline to determine its problems; then, write a two-page explanation of the problems and your suggested solutions.
5. Create a new outline of the material. Note: Keep in mind the demands of your new document type when evaluating and reorganizing the text.
6. Meet with your fellow editors. Discuss with them the best reorganization of the information, given your chosen document type. Produce a final revised outline to indicate the organizational changes.
7. Discuss with your fellow editors the best layout and design for this document. Produce two or three pages or screen mockups, editing the original text as necessary and indicating the types of graphics needed.
8. Write a brief report on your design and organizational editing strategies, including a definition of the problems presented by the original document and the solutions effected by the revised document plan.

    Note: You do not need to rewrite the text or edit it for problems at the sentence level, though you may indicate in your two-page analysis what the major problems are.
9. Prepare a 15-minute presentation of your editing solutions for the class.

    Note: It will be helpful to have handouts or transparencies for your in-class presentation. If you need an overhead projector, let your instructor know well in advance.
10. Submit to the instructor the following documents to represent your editing solution:
    a. an outline of the original text and an outline of the revised text
    b. a brief report (no more than two pages, single-spaced) explaining and justifying your document-editing strategy and solutions
    c. all design mockups, for which you should try to use an electronic page-layout program

POISON IVY	**Identification**
POISON OAK	**Precautions**
AND	**Eradication**
POISON SUMAC	

Many people are accidentally poisoned each year from contact with plants that they did not know were harmful. If they had known how to recognize these poisonous plants, they could have escaped the painful experience of severe skin inflammation and water blisters. Many people do not recognize these plants, although they occur in almost every part of the United States in one or more of their various forms.

Few persons have sufficient immunity to protect themselves from poisonous plants. However, poisoning is largely preventable. You can easily learn to identify plants in their various forms by studying pictures and general descriptions, then train yourself by diligent practice in observing the plants in your locality. Children should be taught to recognize the plants and to become poison ivy conscious.

POISON IVY AND POISON OAK

Poison ivy and poison oak are neither ivy nor oak species. Rather, they belong to the cashew family and are known by a number of local names; actually, several different kinds of plants are called by these names. Poison ivy and poison oak plants vary greatly throughout the United States. They grow in the form of: (1) woody vines attached to trees or objects for support, (2) trailing shrubs mostly on the ground, or (3) erect woody shrubs entirely without support. They may flourish in the deep woods, where soil moisture is plentiful, or in very dry soil on the most exposed hillsides. Plants are most frequently abundant along old fence rows and edges of paths and roadways. Plants ramble over rock walls and climb posts or trees to considerable height. Often they grow with other shrubs or vines in such ways as to escape notice.

Leaf forms among plants, or even on the same plant, are as variable as the habit of growth; however, the leaves almost always consist of three leaflets. The old saying, "Leaflets three, let it be," is a reminder of this consistent leaf character but may lead to undue suspicion of some harmless plant. Only one three-part leaf leads off from each node on the twig. Leaves never occur in pairs along the stem.

Flowers and fruit are always in clusters on slender stems that originate in the axils, or angles, between the leaves and woody twigs. Berrylike fruits usually have a white, waxy appearance and ordinarily are not hairy, but may be so in some forms. The plants do not always flower and bear fruit. The white or cream-colored clusters of fruit, when they occur, are significant identifying characters, especially after the leaves have fallen.

For convenience, these plants are discussed under three divisions—common poison ivy, poison oak, and Pacific poison oak.

Common Poison Ivy

The plant is known by various local names—poison ivy, three-leaved ivy, poison creeper, climbing sumac, poison oak, markweed, picry, and mercury. Common poison ivy may be considered as a vine in its most typical growth habit.

Vines often grow for many years, becoming several inches in diameter and quite woody. Slender vines may run along the ground, grow with shrubbery, or take support from a tree. A plant growing along the edge of a lawn and into the shrubbery may be inconspicuous compared with a vine climbing on a tree. The vine develops roots readily when in contact with the ground or with any object that will support it. When vines grow on trees, these aerial roots attach the vine securely. A rank growth of these roots often causes the vines on trees to have the general appearance of a fuzzy rope.

The vines and roots apparently do not cause injury to the tree except where growth may cover the supporting plant and exclude sunlight. The vining nature of the plant makes it well adapted to climbing over stone walls or on brick and stone houses.

Poison ivy may be mixed in with ornamental shrubbery and vines. Sometimes people do not recognize the plant and cultivate it as an ornamental vine. An ivy plant growing on a house may be prized by an unsuspecting owner. The vine is attractive and sometimes turns a brilliant color in the fall. This use as an ornamental can result in cases of accidental poisoning, and these plants may serve as propagating stock for more poison ivy in the vicinity.

Poison ivy, mixed in with other vines, may be difficult to detect, unless you are trained in recognizing the plant. Virginia creeper and some forms of Boston ivy often are confused with it. You can recognize Virginia creeper by its five leaflets radiating from one point of attachment. Boston ivy with three leaflets is some times difficult to detect. Study a large number of Boston ivy leaves and you will usually find some that have only one deeply lobed blade or leaflet. Poison ivy has the three leaflets. A number of other plants are easily confused with poison ivy. Learn to know poison ivy on sight through practiced observation, then make sure by looking at all parts of the suspected plant.

Common poison ivy when in full sunlight grows more like a shrub than like a vine along fence rows or in open fields. In some localities, the common form is a low-growing shrub that is 6 to 30 inches tall. Both forms usually have rather extensive horizontal systems of rootstocks or stems at or just below the ground level. Under some conditions, the vining form later becomes a shrub. Plants of this type may start as a vine supported on a fence and later extend upright stems that are shrublike. In some localities, the growth form over a wide range is consistently either vine or shrub type. In other areas, common poison ivy apparently may produce either vines or shrubs.

Leaves of common poison ivy are extremely variable, but the three leaflets are a constant character. The great range of variation in the shape or lobing of the leaflets is impossible to describe. The five leaves shown in figure 6 give a fair range of patterns. Other forms may be found. One plant may have a large variety of leaf forms, or it may have all leaves of about the same general character.

Most vines or shrubs of poison ivy produce some rather inconspicuous flowers that are always in quite distinct clusters arising on the side of the stem immediately above a leaf. Frequently, the flowers do not develop or are abortive and no fruit is produced. Poison ivy fruits are white and waxy in appearance and have rather distinct lines marking the outer surface, looking like the segments in a peeled orange.

In some forms of poison ivy, the fruit is covered with fine hair, giving it a downy appearance; however, in the more common form, fruits are entirely smooth. The fruit is especially helpful in identifying plants in late fall, winter, and early spring when the leaves are not present.

Poison Oak

Poison oak is more distinctive than some other types. Some people call it oakleaf ivy while others call it oakleaf poison ivy.

Poison oak usually does not climb as a vine but occurs as a low-growing shrub. Stems generally grow upright. The shrubs have rather slender branches, often covered with fine hairs that give the plant a kind of downy appearance. Leaflets occur in threes, as in other ivy, but are lobed, somewhat as the leaves of some kinds of oak. The middle leaflet usually is lobed alike on both margins and resembles a small oak leaf, while the two lateral leaflets are often irregularly lobed. The lighter color on the underside of one of the leaves is caused by the pubescence, or fine hairs, on the surface. The range in size of leaves varies considerably, even on the same plant.

Pacific Poison Oak

Pacific poison oak of the Pacific Coast States, usually known as poison oak, occasionally is referred to as poison ivy or yeara. This species is in no way related to the oak but is related to poison ivy.

The most common growth habit of western poison oak is as a rank upright shrub that has many small woody stems rising from the ground. It frequently grows in great abundance along roadsides and in uncultivated fields or on abandoned land.

Pacific poison oak sometimes attaches itself to upright objects for support and takes more or less the form of a vine. The tendency is for individual branches to continue an upright growth rather than to become entirely dependent on other objects for support. In some woodland areas, 70 to 80 percent of the trees support vines extending 25 to 30 feet in height.

In open pasture fields, Pacific poison oak usually grows in spreading clumps from a few feet to several feet tall. Extensive growth greatly reduces the area for grazing. It is a serious menace to most people who frequent such areas or tend cattle that come in contact with the plants while grazing.

Low-growing plants, especially those exposed to full sunlight, often are quite woody and show no tendency for vining. These plants are common in pasture areas or along roadsides. Livestock in grazing do not invade the poison ivy shrub. As a rule, these plants spread both by rootstock and seed.

As in other poison ivy, leaves consist of three leaflets with much irregularity in the manner of lobing especially of the two lateral leaflets. Sometimes lobes occur on both sides of a leaflet giving it somewhat the semblance of an oak leaf. The middle, or terminal leaflet is more likely to be lobed on both sides, and resembles an oak leaf more than the other two. Some plants may have leaflets with an even margin and no lobing whatsoever. The surface of the leaves is usually glossy and uneven, giving the leaves a thick leathery appearance.

Flowers are borne in clusters on slender stems diverging from the axis of the leaf. Individual flowers are greenish white and are about one-fourth inch across. The cluster of flowers matures into greenish or creamy white berrylike fruits about mid-October. These are about the size of small currants and much like other poison ivy fruits. Many plants bear no fruit, although others produce it in abundance. Fruits sometimes have a somewhat flattened appearance. They remain on plants throughout fall and winter and help identify poison oak after leaves have fallen.

POISON SUMAC

Poison sumac grows as a coarse woody shrub or small tree and never in the vinelike form of its poison ivy relatives. This plant is known also as swamp sumac, poison elder, poison ash, poison dogwood, and thunderwood. It does not have variable forms, such as occur in poison oak or poison ivy. This shrub is usually associated with swamps and bogs. It grows most commonly along the margin of an area of wet acid soil.

Mature plants range in height from 5 or 6 feet to small trees that may reach 25 feet. Poison sumac shrubs usually do not have a symmetrical upright treelike appearance. Usually, they lean and have branched stems with about the same diameter from ground level to middle height.

Isolated plants occasionally are found outside swampy regions. These plants apparently start from seed distributed by birds. Plants in dry soil are seldom more than a few feet tall, but may poison unsuspecting individuals because single isolated plants are not readily recognized outside their usual swamp habitat.

Leaves of poison sumac consist of 7 to 13 leaflets, arranged in pairs with a single leaflet at the end of the midrib.

The leaflets are elongated oval without marginal teeth or serrations. They are 3 to 4 inches long, 1 to 2 inches wide, and have a smooth velvetlike texture. In early spring, their color is bright orange. Later, they become dark green and glossy on the upper surface, and pale green on the lower, and have scarlet midribs. In the early fall, leaves turn to a brilliant red-orange or russet shade.

The small yellowish-green flowers are borne in clusters on slender stems arising from the axis of leaves along the smaller branches. Flowers mature into ivory-white or green-colored fruits resembling those of poison oak or poison ivy, but they usually are less compact and hang in loose clusters that may be 10 to 12 inches in length.

Because of the same general appearance of several common species of sumac and poison sumac there is often considerable confusion as to which one is poisonous. Throughout most of the range where poison sumac grows, three nonpoisonous species are the only ones likely to confuse. These are the smooth sumac, staghorn sumac, and dwarf sumac, which have red fruits that together form a distinctive terminal seed head. These are easily distinguished from the slender hanging clusters of white fruit of the poison sumac. Sometimes more than one species of harmless sumac grow together.

Introduced Poisonous Sumac and Related Species

The small Japanese lacquer-tree (*Rhus verniciflua*), uncommon in the United States, is related to native poison sumac. Native to Japan and China, it may be a source of Japanese black lacquer. Poisoning has followed contact with lacquered articles. Never plant this tree.

A native shrub or small tree (*Metopium toxiferum*) called poison-wood, doctor gum, Metopium, Florida poison tree, or coral sumac is commonly found in the pinelands and hummocks of extreme southern Florida, the Keys, and the West Indies. It is much like, and closely related to, poison sumac. The shrub or small tree has the same general appearance as poison sumac. However, the leaves have only three to seven, more-rounded leaflets. Fruits are borne in clusters in the same manner as those of poison sumac, but they are orange colored and each fruit is two to three times as large. All parts of the plant are poisonous and cause the same kind of skin irritation as poison-ivy or poison sumac.

When seed heads or flower heads occur on plants, it is easy to distinguish poisonous from harmless plants; however, in many clumps of either kind, flowers or fruit may not develop. The leaves have some rather distinct characteristics.

Leaves of the smooth sumac and of staghorn sumac have many leaflets, which are slender and lance shaped and have a toothed margin. These species usually have more than 13 leaflets. Leaves of dwarf sumac and poison sumac have fewer leaflets; these are more oval shaped and have smooth or even margins. The dwarf sumac is readily distinguished from poison sumac by a winged midrib. Poison sumac never has the wing margin on the midrib.

POISONING

Many people know through experience that they are susceptible to poisoning by poison ivy, poison oak, or poison sumac. Others, however, either have escaped contamination or have a certain degree of immunity. The extent of immunity appears to be only relative. After repeated contact with the plants, persons who have shown a degree of immunity may develop poisoning.

The skin irritant of poison ivy, poison oak, and poison sumac is a nonvolatile phenolic substance called urushiol, found in all parts of the plant. The danger of poisoning is greatest in spring and summer and least in late fall or winter.

Poisoning usually is caused by contact with some part of the bruised plant, as actual contact with the poison is necessary to produce dermatitis. A very small amount of the poisonous substance can produce severe inflammation of the skin. The poison is easily transferred from one object to another.

Clothing may become contaminated and is often a source of prolonged infection. Dogs and cats frequently contact the plants and carry the poison to children and other unsuspecting persons. The poison may remain on the fur of animals for a considerable period after they have walked or run through poison ivy plants.

Smoke from burning plants carries the toxin and can cause severe cases of poisoning.

Children who have eaten the fruit have been poisoned although the fruit when fully ripe is reported as nonpoisonous. A local belief that eating a few leaves of the plant will develop immunity in the individual is unfounded. Never taste or eat any part of the plant.

Cattle, horses, sheep, hogs, and other livestock apparently do not get the skin irritation caused by these plants, although they graze on the foliage occasionally. Bees collect nectar from the flowers, but no ill effects from the use of honey have been reported.

The time between contamination of the skin and first symptoms varies greatly with individuals and probably with conditions. The first symptoms of itching or burning sensation may develop in a few hours or even after 7 days or more. The delay in development of symptoms is often confusing when an attempt is made to determine the time or location when contamination occurred. The itching sensation and subsequent inflammation that usually develops into water blisters under the skin may continue for several days from a single contamination. Persistence of symptoms over a long period is most likely caused by new contacts with plants or contact with previously contaminated clothing or animals.

Severe infection may produce more serious symptoms, which result in much pain through abscesses, enlarged glands, fever, or other complications.

If it is necessary to work among poisonous plants, some measure of prevention can be gained by wearing protective clothing. It is necessary, however, to remember that the active poison can be easily transferred. Some protection may also be obtained by using protective creams or lotions. They prevent the poison from contact with the skin, or make it easily removable by washing with soap and water, or neutralize it to a certain degree.

All measures to get rid of the poison must be taken within a few minutes after contact. A 10-percent water solution of potassium permanganate, obtainable in any drug store, usually is effective if applied within 5 to 10 minutes after exposure.

Many ointments and lotions are sold for prevention of poisoning by chemical or mechanical means. Their use should always be followed by repeated washings with soap and water to remove the contaminant.

Contaminated clothing and tools often are difficult to handle without causing further poisoning. Automobile door handles or steering wheels may, after trips to the woods, cause prolonged cases of poisoning among persons who have not been near the plants.

Decontaminate such articles by thorough washing in several changes of strong soap and water. Do not wear contaminated clothing until it is thoroughly washed. Do not wash it with other clothes. Take care to rinse thoroughly any implements used in washing. Drycleaning processes will probably remove any contaminant; but there is always danger that clothing sent to commercial cleaners may poison unsuspecting employees.

Dogs and cats can be decontaminated by washing; take care, however, to avoid poisoning while washing the animal.

There seems to be no absolute, quick cure for all individuals, even though many studies have been made to find effective remedies. Remedies may be helpful in removing the poison or rendering it inactive, and for giving some relief from the irritation. Mild poisoning usually subsides within a few days, but if the inflammation is severe or extensive, consult a physician.

CONTROL BY MECHANICAL MEANS

Poison ivy and poison oak can be grubbed out by hand quite readily early in spring and late in fall, only if a few plants are involved. Roots are most easily removed when soil is thoroughly wet. Grubbing when soil is dry and hard is almost futile because the roots break off in the ground, leaving large pieces that later sprout vigorously. Grubbing is effective if well done.

Poison ivy vines climbing on trees should be severed at the base, and as much of the vine as possible should be pulled away from the tree. Often the roots of the tree and weed are so intertwined that grubbing is impossible without injury to the tree. Bury or destroy roots and stems removed in grubbing because the dry material is almost as poisonous as the fresh.

Smoke from burning poison ivy plants or contaminated articles may carry the poison in a dispersed form. Take extreme caution to avoid inhalation or contact of smoke with the skin or clothing.

Old plants of poison ivy produce an abundance of seeds, and these are freely disseminated, especially by birds. A poison ivy seedling 2 months old usually has a root that one mowing will not kill. Seedling plants at the end of the first year have well established underground runners that only grubbing or herbicides will kill. Seedlings are a threat as long as old poison ivy is in the neighborhood.

Plowing is of little value in combating poison ivy and poison oak.

Mowing with a scythe or sickle is not an efficient means of controlling poison ivy and poison oak. It has little effect on the roots unless frequently repeated.

Weed burners are also inefficient in controlling poison ivy and poison oak.

CONTROL BY HERBICIDES

Poison ivy and poison oak can be destroyed with herbicides without endangering the operator. One usually may stand at a distance from the plants and apply the herbicide without touching them. Most herbicides are applied as a spray solution by sprayers equipped with nozzles on extensions 2 feet or more in length. The greatest danger of poisoning occurs in careless handling of gloves, shoes, and clothing after the work is finished.

The most satisfactory herbicides for control of poison ivy, poison oak, and poison sumac are: (1) amitrole (3-amino-s-triazole); (2) silvex [2-2.4,5-(trichlorophenoxy) propionic acid]; (3) ammonium sulfamate; and (4) 2,4-D [2,4-(dichlorophenoxy) acetic acid]. These herbicides are sold under their common names and under various trade names.

Any field or garden sprayer, or even a sprinkling can, can be used for applying the spray liquid, but a common compressed-air sprayer holding 2 to 3 gallons is convenient and does not waste the spray.

Use moderate pressure giving relatively large spray droplets, rather than high pressure giving a driving mist, because the object is to wet the leaves of the poison ivy and poison oak and avoid wetting the leaves of desirable plants. High pressure causes formation of many fine droplets that may drift to desirable plants.

Follow the manufacturer's recommendations shown on the container label in preparing the spray solution. Cover all foliage, stems, shoots and bark of poison plants with herbicide spray. Although best results normally are obtained soon after maximum foliage development in the spring, applications may be made up to 3 weeks before fall frost is normally expected under good growing conditions in the humid areas.

Many herbicides used on poison ivy and poison oak will injure most broad-leaved plants. Apply them with caution if the surrounding vegetation is valuable. During the early part of the growing season, the leaves of poisonous plants usually tend to stand conspicuously apart from those of adjacent plants, and they can be treated separately if sprayed with care. Later the leaves become intermingled, and injury to adjacent species is unavoidable. Chemicals other than oil are not injurious to the thick bark of an old tree, and poison ivy clinging to the trunk safely can be sprayed with them. However, cutting the vine at the base of the tree and spraying regrowth may be more practical.

Apply sprays when there is little or no air movement. Early morning or late afternoon, when the air is cool and moist, usually is a favorable time.

No method of herbicidal eradication can be depended on to kill all plants in a stand of poison ivy and poison oak with one application. Retreatments made as soon as the new leaves are fully expanded are almost always necessary to destroy plants missed the first time, to treat new growth, and to destroy seedlings. Plants believed dead sometimes revive after many months. An area under treatment must be watched closely for at least a year and retreated where necessary.

Dead foliage and stems remaining after the plants have been killed with herbicides are slightly poisonous. Cut off dead stems and bury or burn them, taking care to keep out of the smoke.

USE OF PESTICIDES

This publication is intended for nationwide distribution. Pesticides are registered by the Environmental Protection Agency (EPA) for countrywide use unless otherwise indicated on the label.

The use of pesticides is governed by the provisions of the Federal Insecticide, Fungicide, and Rodenticide Act, as amended. This act is administered by EPA. According to the provisions of the act, "It shall be unlawful for any person to use any registered pesticide in a manner inconsistent with its labeling." (Section 12(a) (2) (G))

EPA has interpreted this Section of the Act to require that the intended use of the pesticide must be on the label of the pesticide being used or covered by a Pesticide Enforcement Policy Statement (PEPS) issued by EPA.

The optimum use of pesticides, both as to rate and frequency, may vary in different sections of the country. Users of this publication may also wish to consult their Cooperative Extension Service, State agricultural experiment stations, or county extension agents for information applicable to their localities.

The pesticides mentioned in this publication are available in several different formulations that contain varying amounts of active ingredient. Because of this difference, the rates given in this publication refer to the amount of active ingredient, unless otherwise indicated. Users are reminded to convert the rate in the publication to the strength of the pesticide actually being used. For example, 1 pound of active ingredient equals 2 pounds of a 50 percent formulation.

The user is cautioned to read and follow all directions and precautions given on the label of the pesticide formulation being used.

Federal and State regulations require registration numbers. Use only pesticides that carry one of these registration numbers.

USDA publications that contain suggestions for the use of pesticides are normally revised at 2-year intervals. If your copy is more than 2 years old, contact your Cooperative Extension Service to determine the latest pesticide recommendations.

The pesticides mentioned in this publication were federally registered for the use indicated as of the issue of this publication. The user is cautioned to determine the directions on the label or labeling prior to use of the pesticide.

Farmers' Bulletin Number 1972. United States Department of Agriculture, revised 1978. Stock Number 001-000-03883-4. [Figures in original not included.]

References

Apple publications style guide. 1992. Cupertino, CA: Apple Computer, Inc.

England, Bruce. 1994. *Desktop publishing: What's next?* Unpublished report. San Jose, CA: San Jose State University. Graphic reprinted by permission.

Hofmeister, Marie L. 1989. *Designing screens to help users retrieve and comprehend online information.* Document number TR03-334. San Jose, CA: IBM.

Huckin, Thomas. 1983. A cognitive approach to readability. In Paul V. Anderson, R. John Brockmann, Carolyn R. Miller, Eds. *New essays in scientific communication: Research, theory, practice.* Farmingdale, NY: Baywood.

Killingsworth, M. Jimmie and Michael K. Gilberston. 1992. *Signs, genres, and communities in technical communication.* Amityville, NY: Baywood.

Miller, George. 1967. *The psychology of communication.* New York: Basic Books.

Mirel, Barbara. 1988. Cognitive processing, text linguistics and documentation writing. *Journal of Technical Writing and Communication.* 18.2.

Pugh, A. K. 1978. *Silent reading: An introduction to its study and teaching.* London: Heineman.

Puri, Anjali. 1993. cc:Mail Alpha Help editorial survey. Reprinted by permission.

Rew, Lois Johnson. 1993. *Introduction to technical writing: Process and practice.* 2nd ed. New York: St. Martin's.

Simon, Herbert A. 1974. How big is a chunk? *Science* 183 (February), 482–488.

Williams, Thomas R. 1994. Schema theory. In Charles H. Sides, Ed. *Technical communications frontiers: Essays in theory.* St. Paul, MN: Association of Teachers of Technical Writing.

15 Graphics Editing

Contents of this chapter:

- Defining Key Graphics Terms
- Understanding the Factors Involved in Choosing Graphics
- Developmental Editing of Graphics: Major Questions
- Copyediting Graphics: A Checklist
- Proofreading Graphics: A Checklist

Editors often think of themselves as word experts and see editing as a task of "fixing up" the words in a document. But in the workplace of the 1990s and beyond, words are only one factor in a document's effectiveness, whether that document is on screen or on the pages of a book. Another factor is the presence of graphics: drawings, photographs, graphs, icons, flowcharts, animated sequences, and the like, which allow readers to visualize concepts and relationships without having to construct the visualization themselves.

How important are graphics? S. M. Shelton, a communications consultant, says, "Today people get most of their day-to-day information through graphics images. In fact, some 99 percent of all the information we receive comes through the eyes from *objects* we see" (1993, 617). Shelton's high percentage is more understandable if we recognize that words are also graphic images. Handwriting, and especially calligraphy, remind us of the visual properties of words. But words and what we usually call graphics are different and are, in fact, complementary.

Graphics are important in workplace documents for several reasons.

1. **Graphics help readers learn more.** In a 1982 report on 46 experiments assessing learning from illustrated text, researchers Levie and Lentz found that the groups who read illustrated text learned more—on average 36 percent more—than did the groups who read unillustrated text.

2. **Graphics help readers remember more.** A University of Minnesota study in 1989 "showed that people remember 43 percent more information when visuals are used than they do when visuals are not used" (Shelton 1993 citing Morrison and Jimmerson).

3. **Graphics convey a message quickly.** According to Professor Thomas R. Williams, much of the *meaning* of the information in graphics can be processed by readers "pre-attentively" or automatically. He quotes research that provides "evidence that the gist, theme, or global message of a visual can be processed in a single fixation, a glance seldom exceeding one-third of a second. In contrast," says Williams, "beyond the initial apprehension of word shape, most text processing takes place serially and with considerable expenditure of attentional resources" (1993, 673).

4. **Graphics maintain reader interest in the subject.** Graphics focus the reader's attention, break up large blocks of text, clarify difficult concepts through metaphor or example, and assist international readers by reducing the reliance on language. In fact, graphics have become so common in workplace documents that many readers expect graphics and consider documents without them to be bland and boring.

5. Graphics allow parallel processing of information. "Documents typically communicate through a combination of verbal and visual information. . . . The reason to use graphics instead of text, or vice versa, is that one method can communicate more effectively than the other at any given juncture. . . . Sometimes illustrations can convey complex concepts better than words. Pictures can give substantial shape to ideas, desires, and other information in a way that is beyond the limits of language" (Quimby 1996, 23).

As Quimby's statement makes clear, graphics and words work together, and the editor has a significant responsibility to ensure that (1) the most effective kinds of graphics are used for specific purposes, (2) graphics are included in a document where they can be most effective, and (3) all graphics used are accurate, consistent, clean (uncluttered), complete, clear, referenced, and documented. Editors should check the first two requirements at the developmental level (or alpha draft) and the third requirement at the copyediting level (or beta draft). See sections 15.3 through 15.5 for editing details.

Of the writers and editors I queried in researching this book, 75 percent stated that anyone learning to edit should study graphics principles. As placement director Shirley Krestas said, "Editors need not be experts in graphics, but they need to know the principles that undergird graphics." Sherri Sotnick, managing editor of a company journal, said, "Editors need practice in deciding where illustrations are needed to exemplify or clarify the text. The ability to decide quickly how to present information has been critical in my position." Senior editor Anjali Puri emphasized the editor's responsibility at the beta draft level: "The editor's role is to ask 'Is the graphic clear? Is it essential? Does it help the user perform a task?'"

15.1 Defining Key Graphics Terms

Within government, publishing, and the technical community, there is some ambiguity and overlap in the terminology used to describe graphics. Most commonly, three terms are used to describe the visual form of presenting information: *graphics*, *visuals*, and *illustrations*. This book uses *graphics* as the umbrella term, the word coming from the Greek word meaning *to draw or write*. The word *visuals* is most often used in the workplace to describe the projected pictures, graphs, and so on that accompany a speech, or the picture elements (as opposed to the sound elements) in a film or video. Sometimes, however, the word *visuals* is used as the general term. Since the word *illustrate* means *to make clear*, *illustrations* can be either examples or pictorial representations. In common use, the term *illustrations* usually refers to presentations like line drawings or photographs, but not to graphs, flowcharts, or tables.

Terms particular to software are *screen snapshots*, *screen shots*, *screen captures*, or *screen dumps*, all of which refer to "literal representations of what appears on the user's screen or in a window on that screen. Many computer manuals use no other type of illustration" (Horton 1993, 146). So common are screen shots in such manuals that, as editor Ruth Chase notes, "At my software company, tables and screens are part of the text; we don't consider them graphics." Other companies, however, do consider screen shots graphics.

Traditionally, and especially in publishing, graphics have been divided into two large categories: *tables* (ordered columns of numbers and/or words) and *figures* (all graphics that are not tables). The two categories are often numbered separately; thus, there can be both a Table 2 and a Figure 2. Some companies now simplify this practice by grouping all graphics into one category and numbering them consecutively. The accepted practice is usually defined in a company style guide.

15.2 Understanding the Factors Involved in Choosing Graphics

In order to edit graphics effectively (whether your own or someone else's), you need to understand the key factors that affect the choice of graphics. Those key factors include:

- the purpose of the document or a section of the document (or screen or slide)
- the type and experience of the reader or viewer
- the language and cultural background of the reader
- the content or the kind of information being presented
- the medium (the form in which the information is being presented)
- the time and resources available to produce the graphic

Understanding these factors will help you evaluate what Professor Ralph Wileman calls the "degree of visualization" of any graphic (1993) and decide if the graphic should be more visual or more verbal.

The Purpose of the Document

Just as different kinds of documents are written to achieve specific purposes (for example, to describe, to propose, to instruct, to report), so different kinds of graphics are chosen or designed to achieve specific purposes within any document. In drafting a document, the writer plans and chooses the graphics that he or she deems most effective, but the editor should be a second evaluator. This is true even if you are your own editor. When editing, you need to rethink the purpose of each section of text and the effectiveness of each graphic to fulfill that purpose.

Table 15.1 lists the common purposes for documents or parts of documents, suggests appropriate graphics choices to fulfill those purposes (Rew 1993), and shows where examples appear in this chapter. The pages containing graphics examples are marked with a bar in the margin so you can find them easily. In addition to the examples, each page contains specific editing questions you can ask about that graphics type. At the end of this chapter, sections 15.3, 15.4, and 15.5 provide general questions and checklists you can apply to graphics while developmental editing, copyediting, and proofreading.

The discussion of the factors involved in choosing graphics continues on page 315.

TABLE 15.1 Graphics for Specific Purposes

Purpose	Graphics Possibilities	Example on Page
showing exterior or interior views	photographs	298
	drawings (line, cutaway, exploded)	299, 300, 301
	examples (screen shots, slides, models)	312
displaying part/whole relationships	graphs (pie, segmented, bar)	302
	diagrams (organizational, hierarchical)	304
	maps and charts	305, 307, 308
	photographs	298
	drawings (cutaway, cross-sectional, exploded)	300, 301
	icons, models	313
explaining processes or operations	diagrams (flowcharts, Gantt charts or schedules, block, schematic, wiring)	307, 308, 314, 315
	drawings (action, line)	299
	icons, animated sequences	313
summarizing information	tables	309
	diagrams (organizational, hierarchical)	304
	drawings (cutaway, cross-sectional, exploded)	300, 301
	examples (maps, screen shots)	305, 312
comparing two or more items, showing trends	tables	309
	graphs (line, bar, pictographs)	311
providing examples	tables	309
	graphs	311
	diagrams	304
	photographs	298
	drawings	299, 300, 301
	screen shots, slides, models, animated sequences	312
representing ideas or functions	icons	313
	diagrams (block, schematic, wiring)	314, 315
	symbols (as in schematics)	315

Graphics Examples

Photographs

Photographs, whether black and white or in color, can be effective components of many documents, including reports, manuals, proposals, brochures, annual reports, and newsletters. Photographs can show exterior or interior views, relationships of parts and wholes, and exact situations—such as wear patterns or damage. They are effective in introductions or on covers to give an overview of an object or a piece of equipment. They can be ineffective if they are cluttered with extraneous detail or if they are not properly cropped. *Cropping* means marking the photograph to indicate what should be cut away before the photograph is used. The background can even be completely removed, as it is in Figure 15.1.

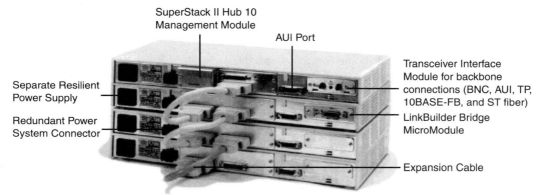

FIGURE 15.1 Photograph to Show Major Components *(Reprinted by permission of 3 Com Corporation)*

Editing Questions

☐ 1. Is a photograph the best choice for this purpose and reader?
☐ 2. Is the photograph properly cropped, enlarged, or reduced? Is it properly oriented, or has it been reversed or inverted?
☐ 3. Is it titled informatively and numbered properly?
☐ 4. Are parts labeled accurately and in such a way that labels can be expanded if necessary in translation?
☐ 5. Is it necessary to show the scale by including a ruler or a common object for comparison?

☐ 6. If the photograph includes people, do they need to be identified? Do you need to seek their permission to use the photograph?

☐ 7. Is the photograph referred to in the text? Is the reference by figure number? Is the photograph placed immediately after the reference?

☐ 8. Does the source of the photograph need to be acknowledged?

Drawings

Drawings can be simplified to show only important features, so they are an effective way to show people, locations, and wholes or parts of objects. Drawings can also show how systems work, how something is done, or the steps in a process. Because they can concentrate on critical features, drawings are often used in manuals for assembly, maintenance, and repair.

Line drawings can be surface, three-dimensional, or shaded perspective, showing an object or place as it is or will be when built. Sometimes a section of the drawing is enlarged for easy reading. (See Figure 15.2.) *Cutaway* and *cross-sectional* drawings show what is under the surface. In a cutaway drawing, the surface has been removed, and relationship of parts is shown beneath that surface. In a cross-section, the object or place has been sliced through the center, revealing the constituent layers or the internal makeup. (See Figure 15.3.) An *exploded* drawing shows the order of assembly of parts. The parts are drawn to scale and spread out along an imaginary line or line of axis showing how they fit together. (See Figure 15.4.)

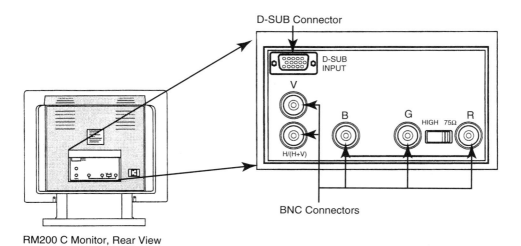

FIGURE 15.2 Enlargement of Drawing to Show Details and Location *(Reprinted by permission of Siemens Pyramid Information Systems, Inc.)*

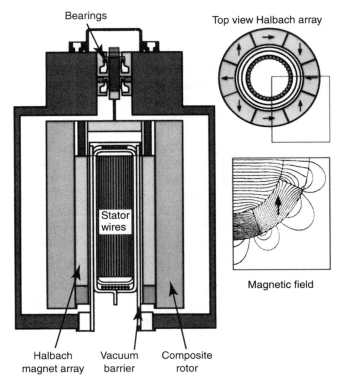

FIGURE 15.3 Cross-sectional Drawing *(Richard F. Post. 1996. A new look at an old idea: The electromechanical battery.* Science & Technology Review *(April) VCRL-52000-96-4.)*

Editing Questions

- ☐ 1. Is this drawing the most effective way to fulfill this purpose and meet the reader's needs?
- ☐ 2. Is the title informative? Is the drawing numbered correctly?
- ☐ 3. Are all the parts included and are they drawn to scale?
- ☐ 4. Are the labels correctly placed? Can the text be expanded in translation? Should there be a key with numbers instead of individual labels?
- ☐ 5. Do callouts require *leader lines* or *arrows* to point directly to the part indicated?
- ☐ 6. Should the drawing include hands to show how something is done?

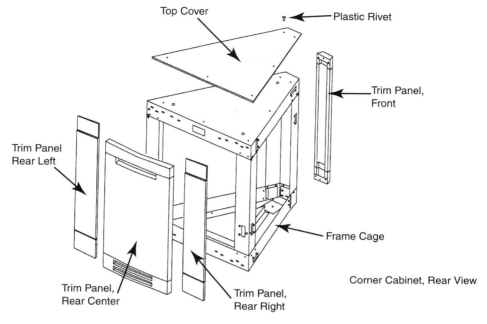

FIGURE 15.4 Exploded Drawing Showing Assembly *(Reprinted by permission of Siemens Pyramid Information Systems, Inc.)*

- ☐ 7. Is the drawing from the perspective of the viewer or user?
- ☐ 8. Is the drawing referenced in the text? Is it placed right after the reference?
- ☐ 9. Does the source of the drawing need to be acknowledged?

Pie Charts and Segmented Bar Graphs

Pie charts (also called circle graphs) and segmented bar graphs clearly show the relationship of a whole and its parts, especially the percentage composition of a whole. Pie charts and segmented bar graphs are good for generalists and managers, and they are used in reports, brochures, and proposals.

A *pie chart* uses a circle that represents 100 percent; the circle is then divided into segments representing the percentage of each part. (See Figure 15.5.) A well-designed pie chart begins with the largest segment starting at 12 o'clock; successively smaller segments move clockwise around the circle. A *segmented bar graph* shows the same relationship, but the segments are stacked vertically in a single bar starting with the largest bar at the bottom. Figure 15.6 shows a segmented bar graph providing the same information that was in the pie chart.

Types of Federal Income Tax Returns in 1996
Telefile 3%
1040PC 6%
Electronic 11%
1040EZ 15%
1040A 16%
1040 49%

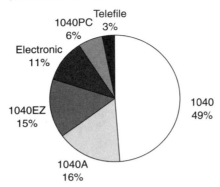

FIGURE 15.5 Pie Chart *(Internal Revenue Service. 1997.)*

Types of Federal Income Tax Returns in 1996
Telefile 3%
1040PC 6%
Electronic 11%
1040EZ 15%
1040A 16%
1040 49%

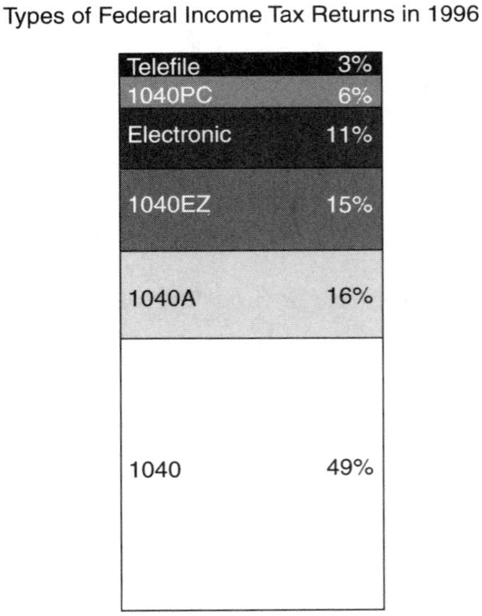

FIGURE 15.6 Stacked Bar Chart *(Internal Revenue Service. 1997.)*

Editing Questions

☐ **1.** For this audience, is this pie chart or segmented bar graph the best choice? (For example, specialist audiences might prefer a table with specific numbers.)

☐ **2.** Is the title accurate and informative? Is the chart or graph numbered accurately?

- ☐ 3. Is the chart or graph simple:
  - ○ Fewer than seven segments?
  - ○ Not three-dimensional or tilted?
  - ○ Not a mixture of decorative types of shading?

  Computer programs frequently clutter these figures; editors should simplify them for easy reading.
- ☐ 4. Do the percentages total 100 percent?
- ☐ 5. Are the parts clearly labeled? Is there room to expand labels in translation?
- ☐ 6. Is the graphic referenced in the text? Is it placed right after the reference?
- ☐ 7. Does the information source need to be acknowledged?

Diagrams, Maps, and Charts

Diagrams, maps, and charts are the figures that work best to show how companies are organized; how ideas or concepts are related; or where cities, stars, or shipping lanes are located relative to one another. These figures are suitable for all reader types, though maps and charts are most often used by operators.

Organizational and *hierarchical* diagrams provide an overview of relationships by simultaneously showing the whole, the parts of that whole, and how the parts relate to one another. The diagrams consist of a series of boxes connected by lines; they are read from top to bottom or left to right. Those boxes on the same line are presumed to be of equal value. (See Figure 15.7.)

Maps and *charts* also show parts and their relationship to the whole, but the whole in this case is on the land or sea or in the sky. Aeronautical and nautical maps are usually called charts. See Figure 15.8 for a map featuring an insert enlargement.

Editing Questions

- ☐ 1. Is this graphic the right one for this purpose and reader?
- ☐ 2. Is the title accurate and informative? Is the figure numbered correctly?
- ☐ 3. In an organizational or hierarchical diagram:
  - ○ Are the same size boxes used for the individuals, groups, or ideas on the same level?
  - ○ Are the lines of authority or relationship clear and accurate?

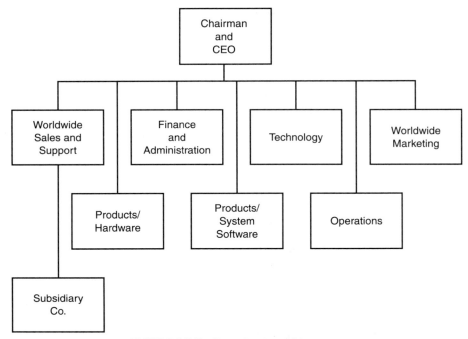

FIGURE 15.7 Organizational Diagram

- Is the concept simple enough, or does the diagram need to be split into sections?
☐ 4. On a map or chart:
- Is north indicated?
- Is a scale included? Is there a legend if needed?
- Is the use of color or shading logical (for example, blue for water)?
- Are the important items included for this graphic's purpose?
- Do other items need to be added?
- Is the graphic simple and easy to read?
☐ 5. Is it referenced in the text? Is it placed immediately after the reference?
☐ 6. Does the source need to be acknowledged?

FIGURE 15.8 Maps Showing Center of U.S. Population: 1790–1990 *(U.S. Bureau of the Census. 1994. Statistical Abstract of the United States: 1994. 114th ed. Washington, D.C.: U.S. Government Printing Office.)*

Flowcharts and Task-Breakdown Charts

Two effective ways to show the steps in a process are flowcharts and the kind of task-breakdown schedules called milestone or Gantt charts.

Technical reports and procedural descriptions often use *flowcharts* to provide an overview, orient readers to particular steps, show how a step branches or loops back to a previous step, and indicate how many steps are involved. Flowcharts can be constructed with a series of labeled blocks, photographs, or drawings. Standard shapes indicate specific information. For example, ovals with the words *start* and *end* may appear at the beginning and end of the procedure; a rectangular box with words inside represents action; a diamond-shaped box represents a decision or question. Arrows show the direction of flow. See Figure 15.9.

Milestone or *Gantt charts* effectively explain schedules of tasks by showing an overall project timeframe and its individual components. These graphics are often used in proposals and progress reports. A Gantt chart lists each major activity on a separate line in sequential order. As in a horizontal bar chart, a line entered on a time-line grid indicates the total amount of time to be spent on the activity. A milestone chart is similar but focuses on deadlines when tasks are to be completed. (See Figure 15.10.)

Editing Questions

- [ ] 1. Is this the best graphic for this purpose and reader?
- [ ] 2. Is the title accurate and informative? Is the figure numbered accurately?
- [ ] 3. In a flow chart:
    - Are the box shapes standard for portraying that step?
    - Do the words accurately state what is occurring?
    - Are the arrows going in the right direction?
- [ ] 4. In a Gantt or milestone chart:
    - Are the beginning and ending points of each action clearly shown?
    - Is the timeline detailed enough so readers can calculate time spent?
    - Is the chart accurate?
- [ ] 5. Is the flowchart or Gantt chart referenced in the text? Is it placed immediately after the text reference?
- [ ] 6. Do information sources need to be acknowledged?

CHAPTER 15 GRAPHICS EDITING **307**

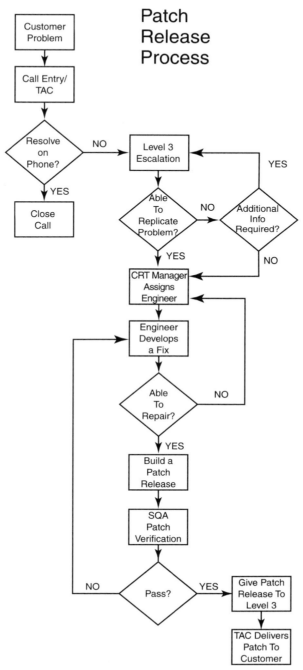

FIGURE 15.9 Flowchart *(Reprinted by permission of FORE Systems, Inc.)*

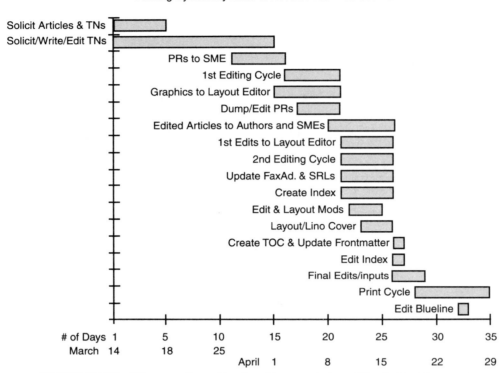

FIGURE 15.10 Milestone Chart *(Reprinted by permission of Newbridge Networks)*

Tables

Tables summarize large amounts of information in an organized manner by displaying it in columns and rows. The data in tables always involves at least two sets of variables. Tables are especially good at providing exact numbers, measurements, and specifications. However, they do not point out relationships, so either the reader must make comparisons, or the text must do so. Tables appeal to specialists who are looking for specifics.

Tables should use a minimum of lines to avoid clutter; most of the separation of items should be by white space. Tables typically use one horizontal line (called a "rule") above and one below the column heads to form a "boxhead" and one rule at the bottom to separate the table from the following text. The first column on the left (called the "stub") contains the list of items. Tables are easier to read if the number of rows and the number of columns are fewer than 10. Table 15.2 shows a table with numbers; Table 15.3 shows one with words.

TABLE 15.2 Table Using Numbers: Magnetic Field Measurements

	1.2" away	12" away	39" away
Microwave Oven	750 to 2,000	40 to 80	3 to 8
Clothes Washer	8 to 400	2 to 30	0.1 to 2
Electric Range	60 to 2,000	4 to 40	0.1 to 1
Fluorescent Lamp	400 to 4,000	5 to 20	0.1 to 3
Hair Dryer	60 to 20,000	1 to 70	0.1 to 3
Television	25 to 500	0.4 to 20	0.1 to 2

Adapted from Gauger 1985.

TABLE 15.3 Table Using Words: AC Power Module LEDs

Label	Color	Indicates...
AC	Amber	The power module is receiving current from the AC power source. If this light is on, but the following two lights are off, a short circuit or other overload has occurred. See Section 6.4.3 for information on resetting the module following an overload.
5V	Green	The power module is supplying +5-volt current to the PowerHub components that need it.
12V	Green	The power module is supplying +12-volt current to the PowerHub components that need it.

Reprinted by permission of FORE Systems, Inc.

Editing Questions

☐ 1. Is a table the best way to fulfill the purpose and meet readers' needs?

☐ 2. Is the table informatively and accurately titled and numbered?

☐ 3. Is the type size big enough to read (at least 8-point type)? See Chapter 16.3 for details on type size.

☐ 4. Is all the information included and is it accurate?
☐ 5. Is the table simple, with adequate separation between items?
☐ 6. Do column heads and stub heads clearly identify the item and the units of measure used?
☐ 7. Are decimal points aligned?
☐ 8. Do footnotes (with lower case letters) provide needed additional information?
☐ 9. Is the table understandable without the text? Is it referenced in the text? Is it placed right after the text reference?
☐ 10. Is the source acknowledged?

Line Graphs, Bar Graphs, and Pictographs

Graphs condense large amounts of information and can compare items; show relationships; and indicate trends, distributions, cycles, and changes over time. Because they provide a visual presentation of numbers, they make a reader's job easier.

Line graphs are plotted on a horizontal and vertical axis. Usually variables of distance, time, load, stress, or voltage are placed on the horizontal axis, and variables of money, temperature, current, or strain on the vertical axis. Scales should start at zero. Plotted points are connected by a line. Specialists and technicians like line graphs for their detail. (See Figure 15.11.)

Bar graphs compare amounts and show proportional relationships. Bars can be vertical or horizontal and should be the same width. Unless tied to time, bars may be arranged in increasing or decreasing length. Generalists and managers like bar graphs for their quick view of relationships. (See Figure 15.12.) A variation of a bar graph is a *pictograph*, which uses a single simple drawing to represent one unit and then repeats that unit to make up a bar. Pictographs add interest, but they are not as accurate as bar graphs.

Editing Questions

☐ 1. Based on what you know about the purpose and the intended reader, is this graphic the best choice?
☐ 2. Is it accurately and informatively titled? Is it accurately numbered?
☐ 3. Are the numbers accurate? (Check against the source information.)
☐ 4. For line graphs:
   ○ Are the X and Y axes properly labeled?
   ○ Is the unit of measurement indicated?

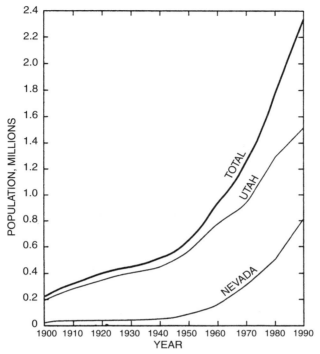

FIGURE 15.11 Example of a Line Graph *(David E. Prudic, James R. Harrill, Thomas J. Burbey. 1995. Conceptual evaluation of regional ground-water flow in the Carbonate Rock Province of the Great Basin, Nevada, Utah, and adjacent states. U.S. Geological Survey Professional Paper 1409-D. U.S. Government Printing Office.)*

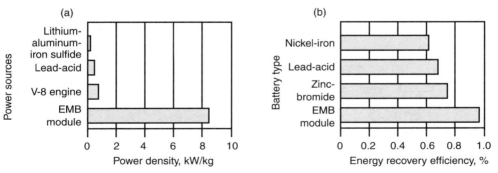

FIGURE 15.12 Example of Bar Graphs *(Richard F. Post. 1996. A new look at an old idea: The electromechanical battery. Science & Technology Review (April) VCRL-52000. 96-4.)*

- Are there enough data points to plot a meaningful line?
- Is the amount of information limited, or should a second graph be constructed?
- Does shading influence the attention directed to any part?
- Is a legend needed to explain items on the graph?
- Are the scales in proportion, or do they distort the data?
- Are the grid marks omitted?

☐ 5. For bar graphs:
- Is the spacing between bars wide enough to isolate the bars visually?
- Is the graph simple? (Three-dimensional bars frequently are hard to read.)
- Is the vertical or horizontal axis labeled?
- Does the graph need a legend?
- Is the graph accurate? (Check against the original data.)

☐ 6. Is the graph referenced in the text? Is it placed right after its reference?

☐ 7. Does the source of data need to be credited?

Examples

Graphic examples help readers understand in the same way that verbal examples do—by making something abstract concrete. Examples include such things as screen shots of computer monitors, EKG monitors, and oscilloscopes that capture for the readers what happened or what the screen looked like. (See Figure 15.13.) Other examples include photographs of microscope slides and of three-dimensional models. A slightly different kind of example is the

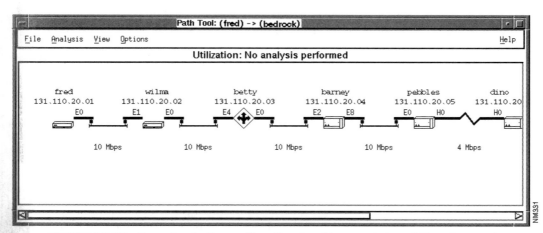

FIGURE 15.13 Example of a Screen Shot *(Reprinted by permission of Cisco Systems.)*

CHAPTER 15 GRAPHICS EDITING **313**

 Caution   Means reader must be careful. In this situation, you might do something that could result in equipment damage or loss of data.

Note   Means reader take note. Notes contain helpful suggestions or references to material not covered in the manual.

 Timesaver   Means the described action saves time. You can save time by performing the action described in the paragraph.

 Warning   Means you are in a situation that could cause bodily injury. Before you work on any equipment, you must be aware of the hazards involved with electrical circuitry and familiar with standard practices for preventing accidents.

FIGURE 15.14   Examples of Icons *(Reprinted by permission of Cisco Systems.)*

icon, a picture that can either symbolize action like "Timed Activity" or a situation like "Warning." (See Figure 15.14.)

Graphic examples are especially helpful for generalist readers, who seek ways to verify what they are reading or doing.

Editing Questions

- ☐ **1.** Based on what you know about the purpose and the intended reader, is this graphic example a good choice?
- ☐ **2.** Is this example titled informatively and accurately? Is it numbered accurately?
- ☐ **3.** For screen shots and photographs of slides or models:
    - ○ Is the graphic reduced to 50 to 75 percent of actual size without sacrificing readability?
    - ○ Is the point obvious by the graphic alone, or does some area need highlighting or a callout?
    - ○ Is the graphic referenced in the text? Is it placed immediately after the reference?
    - ○ Does the source need to be cited?
- ☐ **4.** For icons:
    - ○ Will the meaning of the icon be clear to the intended reader, including international readers?
    - ○ Is the icon culturally sensitive?
    - ○ If necessary, is the icon explained?

Block, Schematic, and Wiring Diagrams

Three different graphics are commonly used to show the relationships among items in complex subjects.

Block diagrams provide an overview by chunking information into blocks or symbols and connecting the blocks to show relationships. Block diagrams are often used in overview sections of manuals or reports. They appeal to all readers. (See Figure 15.15.) *Schematic* and *wiring* diagrams show specialists, technicians, and operators details in manuals of assembly, test, and maintenance. Schematics show the *logical* connections and current and signal flow. Wiring diagrams show the *actual* point-to-point connections. Both diagrams use standardized symbols. (See Figure 15.16.)

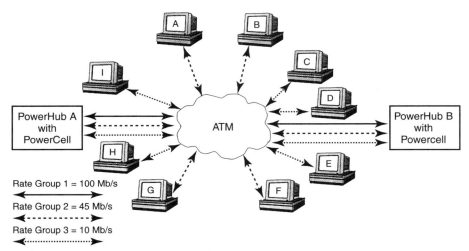

FIGURE 15.15 Block Diagram *(Reprinted by permission of FORE Systems, Inc.)*

Editing Questions

☐ 1. Is this diagram the best graphic for the purpose and the intended reader?

☐ 2. Is it accurately titled and numbered?

☐ 3. Is it accurate? (Either have it checked by an SME or user test it against an actual assembly.)

☐ 4. Is it referenced properly in the text? Is it placed right after the reference?

☐ 5. Does the source of the graphic need to be identified?

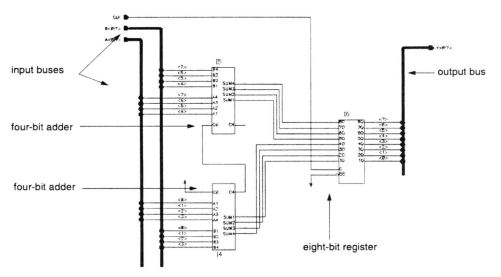

FIGURE 15.16 Schematic Diagram *(Reprinted by permission of Cadence Design Systems, Inc.)*

The Type and Experience of the Reader or Viewer

Editors and writers need to understand *who* the readers of a document will be, *why* those readers are reading, and *what* technical experience or understanding they have.

One convenient way to understand who readers might be is to categorize them in a continuum from least to most technical, as shown in Table 15.4.

Certain kinds of graphics appeal to, and are especially appropriate for, each of these types of readers—giving editors another criterion for judging graphics effectiveness.

Generalists like graphics such as bar graphs, block diagrams, and organizational diagrams, which make comparisons and connections for the reader. Generalists respond well to pie charts, exploded drawings, and maps, which show them relationships. And they constantly seek examples, screen shots, icons, and photographs, to reinforce their understanding of the text. They like pictographs for their visual interest.

Managers look for bar charts that compare, and milestone (Gantt) charts that set schedules. They rely on organizational and hierarchical diagrams to understand relationships. They expect flowcharts and block diagrams that explain processes, and they use tables to understand data summaries.

TABLE 15.4 Reader Types for Graphics Choices

Type	Reader's Purpose	Technical Experience
generalists	seek information, satisfy interest	limited; readers are outside their area of specialization
managers	make decisions	limited to broad; concern is not with the technology as much as with costs
operators	perform an operation, learn "how to do" something	moderate but can range from the novice to the expert
technicians	build, install, maintain, test	moderate to extensive
specialists	design, plan, experiment, develop, research	extensive

Operators use schematic and wiring diagrams and maps to carry out operations. They look for flowcharts to understand processes; exploded, cutaway, and line drawings to see how parts fit together; screen shots, slides, and photographs to see how the results should look.

Technicians, because of their more technical orientation, want line graphs, bar graphs, and schematic diagrams that provide detailed information. Like operators, they look for flowcharts, schematics, exploded diagrams, and wiring diagrams to understand the relationship of processes or parts.

Specialists, who may be more theoretical than any of the other reader types, look for tables, equations, and line graphs—all of which provide details. They appreciate block and schematic diagrams and examples of slides or screens to verify results.

The Language and Cultural Background of the Reader

In this day of extensive international marketing, editors must also consider their readers' ability to read English and their cultural backgrounds. International considerations are explained in detail in Chapter 19, but the following general graphics principles for international readers need to be emphasized in this chapter.

1. Graphics should play a prominent part in documents that will be translated or that will be read by English as a second language (ESL) readers. Graphics—especially photographs, line drawings, and maps—are often more effective than equivalent words. The safety brochures available in airline seat pockets are good examples of communication through graphics.

2. Graphics for international readers must be culturally sensitive, neither relying on negative stereotypes (the Mexican in a big sombrero asleep under a cactus) nor using colors or symbols with unpleasant or negative connotations. (Black, for example, is associated with darkness and evil in India.) Drawings or photographs of hands in symbolic (as opposed to action) gestures are especially problematical; hand positions that may have one meaning in the United States may mean something totally different in another culture.

3. Graphics need to be designed with an international reader in mind. Callouts (word identifications of parts of the graphic) should not be superimposed on the graphic itself but placed in a separate key or in white space around the graphic to allow for text expansion in translation. Graphics that must be read from left to right or from top to bottom must use arrows or numbers to indicate the reading order, especially for people in cultures that typically do not read from left to right.

See Chapter 19 for more information on graphics for international and intercultural readers.

Content

The content, or the kind of information that makes up the document, is another factor influencing graphics choices. Content is closely tied to purpose but also can be different in two documents that have the same purpose. For example, one feasibility report about closing a facility that asks: "Can and should we carry out this action?" might rely heavily on statistical data presented in tables and graphs such as operational costs, projected savings, tax advantages, relocation costs, and so on. Another feasibility report about closing a facility might emphasize location, traffic and parking, proximity to railroads and airports, and the presence or absence of a skilled workforce. This information might be presented through maps, photographs, line drawings, and the like. Any writer must use the collected information in the most effective way when choosing graphics; any editor must respect the limitations of the content and be a second evaluator of how well the chosen graphics function.

The Medium

In considering the medium (the type or form of presentation), the biggest factor that editors and writers must consider is whether to use *online documentation* (a computer screen, a Web page, a video, a film, animation) or *hardcopy* (a book, a manual, a report, a series of slides, a journal article). As technology improves the readability of screens and makes the production of Web pages, CDs, videos, and animation cheaper and easier, more documents are moving from hardcopy to online documentation. Many companies now routinely issue their manuals on CDs and the Web; if customers want hardcopy, they are free to download and print at their location. It is likely, however, that hardcopy documents will persist in the workplace; what these trends mean to you as an editor is that you need the ability to envision effective ways of designing and using graphics for both online use and hardcopy. Chapter 18 considers online editing in detail.

Within the hardcopy medium, editors must be aware of the many choices in the type or form of presentation. Common types include communications (like memos and letters), job-search documents (résumés, application letters), proposals, reports, instructions and manuals, and marketing material. Each form has its own requirements, reader purposes, reader types—and, consequently—requirements for graphics.

Editors Bush and Campbell, for example, say that proposals "should be 50 percent art. Thus, it's a good idea to plan the artwork first and work the text around it, especially because art is a long lead-time item" (1995, 101). Editors can help writers by looking for ways to add relevant graphics that will make proposals more effective. Professor Ralph Wileman notes, "Human fine motor skills . . . are best presented in live-action video or animation; three-dimensional objects . . . are best depicted as models or holograms, while realistic details are best shown through photography" (1993, 29). Proposal editors should also ensure that the graphics comply with any proposal guidelines. Some requests for proposals (RFPs) are very specific in terms of what kinds of graphics can be used.

Proposals are just one example of documents that should be planned around graphics. At the developmental edit, editors can look for key tables and figures around which the text is written.

Time and Resources

Writers have the primary responsibility in evaluating the time and resources available to create graphics, but editors play an important role. Some organizations have graphic designers on staff to create line drawings, graphs, and animated sequences from a writer's rough sketches. In other organizations, the writer must create all the graphics; in still other places, that job falls to the editor.

Software packages for drawing, graph-making, and presentations have made those jobs easier and the results professional in appearance, but graphics preparation can still require enormous amounts of time. Thus, time and cost must be factored in whether you are evaluating graphics (and perhaps suggesting additional or changed graphics) or creating them.

Degrees of Visualization

You can be more effective as an editor if you understand the relationship between verbal images and visual images and their impact on the reader or viewer. Ralph Wileman (1993, 18–26) shows this relationship along a continuum from the purely verbal (even though it may be projected on a slide to accompany a speech) to the purely visual.

Verbal

1. **Purely verbal**—Titles, headings, lists of words.
2. **Emphasized verbal**—Words have selected typographic emphasis: boldface, italics, change of font.
3. **Verbal with visual cues to meaning**—Drawings, icons, and other graphic symbols work with the words to convey meaning.
4. **Verbal/visual balance**—Viewers can receive the message either verbally, visually, or both.
5. **Visual with verbal cues to meaning**—Drawings and graphs provide the primary meaning, aided by words.
6. **Emphasized visual**—Arrows, circles, enlargements, and other emphases draw attention to parts of the visual; no words are used.
7. **Purely visual**—Photographs, line drawings, slides, and the like are used without words, though words may accompany the presentation, as in a video or slide presentation.

Visual

Wileman notes that types 3, 4, and 5 are the most "helpful in terms of learning" (27), but that writers and editors should make an effort to use all seven types. What is important is the degree to which the visual or the verbal (or a visual-verbal combination) is effective in helping the reader comprehend and apply the information.

15.3 Developmental Editing of Graphics: Major Questions

At the developmental or first-draft editing stage, as an editor you should ask the big questions:

☐ 1. **Are the graphics used for specific purposes and is the most effective graphic chosen to fulfill each purpose?** This question is critical whether you are editing your own or someone else's work. If new graphics need to be incorporated or if the type of graphic needs to be changed, this is the time to do it. If something needs to be *shown*, a graphic is probably needed. As editor Judith Goode says, "Suppose text would be more effective as an enumerated list or even a table. If the editor points out such a problem in chapter 1, the writer will have a better way to handle similar material when it appears in chapter 3. Distilling information into tabular form can be difficult, so writing time may not actually be saved. But the time needed to create a table will probably be about the same as drafting copy and then revising, editing, and reformatting it" (1995, 6). While checking for purpose and suitability, you should also check for overuse or misuse of graphics. Graphics in workplace writing should never be merely decorative.

☐ 2. **Are the graphics placed where they can be most effective?** Locating graphics is both an editing problem and a document design problem, and the solutions to the two problems may conflict. Editors need to look at placement from the reader's perspective, which usually means that graphics should appear immediately after their first mention in the text. Readers, even technical readers, are busy and somewhat lazy, and they often will not search through several pages or to the end of a chapter for the relevant graphic. On the other hand, good page design dictates that the number of graphics and the amount of text need to be balanced. The design specifications for a particular document may even mandate that graphics can only be placed at the top or bottom of a page. Thus, graphics may by necessity appear two or three pages after their text reference.

Just such a problem occurs in this chapter. The chapter needs to include examples of and key facts about the major kinds of graphics for those readers just learning about graphics. There are more examples (which take more space) than there is equivalent text. If the examples and explanations were built into the text of this chapter, some of the graphics would not appear immediately after their text reference. Therefore, after consultation, the text designer and I chose to include the examples of graphics and their related text in one section just after the section on *The Purpose of the Document*. The initial text reference comes from Table 15.1, in the column headed "examples." To help readers find the examples easily, the section containing graphics is marked with a bar in the margin like a tab. As the reader of this text, you should evaluate the effectiveness of this graphics placement.

☐ 3. **Are the graphics designed for easy reading?** An editor will ask this question once more while copyediting, but the focus at the developmental edit is, again, on larger issues. For example, is a table so long and with so many columns that it must be printed "landscape" style (that is, lengthwise on the page) forcing the reader to turn the book 90 degrees to read it? Perhaps the table can be redesigned or simplified, so it can be read "portrait" style (vertically on the page). Perhaps important parts of the table can be made into a figure like a bar graph or line graph. For a second example, is a line graph constructed with multiple lines that intersect and force the reader to sort out each one? Perhaps the line graph can be broken into several smaller graphs, each of which will make a single point. For a third example, is a bar graph constructed in three dimensions and with shading and cross-hatching that distort the emphasis? Perhaps it can be redesigned in only two dimensions that use color instead of shading and crosshatching.

☐ 4. **Are the graphics appropriate for the intended reader?** Editors must consider the reader's experience, task, and familiarity with English when evaluating graphics at the developmental stage.

☐ 5. **Are the graphics well rendered?** Are elements in the figures well balanced? Is the choice of photograph or line drawing, line graph or bar graph, and so on appropriate to the content and purpose? Have sex or culture biases been removed? Is the point of the graphic tasteful and tactful?

15.4 Copyediting Graphics: A Checklist

At the copyediting or second-draft level, you owe it to the writer and the reader to scrutinize details. The following checklist lets you systematically approach the copyediting task.

1. **Are the graphics accurate?**
    ☐ Are the figures up-to-date, showing the current or most recent product or release?

- ☐ Do all the numbers add up?
- ☐ Is the chronology correct?
- ☐ Does the information in the graphic agree with the information in the text?
- ☐ Is the terminology consistent in the graphics and the text?
- ☐ Are the objects or parts shown in proportion?
- ☐ Are relative sizes obvious, shown by a scale, or shown by comparison with a common object such as a coin?

2. **Are the graphics consistent?**
   - ☐ Within a document, a document set, or a series of screens, are similar types of graphics designed in the same way: in placement of titles, figure and table numbers, keys, and callouts?
   - ☐ Should scales be the same (for example, not shifting from a 0 to 50 percent scale to a 0 to 100 percent scale)?
   - ☐ Are graphics approximately the same size, given their design differences?
   - ☐ Are the type fonts, type sizes, and type style (bolding, italics, underlining, color) used in the same way for a series of graphics?
   - ☐ Is capitalization used the same way in a series of graphics?

3. **Are the graphics clean and uncluttered?**
   - ☐ Is the graphic itself and every part of it necessary to convey the information? (No cute or decorative touches.)
   - ☐ Are extra lines removed from tables, with white space used as a separator?
   - ☐ Are distracting grids removed from line graphs and bar graphs?
   - ☐ Are photographs cropped to show only relevant items and to omit distractions?
   - ☐ Are extraneous lines omitted from line drawings, cross-sections, and cutaway drawings?

4. **Are the graphics complete?**
   - ☐ Does each graphic have a number and a meaningful title?
   - ☐ Will the title be meaningful without the graphic, for example, in a list of figures or tables?
   - ☐ Are the titles of graphics parallel in structure?
   - ☐ Are all the columns filled in tables?
   - ☐ Do tables have column heads and, if necessary, a key?
   - ☐ Is each part or step labeled so the graphic is understandable even without the textual reference?
   - ☐ Are people and objects in photographs identified in a caption?

5. **Are the graphics clear?**
   - ☐ Are the graphics that need reduction or enlargement clearly marked?
   - ☐ Will the graphic be completely legible after being reduced or enlarged to fit the page layout?

☐ Will the graphic be clear after printing or photocopying? Shading is particularly troublesome.
☐ Is there enough white space around each item for clear separation?
☐ Are significant items highlighted by color, shading, bold lines, and the like?
☐ Are labels and headings horizontal whenever possible?
☐ Are icons, acronyms, and abbreviations obvious to the intended reader or explained in a key?
☐ In screen shots, do special effects like pointers, lines, or combined images appear where they should? Can you see everything clearly (for example, no cropped scroll bars or cut-off words)?
☐ Does each graphic make a coherent point even without the text reference?

6. **Are the graphics referenced?**
☐ Does the text reference the graphic by table or figure number (not by location)? Is referencing consistent (by parentheses, as part of the text, and so on)?
☐ Does the text add information to the graphic itself by explaining it, highlighting portions of it, or relating the graphic to the point being made in the text?

7. **Is the source of each borrowed graphic properly cited?**
☐ If the graphic is borrowed from another source, is the proper credit line attached? (And if the document is published, is there a signed permission for publication?)
☐ If the source is a government document, is the specific source acknowledged?
☐ If the graphic is designed in-house using source material from elsewhere, is that source material acknowledged?

15.5 Proofreading Graphics: A Checklist

At the final or proofreading edit, you again look at details, but this time your primary aim is to ensure that previously marked graphics changes have, in fact, been made. If time permits, you should again check for accuracy, continuity, and completeness.

1. **Have all copyediting changes been made to the final copy?**
☐ Do proof sheets and corrected drafts agree?
2. **For graphics that have been moved:**
☐ Are table and figure numbers in the correct order?

☐ Do text references agree with the new numbers?
☐ Are graphics located immediately after their text reference?
3. **Are all graphics properly oriented on the page?**
☐ Have photographs and line drawings been reversed or inverted? (Check against the original.)
☐ Are all columns and rows present in a table? (Check against the master copy.)

EXERCISE 1

Choosing Effective Graphics

For each of the following purposes, choose two possible graphic types from those listed in Table 15.1 and be prepared to explain why they would be effective. Selected answers appear in the Appendix.

1. To persuade the reader that Airix Chemicals has the resources and experience to carry out the projected task
2. To teach the reader how to assemble the transmission
3. To help the reader choose from among three alternative testing devices
4. To show the reader what happens to sawdust from the time it enters the paper mill until it emerges as the final product
5. To show the reader the timber loss in each of the last 10 years from gypsy moth infestation

EXERCISE 2

Evaluating Graphics in Development

Evaluate each of the following graphics for ease of reading (U.S. Bureau of the Census 1994). Indicate in writing what the problems are and what developmental editing changes you would suggest.

CHAPTER 15   GRAPHICS EDITING   **325**

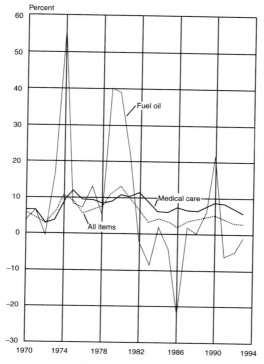

FIGURE A   Line Graph

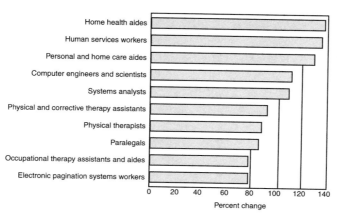

FIGURE B   Bar Graph

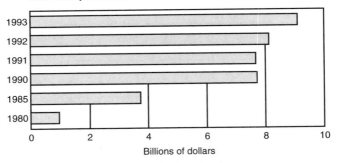

FIGURE C Bar Graph

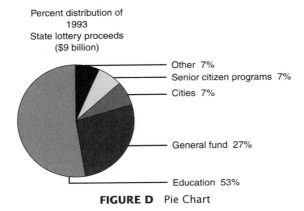

FIGURE D Pie Chart

EXERCISE 3

Graphics Applications in the Workplace

Interview a writer or editor in your local area, asking the following questions and others:

1. What role does he or she play in graphics choice and production?
2. What guidelines are used to decide what graphics to use?
3. How are graphics created?
4. What software is used in graphic production, and how good is it?

Report to the class and your instructor orally and in writing.

EXERCISE 4

Creating Graphics

For each of the following paragraphs of information, suggest and sketch two possible graphics. Specify the purpose, reader, and main point for each one, and tell why that graphic type would be appropriate.

1. In 1993, the total population of the United States was 258 million. Of the total, 24 percent lived in the Midwest, 35 percent lived in the South, 22 percent lived in the West, and 20 percent lived in the Northeast.
2. In 1992, according to the U.S. Dept. of Agriculture, per capita consumption of noncitrus fruits was 74.4 pounds; per capita consumption of citrus fruits was 24.3 pounds. The breakdown by type of fruit in pounds was as follows: bananas 27.3, apples 19.3, grapes 7.2, nectarines and peaches 5.9, pears 3.1, strawberries 3.5, pineapples 2.0, plums and prunes 1.8, oranges 12.9, grapefruit 5.9, other 9.8.
3. Microcomputer software sales in 1993, as reported by Software Publishers Association, totaled (in millions of dollars) 1022 for word processors, 801 for spreadsheets, 476 for databases, 410 for entertainment, 409 for finance, and 185 for desktop publishing.

EXERCISE 5

Evaluating Workplace Graphics

1. Clip or copy an example of a table from a workplace document. In writing, evaluate its strengths and weaknesses. Suggest and sketch another way to present the information.
2. Clip or copy two different examples of effective figures in workplace documents. In a paragraph, analyze why they are effective.

References

Bush, Donald W. and Charles P. Campbell. 1995. *How to edit technical documents*. Phoenix, AZ: Oryx.

Goode, Judith. 1995. Using editors when and where it counts. *Intercom* 48.8 (October).

Horton, William. 1993. Dump the dumb screen dumps. *Technical Communication* 40.1 (January).

Levie, W. H. and R. Lentz. 1982. Effects of text illustrations: A review of research. *Educational Communications and Technology Journal*. 30: 195–232.

Morrison, Claire and William Jimmerson. 1989. Business presentations for the 90s. *Video Manager* (July), 18.

Post, Richard F. 1996. A new look at an old idea: The electromechanical battery. *Science & Technology Review* (April) VCRL-52000-96-4. For Figure 15.3.

Prudic, David, James Harrill, Thomas J. Burbey. 1995. *Conceptual evaluation of regional ground-water flow in the carbonate-rock province of the Great Basin, Nevada, Utah, and adjacent states*. Washington: U.S. Geological Survey Professional Paper 1409-D. For Figure 15.11.

Quimby, G. Edward. 1996. Make text and graphics work together. *Intercom* 43.1 (January).

Rew, Lois Johnson. 1993. *Technical writing: Process and practice*. 2nd ed. New York: St. Martin's.

Shelton, S. M. 1993. Visual communication: Introduction. *Technical Communication* 40.4 (November).

U.S. Bureau of the Census. 1994. *Statistical abstract of the United States: 1994*. 114th ed. Washington, DC: U.S. Government Printing Office.

Wileman, Ralph E. 1993. *Visual communicating*. Englewood Cliffs, NJ: Educational Technology Publications.

Williams, Thomas R. 1993. What's so different about visuals? *Technical Communication* 40.4 (November).

16 Document Design

Contents of this chapter:

- Defining Document Design Terms
- Understanding the Factors That Influence Document Design
- Evaluating the Document Design and the Page Design
- Developmental Editing of Document Design: Major Questions
- Copyediting Document Design: A Checklist
- Proofreading Document Design: A Checklist

Editors—as well as writers—have a responsibility to the reader for good document design. To understand the importance of good design to the workplace reader, we need only to think how we proceed when we read. "When we're confronted with a page of solid, undifferentiated text, what do we do? We use a colored highlighter. In highlighting, we create visual landmarks. We mark key points that we want to remember and refer to, that help us mentally follow the structure of the information, increase our understanding, and make it easier to recall later or to go back for reference" (Keyes 1993, 639).

Good document design, by revealing the organization, provides a framework (or schema) that readers can use to sort and comprehend information. Document design pre-processes the content by providing the visual landmarks that readers need, and it thus leads the reader through the document in the way that the writer intended.

Computers, software programs, and high-resolution printers give workplace writers almost unlimited choices in type sizes, styles, and font; in placement of text and graphics; in use of color and white space; and in boxes, rules, and screens. However, as Professor Mary Lay notes, ". . . with the freedom to arrange text and graphics on each page—and even on each computer screen—comes the responsibility to assess the effect of that arrangement" (1989, 72). The responsibility for that assessment often falls to the editor, who needs to be the "re-viewer," the person who rethinks and evaluates the chosen design.

Large and well-established companies usually have a comprehensive plan for document design so that all their documents of a certain type have the same look. They may even have computer templates governing individual page layout, size and location of headings, and so on. As Jeffrey Vargas, a learning products specialist at a computer manufacturer, noted, "Most companies have a pre-defined document design; there's not much impact individuals can have in those cases."

Nevertheless, more than half of the 50 writers and 14 editors I surveyed in researching this book indicated that they were responsible for editing document design. Yvonne Kucher, who edits at an internetworking company, said, "As an editor, I participate in developing design standards." In small and start-up companies, editors may actually design the documents. Chrystal Francois, who works primarily with the medical and drug industries, said, "I've found that document layout has become very important in consulting." And Elizabeth Van Houten, a freelance editor, said, "For a well-written 20-page business document that I have to design and edit, I probably spend 70 percent of the time doing page layout and 30 percent of the time reviewing."

Because editors are so often involved in document design (either as creators or reviewers), this chapter provides the basic information about design that editors need to know. It also provides editing questions for the developmental, copyediting, and proofreading stages of editing.

16.1 Defining Document Design Terms

The four general terms used in this field are document design, format, page design, and layout. *Document design* is the umbrella term; it refers to the process of determining the way the manual, report, proposal, or other document looks. Through the *format*, or the arrangement of information on every page, the designer can emphasize the document's organization. The terms *page design* and *page layout* refer to the look of each page based on spacing, specifications of type, location of graphics, and so on. Online documents also have format and screen design specifications, which are explained in Chapter 18.

16.2 Understanding the Factors That Influence Document Design

Sometimes as an editor you will be charged with designing the document yourself. At other times, you will be a reviewer or evaluator. The four factors that should influence the design choices in either case are the same: the purpose of the document, the needs of readers, the way readers will probably read this document, and the type of document.

Purpose

Just as purpose shapes the content of the document, so purpose influences the way that content should be presented to the reader. The writer presumably has already determined the purpose: what he or she wants readers to do after reading the information. Your job is to rethink the purpose and ensure that it is clear. Does the document inform? Instruct? Propose? Recommend? Does the content support that purpose? How should the design reinforce the content?

Needs of Readers

As you learned in Chapter 14, specific readers have different interests and needs, and you should learn as much as you can about those needs. In addition, it helps to know how old readers are and what their general level of experience is.

Sometimes you and the writer will know specifically who the intended reader is: for example, a system administrator with experience in previous versions of the product or an expert engineer familiar with computer-aided-design (CAD) systems. More often, reader experience and needs will be harder to define. Writers and editors can learn about intended readers by analyzing responses from any of these sources:

- reader response cards inserted in a manual or journal
- Web sites or Internet communication about the product or service
- questions or problems fielded by help centers or product support
- results of usability tests, conducted either formally or informally
- meetings or interviews with selected readers or users

In the workplace, few readers are reading for pleasure. They are under time pressure and are either reading to learn something new, to do something, or, as researcher Janice Redish notes, "to learn to do" (1993, 21).

The Way Readers Will Read the Document

Related to reader needs in evaluating document design is the concept of the *way* readers read. For example, is the reader seeking information to make a decision? Does the reader need to repair a machine? To learn a new process?

As explained in Chapter 14, most busy workplace readers will read using one of three methods:

1. *skimming:* reading for the general drift of the passage
2. *scanning:* reading quickly to find specific bits of information
3. *search reading:* scanning and then studying specific items

These methods of reading influence many of the techniques for designing documents. For example, skimmers need visually prominent headings, abstracts or summaries set off by white space, and checklists that condense and apply previous information. Scanners rely on size of type, bolding, and white space to signal key information. Search readers expect short paragraphs of chunked information for easy information access. These design techniques and others are explained in section 16.3.

Document Types and the Conventions That Influence Them

As workplace writing has evolved, certain document types have come to be designed in conventional ways. Both writers and editors need to be aware of these conventions because readers have already-developed expectations (or schemata) for these documents.

Some conventional document designs come from tradition: the form of the standard business letter—framed on the page and containing a salutation and complimentary close—has historical roots. The structure of the scientific research report is governed by the profession; scientist readers expect the IMRAD structure—introduction, materials and methods, results, and discussion. These readers expect that the content structure will be made apparent by the design. On the other hand, the design of instructions may be influenced by the company style guide. When starting an editing project, you should find out if a conventional form for this document exists, or if the form is set by the company's or the profession's style guide.

16.3 Evaluating the Document Design and the Page Design

Whether evaluating the design of the whole document or of individual pages, an editor needs always to assume the perspective of the reader. Good design is the result of many choices, but the impact of that design must be measured by how the reader responds. Although you can never be sure of an individual reader's background or motivation, you can simplify the editing task by asking the following six questions:

1. How effective is the design of the whole document?
2. How easy is the document to read?
3. Can the reader find information easily?
4. Can the reader understand the text easily?
5. Will the reader be able to retain and retrieve information?
6. For instructions and manuals, can the reader use this document to perform a task?

How Effective Is the Design of the Whole Document?

People who read novels are highly motivated. Because they are reading for pleasure and at leisure, they do not mind pages densely packed with text and with the only white space from paragraph indentations and margins. But readers in the workplace do not read for pleasure. Because they are busy and pressed for time, they read only because they need information, instruction, or advice. As Jan V. White, an expert in document design notes, "One of the myths in the field of publication making is that readers are readers. In actuality, they start out as viewers. They scan, they hunt and peck, searching for the valuable nuggets of information. Reluctant to work, saturated by media, and a bit lazy, they literally need to be lured into reading" (1990b, 17). A document that looks easy to read and use will help "lure" those readers into the actual reading.

Packaging

One way to make a document effective is to fit the "packaging" of the document to the purpose and the reader. For example, consider a mechanic's manual, which needs to be read in the shop where the lighting is poor and the reader's hands are greasy. Design packaging should perhaps include a spiral binding, so the manual will lie flat; laminated pages that can be wiped off; large font size and generous white space to separate information items. Alternatively, consider a proposal, which will compete with 50 others being evaluated in two days. Design packaging should provide a clear table of contents, summaries at the front, and bold headings that direct the reader's eye.

Packaging includes cover material and design, binding, quality of paper, use of color, and the relationship of text with graphics and white space. Design standards for these elements will also ensure that related documents will look like they belong together—that they are part of a set.

Page-Level Considerations

At the page level, various elements can be evaluated as a unit. As publishing design expert Roger Parker notes, ". . . success is determined by how well each piece of the puzzle relates to the pieces around it. For example, proper headline size is determined partly by its importance and partly by the amount of space that separates it from the adjacent borders, text, and artwork. A large headline in a small space looks cramped" (1997, 11).

On hardcopy documents, pages should be evaluated as a two-page spread since that is how the reader will view them—even before reading. To see if a two-page spread is effective, you can print it out full size and view it from a distance. Or, if your software permits, you can print *thumbnail* proofs, facing pages reduced in size and on the same sheet of paper. Parker notes that "Thumbnail proofs let you see where good design has been sacrificed for expediency" (1997, 25). See Figure 16.1 for examples of thumbnail proofs.

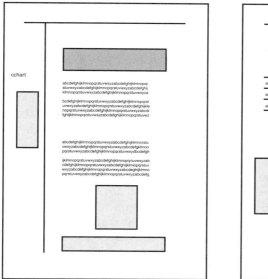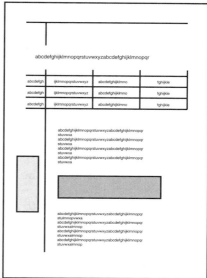

FIGURE 16.1 Thumbnail Proofs: Reduced-size Pages for Editorial Review

Principles of Page Design

Understanding four principles of page design—unity, balance, emphasis, and sequence—will help you evaluate how well the various elements on pages work together.

The first principle is *unity*. Graphics, sections of text, headings, and subheadings must look like they belong together. You can promote unity by placing those elements in close proximity (with more white space around the elements than between them), by repeating the general shapes of graphics or sections of text, and by using a single color. According to Jan V. White, "A single color used throughout a document provides unity, continuity, and character. Color consistency gives the product an identity and places it in a corporate context" (1990a, 44). In designing screens, the same type of information should be located in the same place on successive screens. (See Chapter 18 for more information on screen design.)

Pages should also be evaluated for their *balance*, which can be symmetrical and formal or asymmetrical and informal. Professor Mary Lay says, "Large, unusual, or colored elements weigh more than small, ordinary, or black and white elements. A large area of white space also appears heavy. If elements on a page or screen are not in balance, we may feel a need to rearrange" (1989, 77). But too much symmetry on pages may be boring and may, in fact, make the text uninviting. Asymmetrical balance, with two or even three columns of unequal width, can be achieved by balancing headings and graphics with blocks of text and white space. Asymmetrical formats also are easier to scan. Figure 16.2 shows how asymmetrical balance might look.

Emphasis and *sequence* are the third and fourth principles of page or screen design; they help readers who skim, scan, or search-read by providing visual cues about what is important. Emphasis can be achieved with color, bold text type or headings, screens, rules, or boxes. Emphasized areas influence sequence, or the direction the readers' eyes travel over the information. Outdented headings, the proximity of sections of text, and the use of white space all contribute to good sequencing.

Decisions about the placement of text, headings, graphics, and white space are influenced by what we know about how readers read. Readers of English read from left to right, their eyes scanning a page in a Z pattern, as shown in Figure 16.3. This pattern means that important elements, like key graphics or crucial text, should be located in places that the eye settles on—the upper left and lower right sections, for example.

Page layout is also influenced by the type of document and the design conventions for that type. For example, a technical report within a particular profession might be conventionally designed in single-column format; pages for a manual in the same profession might be designed in two columns.

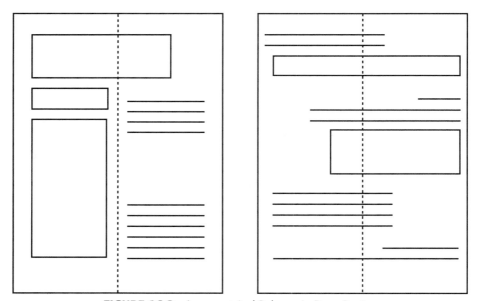

FIGURE 16.2 Asymmetrical Balance in Page Design

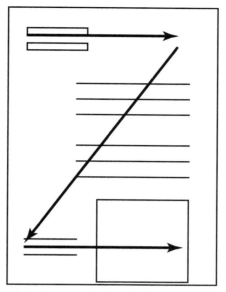

FIGURE 16.3 Scanning Pattern for Readers of English

How Easy Is the Document To Read?

Of all the questions an editor can pose to assess design effectiveness, none is as important as "How easy is it to read?" Within the technical community, a useful guideline is this: The more difficult the technical concept, the easier the prose should be to read.

According to *The Art and Technology of Typography*, readability is "the level of comprehension and visual comfort when reading printed material. Readability is concerned with how the type is arranged on a page. Readability is affected by line length, word spacing, letter spacing, and leading" (Agfa 1988, 8). As an editor, you need to understand the basic guidelines that help ensure good design for each of these elements.

Line Length and Grid Divisions

Decisions about line length and page layout can be simplified by sketching trial designs with a pencil and paper. If you are the designer, try at least three different sketches before making a decision. If you are editing someone else's design, sketch an alternative arrangement if something doesn't look right. (But be sure you can justify your new design with more than "It looks better this way.")

Once you have the basic idea, you can use a grid to determine the number of columns, margin size, heading placement, text placement, and so on. Most computer design programs allow you to establish formats based on grids that will be maintained from page to page.

The columns of a grid layout provide the vertical structure. A single column on a page will allow a great deal of text, but the lines may be so long that they are hard to read. David Matis, who researched the literature on text design, found that "Ten to twelve words per line, or about 50 to 70 characters, is the optimal line length for most conventional type sizes" (1996, 23). Thus, if a single column is used, substantial white space is needed.

Two-column pages can provide comfortable line length, but they may be boring unless they are broken up with subheadings or illustrations. Three columns can provide more interest but may be confusing. In general, column size is related to type size: the narrower the column, the smaller the type size; the wider the column, the larger the type size. In addition to the vertical grid, a horizontal grid of six to eight divisions will help in placement of graphics, text chunks, and headings. Consult a design program for details in setting up grids.

Alignment

Related to column width and line length is the decision about alignment: whether lines of type are "squared off," or fully justified, at the right and left margins. While many commercially published books are fully justified, manuals and reports in the workplace frequently are unjustified on the right margin (a technique also called "ragged right"). Readability studies show that ragged right text is somewhat easier to read. In ragged right text, no extra spaces have been inserted to even out the line, and the tighter word spacing helps readers recognize word groups. Because words can wrap to the next line, hyphenation also can be eliminated, which helps readability. In addition, the ragged right margin gives the page a more open and informal look. See Figure 16.4 for examples of the look of justified and ragged right pages.

Type Size and Style

In evaluating the writer's choices of type size and style, you should look for ease of reading. Jan White says it well: "Functional typography is invisible because it goes unnoticed. The aim is to create a visual medium that is so attractive, so inviting, and so appropriate to its material, that the process of reading (which most people dislike as work) becomes a pleasure" (1988, 2). You need to choose or review type based on the following factors:

1. number of fonts
2. point size of text type

FIGURE 16.4 Justified and Unjustified Pages

3. point size of headings and subheadings
4. uppercase or lowercase type
5. serif or sans serif font
6. highlighting techniques: bold, italic, underlining

Number of fonts. A *font* (also called a typeface) is a complete set of type of a distinctive design, for example, Helvetica or Times Roman. In choosing fonts, the experts recommend restraint and simplicity—using one or two fonts is more effective than using many.

Point size of text type. Experts recommend a character size that is from 8 to 12 points for text type. A *point* is a measurement of type, with 72 points to the inch. Point size is influenced by column width and by the distance from which readers will be reading. For example, the farther away the type will be from the readers' eyes, the larger the point size should be.

Point size of headings and subheadings. To help readers scan the text, headings and subheadings need to be visually prominent. Parker says, "Generally, headings are 1.5 times larger in point size than the body text and one full level greater in boldness. For example, if you select 12-point body text, then you'll use 18-point size for headlines" (1997, 71). Figure 16.5 shows the relationship of font size in headings and text type.

> ## International Icons
>
> Document designers must keep in mind the culture of their target audience when creating icons. Common symbols in America may be unrecognized in other countries.

FIGURE 16.5 Relationship between Heading and Text Type. Heading is 18-point font; text is 12 point.

Upper-case or lower-case type. Experts recommend using a mixture of upper-case and lower-case type instead of all capitals for both text and headings—unless the headings are very short. Researcher Matis says, "Text set in lower-case letters is read 10 percent faster than similar material set in all capitals. Text set in all capitals is read about 19 percent slower than text set in mixed case" (1996, 23).

Serif or sans serif font. Serifs are the little extensions or "legs" on type that tend to join letters together as the eye moves over them. A sans serif font is one without the extensions. A common serif font is Times Roman; a common sans serif font is Helvetica, as shown in Figure 16.6.

> **Perception: What It Is**
> Seeing something is a physiological act. The eye sees something, then transmits the information to the brain where the brain interprets the information. What the eye sends to the brain is generally filtered through sensory data filters.
> Information that passes the sensory data filters and makes its way to the mind then becomes interpreted. The interpretation of the something involves recognizing a pattern, processing the pattern, and organizing it. (Ishizaka 1997)
>
> **VRML Makes 3-D Graphics a Reality**
> Virtual Reality Modeling Language, or VRML (pronounced "vermal"), is a language used to create interactive three-dimensional worlds and objects. Just as Hypertext Markup Language (HTML) has brought to life previously flat screens of text, VRML has pumped life into two-dimensional text and graphics by allowing users to enter the virtual world and become part of what they see. As with HTML, VRML publishers can distribute their work on CD-ROM, the Internet, or on a company's intranet. (Saunders 1997)

FIGURE 16.6 Serif and Sans Serif Fonts

Although the research results on ease of reading of serif or sans serif fonts are mixed, American readers tend to find serif fonts easier to read in body text. Sans serif fonts are often used for headings, captions, online documents, and visuals for transparencies that will be projected on an overhead projector. The best way to choose fonts is to print out a page or two in the column width you have chosen to see how the printed lines look on the page.

Highlighting techniques. The highlighting techniques of boldface, all capitals, italics, and underlining are also called type style and emphasis. Experts agree that boldface provides a clear distinction; italics set off a word but do not emphasize it; but underlining (a leftover from the days of typewriters) is generally not effective. Parker says, "Portions of the descenders [the parts of letters that drop below the line] often become lost in the underlining, making letters harder to identify and words harder to read" (1997, 305). The key to any effective highlighting technique is to use it with restraint. For example, italics in large chunks are hard to read, especially in online documents.

Spacing. Spacing includes all the uses of white space including margins and space around graphics, but it generally refers to the space between letters, words, lines of type, and paragraphs. Spacing can be an effective design tool: "The space between lines of type or around blocks of copy is like the space between and around your living room furniture. The addition or subtraction of space between a coffee table, bentwood rocker, and sofa can create extreme openness or cozy comfort" (Agfa 1988, 32).

Kerning is the adjustment of space between selected pairs of letters. Kerning can effectively reduce the space in headings that have an uppercase "W" next to a lowercase "a," or an uppercase "T" next to a lowercase "o." In editing, you can examine such combinations to see if kerning is needed.

Word spacing is the amount of space between words. If spacing is reduced, you can fit more words in a line, but the text may be difficult to read. Most computer programs set word spacing at 20 to 25 percent of type size, and you can use this default setting for most documents. If two spaces have been left at the end of each sentence (as typists customarily did), the page of type may have distracting white space "rivers" running through it. Therefore, the new convention is to have only one space between sentences.

Leading (pronounced "ledding") is the term for the space above and below a line of type. Experts say that text type reads most easily with one or two points of leading. For example, 10-point type may be set on a 12-point line. This is less spacing than what we call "double spacing," but it does open up the text. Headings, on the other hand, are often "set solid"; that is, they have no extra white space between lines.

Paragraph spacing adds to readability by opening up the page and making each paragraph a discrete unit. By using the paragraph spacing command on your desktop publishing program, you can increase paragraph spacing without adding too much space. When editing, this is an important item to check. Paragraphs can also be set off by tabbing about 0.5 inch (standard indentation) and setting the paragraphs solid.

Some design experts say that at least 50 percent of a page should be white space. "Space is as important in printing as rests are in music" (Linotype 1988, 17).

Boxes, Rules, and Screens

Graphic elements like boxes, rules, and screens help readers by emphasizing, separating, or grouping information. *Boxes* are made of thin lines that surround a block of text or a figure. Boxed copy is usually related to the main text but can also be read and understood by itself. Sometimes boxed copy is set in a font one size smaller than the text copy. For generalist readers, boxes can set off explanatory material in what is called a *sidebar*. For readers who are operators, boxes can be used to emphasize warnings and cautions in instructions.

Rules are horizontal or vertical lines that can emphasize or link two related columns or isolate the text below or next to them, thus guiding the reader's eye. Consistent use of rules can help the reader navigate through the text.

Screens are the portions of a page set off by pale dots over which type is printed. Screening adds emphasis and interest, but should be used carefully. Jan White says, "Use very pale screens to distinguish areas on the page; 20 percent is usually as dark as it should be" (1990b, 20). See Figure 16.7 for examples of boxes, rules, and screens.

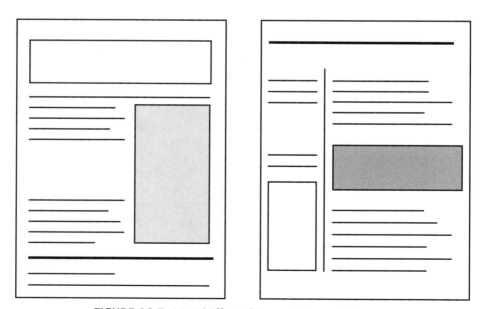

FIGURE 16.7 Visual Effect of Boxes, Rules, and Screens

Color

Consistent use of color can also help guide the reader's eye, especially when used for first-level headings, rules, or margin annotations. But the main text should be set in black. Colin Wheildon, who conducted an extensive study of type and design on reading comprehension, says, "Even copy set in deep colors was substantially more difficult for readers to understand. Seven times as many participants in the study demonstrated good comprehension when text was black as opposed to either muted or high intensity colors" (1995, 101).

Can the Reader Find Information Easily?

Finding information easily depends on what are called organizers or *access aids*—those elements in the whole document and on individual pages that help readers who are skimming or scanning. Access aids for the whole document are so important that they are treated separately in Chapter 17. There you will learn how to edit access aids like the table of contents, index, glossary, menus, and abstracts.

In this section, the emphasis is on headings and subheadings, the access aids that help readers find information on a page. Guidelines for effective headings and subheadings include the following:

- Headings should be short and succinct and set in a font larger than the text type (often as much as 4 points larger).
- A different typeface can set off the headings (such as sans serif heads over serif text).
- Long headings are more effective set with initial caps only or in lowercase—not all caps. Jan White says, "Set just like a normal sentence, differing only in type size and boldness, downstyle heads [the first word of the heading is all that is capitalized] follow the patterns readers are used to. Besides, they are more economical of space, since capitals take up much more space than their lowercase equivalents. Furthermore, proper names and acronyms stand out because they are capitalized in the normal way and thus aid comprehension" (1988, 96).
- Headings set bold and in color are effective.
- Flush-left headings are easier to scan than centered headings. "As people read, they need to return to a vertical axis at the left, where the next line is logically expected to begin" (White 1988, 96.) Headings also scan effectively when they protrude into the left margin, a technique called *hanging-indent* heads or *outdenting*. See Figure 16.8 for an example.
- Headings and subheadings should be self-contained and independent of the text. Readers should be able to determine the document's structure by reading the headings and subheadings.

FIGURE 16.8 Hanging Indent Heads

- Subheadings can be placed within the text or next to the text in the left margin.
- A subhead should be separated from the text above it by white space, twice as much space above the head as below it.
- Consistent use of style and size for headings and subheadings throughout the document will help the reader find information easily.

Can the Reader Understand the Text Easily?

Comprehension is the reader's ability to understand information, and that ability depends on three things: (1) the reader's short-term memory, (2) the reader's long-term memory or activated schemata, and (3) the way the document is organized and designed.

Short-term or working memory, as Chapter 14 explains, allows readers to remember items of information long enough to associate them with new information. However, the limit of short-term memory is approximately seven items.

Long-term memory stores information and experiences in schemata (frameworks or structures of related knowledge). Comprehension occurs when the reader can insert new information into an existing schema or can compare an existing schema to a new experience. (See Chapter 14 for more information on long-term memory.)

In editing document design, you can help the reader comprehend information or instructions by examining the effectiveness of the following related techniques:

- chunking
- queuing
- filtering
- visually prominent headings
- use of lists
- use of color
- figures and tables

Chunking

The technique called *chunking* groups information into manageable units like sections or paragraphs; thus, it reduces the load on short-term memory (Nord 1993). Those chunks are made obvious to the reader by white space for separation, by grouping of related chunks, and by the use of headings in larger and bolder type fonts over chunks of information. Figure 16.9 shows the visual impact of chunking.

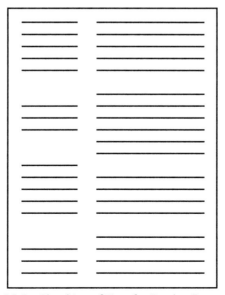

FIGURE 16.9 Chunking of Text for Reader Comprehension

Queuing

Within a chunk, the relationship of items of information can be shown to the reader by the layering technique called *queuing*. In an outline, the best way to show the relationship of one item to another is by the amount of indentation. Outlines, in fact, can be perfectly clear even without any numbering system if items of the same level of importance are indented the same amount. Queuing emphasizes the hierarchy by using headings of proportional size and density, by using horizontal rules, and by using decreasing line spacing as the reader moves down the hierarchy.

Filtering

According to Elizabeth Keyes, professor of graphics, "Filtering creates layers of information within the hierarchy. Filtering visually identifies and differentiates various types of information, so that readers can find what they need" (1993, 641). Examples of the types of information that can be visually identified include instructions that are numbered with outdents (numbers protruding into the left margin); warnings or cautions that are boxed, bolded, or with the word "Warning" outdented; captions that are in different type; overviews that are set in one wide column over two narrower columns of text type.

Visually Prominent Headings

Reader comprehension is increased by headings that are visually prominent, a technique that helps the reader to understand the hierarchy and the chunking. The type of heading can also increase understanding. Researcher Philippa Benson says that readers who are in a hurry or who have a low skill level benefit from headings that are full statements or questions (1985).

Use of Lists

Readers read information in a list easily because the list is set off by white space; they understand information in a list because of the chunking effect and the visual cues accompanying the list.

Numbered lists show sequence or the level in the hierarchy. They can also emphasize the number of items to be considered ("the three steps"). Numbered lists are most effective if the total number of items is fewer than 10. To aid short-term memory, longer lists should be chunked. An editor should ensure that the list numbers are outdented, so they are visually prominent.

Bulleted lists are effective when the sequence is not important. Technical writer Saul Carliner suggests keeping bulleted lists to fewer than five items. "The longer your lists, the more difficult it is for readers to process and use the information, even allowing for the value of chunking . . . [W]hen lists grow beyond five items, it is time to further chunk the information into main points and subordinate points" (1987, 220). Figure 16.10 shows an example of a bulleted list and a numbered list.

Checklists have a small box or line preceding each list item that readers can check off after completing that step. Checklists are useful for procedures that require completion of each item or for summary questions at the end of a chapter. See the end of this chapter for a checklist.

To meet the demands of multimedia, writers should
- be good script writers and story tellers
- work well as team members
- be able to think visually and non-linearly
- understand the medium by playing CD-ROMs, exploring the Web, playing multimedia games, attending conferences, and taking classes
- acquire technical skills such as graphics, authoring or programming, and multimedia software. (Hoskins 1997)

The steps involved in performing a "discount" usability evaluation of a document are:
1. Pick a document (or a particularly important part of a document).
2. Develop a set of usability principles.
3. Develop evaluation questions based on those principles.
4. Select usability evaluators.
5. Have the evaluators review the document based on the usability principles.
6. Schedule a post-evaluation debriefing session with the evaluators and writer.
7. Revise the document based on the evaluators' comments.
8. Have the document evaluated again if possible. (Daugherty 1997)

FIGURE 16.10 Listing to Aid Reader Comprehension

When editing, examine the type of list (numbered, bulleted, or checklist) for its applicability to purpose; examine the way the list is organized and designed for its effectiveness. You should also edit the punctuation. Lists are often introduced by a clause that ends with a colon ("In designing the document, follow these three steps:"). List items that are words or sentence fragments begin with a lowercase letter and are unpunctuated. List items that are complete sentences begin with a capital letter and end with a period.

Use of Color

Color helps reader comprehension by focusing attention. Graphics Professor Keyes says, "We have noted that typographic cueing has limits that are easily exceeded. People have a cognitive limit to the amount of visual cueing they can effectively absorb, process, and utilize before the cues become distracting overload. Color can *extend* this limit by creating a visual layer that we separate perceptually from monochromatic typographic and spatial cues" (1993, 646). Color focuses attention, groups objects, and creates "information targets" that the reader sees first, before seeing the relationship of what is in color to the rest of the text. Therefore, color should be used for the type of information readers need to access independently of the surrounding text. Such uses might include warnings in instructions, running heads, graphics, section dividers and chapter openers, cross references, and hints (Keyes 1993).

Figures and Tables

International readers, those with poor reading skills—in fact, most readers—can often comprehend information more easily when it is graphically presented. When you edit text, you need to ask if this information could be more effectively presented in a table or some kind of figure. See Chapters 15 and 19 for more detail.

Will the Reader Be Able to Retain and Retrieve Information?

Four related words affect the reader's ability to use information once it has been read and understood. They are retain, recall, remember, and retrieve, and all four have implications for document design. *Retain* means "to hold back" or "to keep in mind." To use a computer analogy, to retain is to store information in memory. *Recall* and *remember* are related; recall means "to bring back by a voluntary effort," and remember means "to think of again." *Retrieve*, however, means "to find again." To use another computer analogy, retrieve means to obtain data stored in the memory.

What can document design do to help the reader retain and retrieve information? Researchers present four possibilities:

1. chunking
2. presenting information both verbally and visually
3. using repeated patterns of page layout
4. presenting questions at the end of an information unit

Chunking

Just as chunking information into related groupings helps understanding, so chunking helps retention. Education researchers McBride and Dwyer report that "The demands of the information to be processed affect how much can be stored." Chunking ". . . results in superior performance because it allows a subject to 'think about' more information in his [sic] limited working memory by allowing information to be integrated in effectively larger chunks" (1985, 149).

Presenting Information Both Verbally and Visually

A second technique for helping readers retain information is to reinforce the verbal with graphics or visual presentation. For example, in addition to a verbal explanation of a process, a flowchart can *show* the process. With good page design, both the verbal and visual memory systems of the reader can be triggered. Other examples of visual presentation are text boxes and *pull quotes* (key statements pulled from the text and set off in larger type). Pull quotes emphasize the text and break up the page.

Using Repeated Patterns of Page Layout

Good hierarchical organization—that is enhanced by headings, white space, and repeated layout patterns—also increases retrieval of information. McBride and Dwyer report on research that says, ". . . what is stored is determined by what is perceived and how it is encoded, and what is stored determines what retrieval cues are effective in providing access to what is stored" (1985, 149). Thus, chunking and hierarchical ordering assist retrieval.

Presenting Questions at the End of an Information Unit

While not strictly a design issue, another technique for retrieval and retention is to pose questions at the end of an information unit. Those questions can be made more visually prominent by good design (such as a checklist), which will encourage the reader to apply them. The research indicates that "questions work in a backward manner, organizing and repeating previous prose content" (McBride 1985, 149) and thus help the reader retain the information.

Can The Reader Use This Document to Perform a Task?

Instructions—and the manuals and screens that contain them—are among the most frequently written documents in the workplace. Those documents should be designed to be inviting, readable, and comprehensible, but they also need to be usable. *Usability* means that human factors have been taken into account and that the instructions are task oriented; that is, they are written and designed to help the reader perform a task.

In instructions, chunking and modular design can help the reader. *Modular design* means that each step appears on a single screen or page or on two facing pages. When possible, no text or figure is continued to another page, and when the parts of a step fill less than a whole page, the rest of the page is left as white space. This modular design helps the reader isolate information and process the steps one at a time.

This chapter has provided an overview of the factors that influence document design and the general questions you can ask when editing to evaluate the effectiveness of the design. In addition to these general questions, there are specific questions you can pose at each of the major levels of edit—developmental editing, copyediting, and proofreading.

16.4 Developmental Editing of Document Design: Major Questions

In addition to the six questions posed and explained in section 16.3, when editing at the developmental level you should ask the following questions.

☐ 1. Is there a design style specification sheet—with sample pages—that indicates at a minimum the choices made for type styles and sizes, column structure, margins, and horizontal grid?

☐ 2. Has the writer done rough sketches of layout for facing pages?

☐ 3. Does the design consider line length? Number of columns? Placement of graphics? Length of the whole document?

☐ 4. Does the design (white space, indentation, type size and style) reinforce the structure of the information?

16.5 Copyediting Document Design: A Checklist

At the second-level or copyediting stage, you can examine the draft of the document. Here are some questions you should ask:

- ☐ 1. Does each page have a focal point?
- ☐ 2. Does the design of each page lead the reader's eye through the information?
- ☐ 3. Are page elements with dark areas (like bold headlines, photographs, blocks of text) balanced with light areas? One way to check is to turn the page upside down, so you look at the blocks instead of the words. Another way is to put the page on the floor and look at it from a distance.
- ☐ 4. Do the designs of facing pages work together as a unit? Look at reduced-size pages printed as a unit.
- ☐ 5. Does the visual reinforce the verbal? For example, do sizes of type and placement of headings reflect priorities?

16.6 Proofreading Document Design: A Checklist

At the final or proofreading edit, ask these document design questions. It is best to check all consecutive pages for one item before checking the next item.

- ☐ 1. Are the type sizes and styles at the various levels of headings consistent?
- ☐ 2. Is spacing consistent? Is indentation consistent?
- ☐ 3. Do headings have at least three lines of type above them or below them on a page or column?
- ☐ 4. Has the spacing decision avoided rivers of white space running through the text blocks?
- ☐ 5. Is the text free of widows (lone words at the top of a page or column) and orphans (lone words at the bottom of a page or column)?

EXERCISE 1

Redesigning a Table of Contents

Analyze the following table of contents for its design effectiveness. Then redesign, keeping the information and the numbering system the same. Write a paragraph in which you explain the design problems and tell how you have solved them.

CONTENTS

Section   Page

1.0	PURPOSE AND SCOPE	1
1.1	Purpose	1
1.2	Scope	1
1.3	Definition of Terms	1
2.0	CONTRACT AND TECHNICAL DIRECTION	2
2.1	Contractual Direction	2
2.2	Technical Direction	2
3.0	APPLICABLE DOCUMENTS	3
3.1	Government Documents	3
3.2	MCC Documents	3
3.3	Precedence	4
4.0	MANAGEMENT TASKS	4
4.1	Program Management	4
4.2	Program Management Plan	4
4.3	Program Reviews	4
4.4	Reporting	5
4.4.1	Master Milestone Schedule	5
4.4.2	Status Reports	5
4.5	Red Flag Report	6
4.6	Configuration Management	6
4.7	Design to Unit Production Cost (DTUPC)	6
4.8	Standardization Plan	9
4.8.1	Nonstandard Parts	9
5.0	COST AND SYSTEM ANALYSIS TASKS	9
5.1	Design Verification	10
6.0	SYSTEM EFFECTIVENESS	10
6.1	Reliability	10
6.2	Maintainability	10
6.3	Failure Mode and Effects Report	11
6.4	Critical Items List	11
6.5	Safety Hazards	11

CHAPTER 16 DOCUMENT DESIGN 353

EXERCISE 2

Redesigning Instructions

Analyze the following instructions for their effectiveness of design. (1) Write a page of analysis in which you comment on the strengths and the weaknesses of the design. (2) With another student, redesign the instructions.

<u>Instructions for logging into a workstation</u>

<u>at the CS Computer Lab</u>

Students who have computer accounts in the Department of Computer Science should use the following procedures to log into a terminal/workstation at the Computer Lab.

<u>WARNING:</u>

FAILURE TO FOLLOW THE INSTRUCTIONS GIVEN BELOW MAY CAUSE A SERIOUS MALFUNCTION IN THE WORKING OF THE WORKSTATION AND MAY EVEN LEAD TO A NETWORK BREAKDOWN.

<u>What you need:</u>   1. Student ID
2. Floppy Disk(s) to save your files.

<u>Login Procedure:</u>
1. Present your Student ID to one of the lab assistants.
2. Get the boot floppy disk from the lab assistant and insert it in the A: drive of the workstation indicated on the floppy disk. For example, use boot floppy E-1 for workstation E-1. Note: The 'boot floppy' is a read-only floppy which is used to load the operating system onto the computer and bring up the network.
<u>CAUTION:</u> Use only the boot floppy of the associated workstation to avoid login problems.
3. Switch the computer on and wait for the memory checks to be completed.

4. When the login prompt appears (see Fig. 1), type in your social security number and press <Enter>.

5. Type in your password at the password prompt and press <Enter>. If you are a first-time user, your password will be the same as your social security number. Please change your password to a 8 character password with at least 2 digits and 1 special character, to ensure better security for your files.

```
WELCOME TO THE UNIVERSITY
COMPUTER NETWORK
Monday, October 10, 1999
Login: 999-99-9999 <Enter>
Password: 999-99-9999 <Enter>
```

FIGURE 1 Login Screen: Note the Password Does Not Appear on the Screen

6. Type 'menu' at the command prompt to see the options available to you on the screen.

<u>Logout procedure:</u>

1. After your work is done, save any files you need in your floppy disks.

2. Insert the boot floppy in the A: drive, select the 'Quit' option from the main menu with the mouse.

3. Type 'logout' at the command prompt to log out of the system.

4. Wait for the 'bye' to appear on the screen and remove the boot floppy from the A: drive.

5. Return the boot floppy to the lab assistant and retrieve your student ID card.

CHAPTER 16 DOCUMENT DESIGN **355**

EXERCISE 3

Reorganizing and Redesigning Instructions

This is both an organizational and a design exercise. Study the following safety instructions written for students in an aviation laboratory. Determine its organizational and design strengths and weaknesses. Redesign the document, applying everything you have learned about good organization and design.

AVIATION LABORATORY SAFETY INSTRUCTIONS

The purpose of these instructions is to protect you and your classmates from injuries. These safety instructions have been developed from lab student experiences and common sense. Please read and review them carefully, especially when working in special safety areas. Please forward any recommendations, revisions, deletions to your advisor or any Aviation Department instructor. These instructions are to be followed when you are in the lab area. Refer to special safety instructions when working in special areas.

1. Wear safety glasses at all times when performing such operations as grinding, power filing, drilling, sawing, chipping or cutting. See Figure 1.
2. When working with acids, solvents, or chemical salts in the plating room, wear face and skin protection. See Figure 2.
3. Be careful not to spill harmful liquids on any part of your body.
4. Be familiar with the location of the emergency showers and eye wash fountains and call for help if unable to see your way.
5. When removing a cap from any container, release pressure slowly, and hold your face to one side. Containers may be sufficiently pressurized to expel their contents. See Figure 3.
6. Never attempt to lift any weight beyond your capacity; get additional help when required. Use adequate equipment to lift heavy loads. See Figure 4.
7. When lifting, keep your back nearly straight, bend your knees and use your thigh muscles instead of your back muscles.

8. Floors must be free of slippery or hazardous materials at all times.

a. Wipe out immediately when oil is spilled on the floor.

b. Mop up immediately when water is spilled.

9. No object must be permitted to protrude beyond the edge of a workbench. Objects which must protrude must be barricaded or padded.

10. Be aware of the location of aircraft when moving around the laboratory or ramp. The trailing edge of a light airplane is between five and six feet above the ground level and is a frequent cause of head injuries.

11. When using a ladder, be sure it has a firm footing and that the ladder top is against a substantial rest.

a. Check all steps for soundness, and ensure that the rails are not split. A wobbly ladder is never safe.

b. Never use boxes, chairs, and stools in place of a ladder or maintenance stand. See Figure 5.

12. When using electrical instruments or working on electrical circuits, be sure you understand them completely. Remember that you may damage very expensive equipment and receive lethal voltage discharges.

13. When using portable electrical equipment make sure that the cord and plugs are in a safe condition and grounded.

14. When using sharp tools, make all cuts away from your body.

a. Do not place screwdrivers, scribes, and other sharp tools in trousers or shirt pockets.

b. Never use a file without a handle. It may puncture your skin.

15. Report any defective equipment to the lab instructor at once. Do not attempt to use or repair it.

16. When working with any kind of machinery, keep your eyes and mind on your work. Discontinue your work if you are distracted by anyone talking to you or for any other reason.

17. Never work alone in hazardous areas or when using power equipment.

18. Do not use a tool or machine for any purpose other than that for which it is intended.

19. Read all precautions and warnings associated with each particular item of equipment before using it. Don't ever play rambo or try to be smart while in lab. Ask directions from your instructor if you have questions. See Figure 6.

20. When working in the lab or ramp area, don't wear rings, watches, neckties, and loose clothing. Remove all these items, and wear coveralls during the lab session. See Figure 7.

21. Secure the coverall's sleeves so they don't become entangled in any equipment.

22. Never work with a long hair hanging on any operating machinery. Before entering the working area long hair must be tied back or constrained.

23. Use a respirator when spray painting, using liquids with harmful vapors, or working in an environment of harmful dust. Use exhaust fans where installed.

24. Do not use compressed air for blowing chips from machinery or for removing dust and other items from the body or clothing.

25. Remain clear of any objects being lifted with the hoists. See Figure 8.

26. Always practice good housekeeping procedures during and after each operation. Remember that cleanliness reflects your attitude towards the job you are performing.

27. Always refer to the special safety instructions posted in the main entrance of each of the following areas: welding, grinding, sand and grit blasting, drill press, lathe, soldering, sheet metal, rip and cut-off saw, band saw, cleaning room, batteries, abrasive cut-off saw, and plating rooms. See Figure 9. With your cooperation we will succeed in avoiding severe injuries to you and our fellow students. So far there have not been fatal injuries in the lab area. DO NOT be the first.

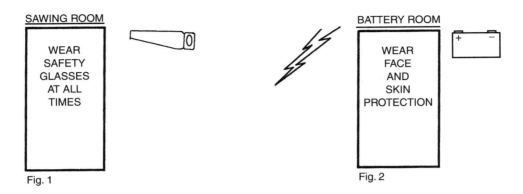

Fig. 1

Fig. 2

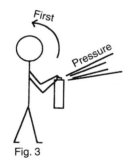

Fig. 3

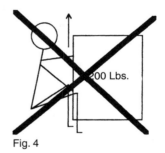

Fig. 4

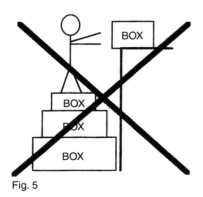

Fig. 5

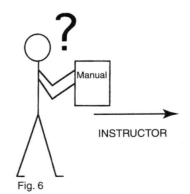

Fig. 6

Fig. 7

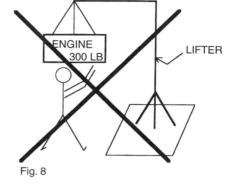

Fig. 8

Fig. 9

EXERCISE 4

Reorganizing and Redesigning Procedures

Study the following employee procedures. Analyze the document for organizational and design problems. Then, with another student, reorganize and redesign.

TRAVEL AND EXPENSE REPORTING

Employees traveling on company business will be reimbursed for all air fare, ground transportation, car rental, hotel, telephone calls, meals, entertainment, and advances, upon proper written approval. This travel policy will provide employees with an uniform method of reporting travel and other expenses.

Any interpretations of this policy's provisions will reside with the manager. Any deviations from this policy must be approved by the manager or a designated substitute. The manager (approver of the travel expense report), as well as the employee, is responsible for complying with all provisions of this policy and procedures with adequate and reasonable explanations.

All travel must be pre-approved. A numbered Travel Request/Authorization/Advance (TR) Form must be filled out by the employee. The TR Form must also be signed by the department manager. This TR Form will then go to the Travel Coordinator. Tickets will not be purchased without a fully completed or approved TR Form. You can obtain a TR Form from the Travel Coordinator.

All travelers must fill out an expense report once they have returned from their trip. The expense report must be signed by the manager and routed to Accounts Payable for processing. Expenditures other than travel expenses should not be reported on the "Travel Expense Form", but a "Check Request" Form should be filled out.

Expenditures that are not considered travel expenses include the following: software purchases, professional books, seminar fees, office supplies, subscriptions, memberships, copies, and all other expenses not related to travel.

Expenses not paid by the Company
The following is a list given as a guide and not necessarily a list of all expenses which are NOT reimbursable: airline or other travel insurance; baby-sitter fees; barbers and hairdressers; pet care; golf fees (except when part of customer entertainment); personal property insurance; annual fees for credit cards; transportation to and from the office on regular work days; shoe shine; movies; magazines; books and newspapers (except if directly related to Company business); parking or traffic tickets; etc.

Expenses incurred by employees traveling on company business will be reimbursed as follows:

Travel Expense Report

- Complete this form for each trip. Make sure Travel/Request/Authorization (TR) number appears on top right of expense report form so it will be processed by the accounting department.
- Sign the date and certification part of the report, and receive approval from management before routing form to accounting. This should be done within a week after returning from your trip.
- Attach receipts for all airline tickets, hotel, and entertainment bills. Receipts are required for all expenditures over $25.00.

Allowance

- Obtain the allowance rate from Accounts Payable before traveling. A dollar amount allowance will be given to each traveler, which will cover hotel and meals. Try to keep the combined costs of hotel and meals within this rate.
- Do not exceed your allowance; you will only be reimbursed if the expense is reasonable.

Travel Agent

- Use the Corporate Travel Agency to book your tickets, hotel, and car reservations so the accounting department will be able to keep updated corporate travel data. The travel agency will only accept orders from the designated Travel Coordinator for the Company.

Air Fares

- Make reservations far in advance to take advantage of discounts.
- Use only coach (Y fare) or lower fares for company traveling. Any upgraded fare will be paid by the employee.
- Do not combine personal and business expenses without prior approval. If approved, only business expenses will be reimbursed.
- Keep airline ticket stubs and attach them to the expense report.
- Return any unused tickets (or parts of ticket not used) for proper credit.

Ground Transportation

- Note separately on expense report taxicab fares in excess of $5.00 with destination and purpose of trip. You will be reimbursed (price plus tip) for airport limousines, buses, or taxicabs if used when traveling to and from airports.
- Use a rental car only when it is more economical.

Car Rental

- Use companies that offer corporate rates to the company. You will be reimbursed only for these companies.
- Use only compact rental cars, unless there are two or more people or equipment being transported.
- Save gasoline and parking lot receipts for reimbursement when you are using a rental car.

Hotels

Give preference to hotels that have corporate rate agreements with the company.

- Try to keep hotel expenses within allowance limits.
- Do not use luxury or special accommodations. They will not be reimbursed.
- Itemize day-to-day receipts as a back-up for reimbursement.
- Itemize separately all charges to hotel bills other than room expenses and taxes.

Telephone

- Log all business calls to obtain reimbursement.
- Keep personal calls home to a minimum; no other calls will be reimbursed.
- List separately all telephone calls charged to hotel bills on expense report form.

Meals

- Keep all meal receipts during traveling days for reimbursement. You will be compensated at actual cost, within reason.
- Include gratuities and refreshments in the meal expense category.

Entertainment

- Do not include unreasonable entertainment expenses; they will not be reimbursed.

- Request an entertainment reimbursement if the expenditure precedes, includes, or follows a substantial and bonafide business discussion. A business benefit is expected to derive from the entertainment.
- Verify, in detail, all expenses incurred: who was entertained, their business relationship, where the entertainment took place, and the purpose of the event, as regulated by the IRS.
- Explain the lack of receipt and the expense.
- Include alcoholic beverages with entertainment expenses.

Advances

- Obtain a reasonable cash advance to cover anticipated expenses for international travel.
- Submit a request for cash advance on the Travel Authorization (TR) Form.

If you need any further information, please contact the Travel Cooordinator at Ext. 2222.

Reprinted by permission of Atmel Corporation.

EXERCISE 5

Evaluating a Workplace Document

Clip or copy two facing pages from a workplace document such as an annual report, a marketing brochure, a manual, a newsletter. Write a brief report in which you comment on the strengths and weaknesses of the design. Apply each of the six questions in section 16.3 to the document and record your answers.

EXERCISE 6

Critiquing Document Design

Bring to class a workplace document or a published document like an annual report, journal, or fact sheet. In small workshop groups, critique the design of these documents, applying the checklists in sections 16.5 and 16.6. Be prepared to tell what works and what doesn't work in the document.

EXERCISE 7

Design Application in the Workplace

Interview a local writer or editor about document design responsibilities. Ask who plans the design, who sets the standards, who edits the design. Ask about design style sheets or templates and how much variation is allowed. Report back to the class and compare your findings.

References

Agfa Corporation. 1988. *The art and technology of typography*. Wilmington, MA: Agfa Corp.

Atmel Corporation. 1997. Travel and expense reporting from the *Atmel Supervisors Manual*. Reprinted by permission. San Jose, CA.

Benson, Philippa. 1985. Writing visually: Design considerations in technical publication. *Technical Communication* 32.4 (November), 35–39.

Carliner, Saul. 1987. Lists: The ultimate organizer for engineering writing. *IEEE Transactions on Professional Communication* PC 39.4.

Daugherty, Shannon. 1997. *Usability testing for technical communication*. Unpublished paper. San Jose State University.

Hoskins, Linda. 1997. *A writer's introduction to multimedia*. Unpublished paper. San Jose State University.

Ishizaka, Janice. 1997. *What technical communicators need to know about perception and why*. Unpublished paper. San Jose State University.

Keyes, Elizabeth. 1993. Typography, color, and information structure. *Technical Communication* 40.4 (November).

Lay, Mary M. 1989. Nonrhetorical elements of layout and design. In Bertie Fearing and W. Keats Sparrow, Eds. *Technical writing: Theory and practice*. New York: MLA.

Linotype Ltd. 1988. *The pleasure of design: A practical guide to layout and typography in desktop publishing*. Cheltenham, U.K.: Linotype.

Matis, David. 1996. The graphic design of text. *Intercom* 43.2 (February).

McBride, Susan and Francis Dwyer. 1985. Organizational chunking and postquestions in facilitating student ability to profit from visualized instruction. *Journal of Experimental Education* 53.

Nord, Martha Andrews and Beth Tanner. 1993. Design that delivers—Formatting information for print and online documents. In Carol Barnum and Saul Carliner, Eds. *Techniques for technical communicators*. New York: Macmillan.

Parker, Roger C. 1997. *Looking good in print*. Research Triangle Park, NC: Ventana.

Redish, Janice C. 1993. Understanding readers. In Carol Barnum and Saul Carliner, Eds. *Techniques for technical communicators*. New York: Macmillan.

Saunders, Greg. 1997. *VRML makes 3-D graphics a reality*. Unpublished paper. San Jose State University.

Wheildan, Colin. 1995. *Type and layout*. Berkeley, CA: Strathmoor Press.

White, Jan V. 1988. *Graphic design for the electronic age*. New York: Watson-Guptill.

———1990a. Consistency: The key to solving the EP&P color puzzle. *Electronic Publishing & Printing*, March.

———1990b. *Great pages: A common-sense approach to effective desktop design*. El Segundo, CA: Serif Publishing Co.

17 Access Aids in Print and Online Documentation

Contents of this chapter:

- Tables of Contents and Menus
- Indexes
- Glossaries
- Navigational Aids
- Advance Organizers
- Citation of Sources
- Developmental Editing of Access Aids: Major Questions
- Copyediting Access Aids: A Checklist
- Proofreading Access Aids: A Checklist

"If readers can't find information, the document is not successful. It's not the size of a document that frustrates readers; it's their inability to find information." Freelance editor Jo Levy is talking about one of the major problems readers face in workplace writing: accessibility of information.

Accessibility depends on many factors, including organization and document design. However, at the document level, writers and editors can boost information retrieval by incorporating what are called *access aids*, *orienters*, *navigational aids*, or *advance organizers*. These terms describe ways writers and editors can help readers find the information they seek. These types of access aids are used both in hardcopy (printed pages) and in online documentation, though their organization and appearance may differ in the two modes. Writers are often responsible for all the access aids in a document, but sometimes editors must write some or all of them. Whoever writes them, editors must evaluate how well the access aids work. Six common types of access aids are

- tables of contents and menus
- indexes
- glossaries
- navigational aids
- advance organizers
- citation of sources

17.1 Tables of Contents and Menus

The *table of contents* (TOC) shows in outline form the major ideas of the document, lists the supporting points for those ideas, and shows the relationship of each point to the whole outline and to the other points. Because it provides an overview of the whole document, a table of contents is often the first place readers search for information. Readers browse the table of contents to see what information the document contains. If they find something useful, the table of contents also helps them navigate through the document. A hardcopy table of contents appears at the beginning of the document and is complete—often, in fact, showing three or four levels of hierarchy. A long book (like this one) may even have two separate tables of contents: a one-page TOC showing only the main divisions or chapters for a quick overview, followed by a detailed TOC extending over many pages.

Entries in the TOC should be compiled from the headings and subheadings in the document. Thus, when you examine the TOC, you can assess how well the headings and subheadings work. Does each entry contain key content words (telling the subject of the entry) and organizational indicators (telling the purpose, such as instruction, evaluation, description)? Is each entry at one level of the hierarchy syntactically parallel with the other entries at that level? Is the type size and style consistent at each level? Are the levels of the hierarchy shown by indentation? See Chapters 14 and 16 for more on parallelism and headings.

The TOC in an online document is usually in the form of a pulldown or dropdown *menu*. If the document has more than a few screens of information, the menu "may have to be layered, in other words given to the reader in stages. [Whereas] paper tables of contents appear at the beginning of the book and are complete, online tables of contents are almost always accessible from every screen and are usually used to navigate the document. . . . Layering is necessary because computer screens are smaller and harder to read than paper documents" (Hibsher 1995).

Figure 17.1 shows a portion of the hardcopy table of contents from the *Lotus cc:Mail Release 8 Administrators' Guide*. Notice that three levels of hierarchy are shown. Figure 17.2 shows the first screen of the online Help Topics menu, which shows only two levels. However, clicking on any of the level-two entries brings up the third level, as shown in Figure 17.3. The online Help Topics menus thus effectively layer the information and present it to the reader only a few items at a time.

You can ask the same questions of menu entries as of headings in a table of contents, but you also need to make sure that each layer is succinct and has three to five (definitely fewer than 10) choices—to meet the limits of short-term memory. Entries should fit together and be in a logical order. Menu items that can be clicked for another layer of information should be visually highlighted with color or underlining.

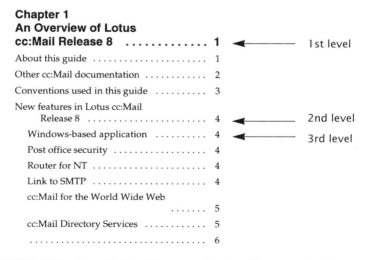

FIGURE 17.1 Paper Table of Contents Showing Three Levels of Hierarchy

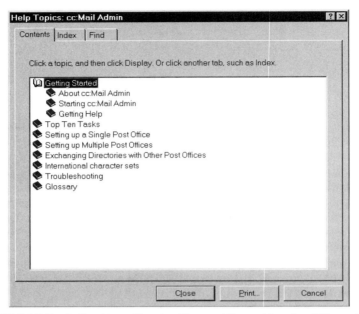

FIGURE 17.2 Online Menu Showing First and Second Levels of Hierarchy

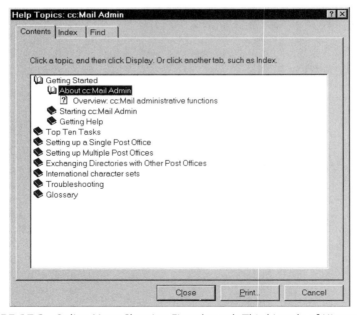

FIGURE 17.3 Online Menu Showing First through Third Levels of Hierarchy

17.2 Indexes

An *index* is an alphabetical and structured hierarchy that allows fast, random access to the contents of a document. As Janice Redish and her associate researchers note, "In work settings, . . . almost all documents longer than letters and memos are used primarily for reference. The reader goes to the document looking for the answer to a particular question, for a particular piece of information, or for instructions on how to do a particular task. The reader's goal is to get in, get the answer, and get out as quickly as possible" (1985, 139). Thus, all four of the general purposes of an index are related to fast and easy access.

The four purposes of an index are as follows:

1. **To help the reader find an item of information easily.** The word *index* comes from a word meaning "pointer." An index assumes that the reader does not want to read the whole document, so it points the way to small, discrete items.
2. **To arrange the information by another organization.** Most documents have a table of contents, which is arranged hierarchically by the author's organization and which groups information in relatively large chunks. The index helps the reader by providing a second organization of information, which is arranged sequentially by the alphabet and which groups the information in much smaller chunks.
3. **To disclose relationships among items.** The second and third levels of an index are hierarchical, so related items are grouped under a main entry. Browsing should, therefore, help the reader locate information.
4. **To indicate omissions.** Presumably, all key information in a document will be indexed. Thus, if the information is not in the index, the reader should assume it is not in the document. Indexes have a broader audience than does the document itself, which puts extra responsibility on the indexer to ensure that all information items are listed in the index.

Because readers look for "what something is," indexes use nouns (for names, concepts, ideas, functions, procedures) or gerunds—the "ing" form of nouns made from verbs (for tasks, subtasks, and restrictions). Figure 17.4 shows examples of types of index entries.

Most indexes will have three levels of entries, moving to another level when there are more than three entries in a list. Some indexes even have four levels in the hierarchy. Here is an example of a three-level entry:

> finite-state machines (FSM)
>   extracting, 10–14
>   assumptions, 10–23
>   process, 10–6

nouns:	clock	[broad term]
	clock DR signal	[name]
	insert scan	[narrow term]
	ASIC vendor	[acronym]
	Custom test vector (CTV) interface formats	[narrow term]
gerunds:	backtracking	[general task]
	creating directories	[specific task]
	formatting test patterns for ATC	[very specific task]

FIGURE 17.4 Types of Index Entries

Indexes should also have *cross references*, which are references to preferred terminology, synonyms, and added information. A *See* reference sends the reader to a synonym or the standard (or preferred) company term. A *See also* reference sends the reader to additional or related information in another entry. Figure 17.5 shows examples of *See* and *See also* entries taken from a variety of indexes.

Another way to cross reference an item is to reverse the terms in a double entry. For example, *devices, input-output* should also be indexed under *input-output devices*. Whether you are the writer or the editor of an index, you should check to ensure that every index item can be found in at least two places—more is better. Sometimes that means you must brainstorm for nontechnical terms that the reader might use or terms that the competition might use as well as actual synonyms. For example, *concatenating files* could also be indexed under *merging files, joining files, appending files, combining files*.

The types of information to be indexed will vary from industry to industry, but you can set up general categories. Figure 17.6 shows the major categories that should be indexed in computer-related documents.

Finally, there should be approximately one index entry for each 100 words of text. If the document is chunked, there may be more entries because key words associated with the headings will be indexed. Index entries should answer readers' questions, and there should be no more than three undefined page entries. *Undefined page entries* are page numbers listed after an index term without any further identification. (For example, bootstrap, 15, 31, 33, 35, 45, 54, 98.) To find specific information, the reader must look under each page number, and since most readers are impatient, they will not search far before quitting.

See cross references

FSM. *See* finite-state machine, 10-2	[explanation of an acronym]
power-programmable cells. *See* parameterized cells, 4-26	
	[preferred term]
drop strobe. *See* strobe	[broader term]
software. *See* Cell FIT	[specific term]
fluid filter. *See* system fluid filter	[narrower term]
tips. *See* pipette tips	[narrower term]

See also cross references
state table design format. *See also* state tables, E-1

[additional information]

beads. *See also* control beads [specific topic]

FIGURE 17.5 Examples of Cross Referencing

What Should Be Indexed

1. measurements, controls, functions, procedures, interactions between products
2. screen selections and menu options
3. command names and parameter names
4. concepts, terms, definitions
5. tasks or utilities
6. alternate names and synonyms
7. acronyms and abbreviations
8. special characters
9. part names and numbers
10. proper names
11. keys
12. terms in figures and tables
13. hardware and software specifications
14. error conditions

FIGURE 17.6 Categories of Items That Should Be Indexed in Computer-Related Documents

The Process of Indexing

Because editors are often assigned to be indexers, you should learn an efficient general process of indexing. Both word processing and page-layout programs include indexing software, but you cannot rely on the computer to make indexing decisions for you. Indexing is a writing task, and it demands the same kind of thoughtful planning and attention to detail as does any other writing task. The software will save you time by automatically alphabetizing the index, setting the hierarchy, and updating the page numbers in subsequent drafts. It will not make the decisions about what should be indexed.

If you both write and edit the index, you will need to follow all six of the following steps. If you edit only, you can concentrate on steps four through six.

1. **Determine the purpose of the document and the intended audience.** Ask: What is the primary task? Is it, for example, installation? Operation? Diagnosis? Programming? What is the level of knowledge and jargon of the primary audience? Is there a secondary audience?
2. **Go through the text on hardcopy printout or on screen, flagging index terms or listing them before the paragraph.**
   - Mark terms in headings and subheadings first.
   - Go through the overview for pertinent terms.
   - Find keywords in the text. Turn verbs into gerunds (*process* into *processing*). Use the singular form unless the plural is the commonly accepted term.
   - Determine the level of the entry. Some indexing programs will do this for you.
   - In compound or multiword terms, also reverse the terms for cross referencing (*primary display adapter; display adapter, primary; adapter, primary display*).
   - Add synonyms or words that clarify or modify.
   - Add *See* or *See also* references.
3. **Print a copy for editing.**
4. **Analyze index entries.**
   - Analyze level-one entries first. Can any entries be consolidated? Should any level-one entries become level-two entries? If your application supports third-level entries, should you add a third level?
   - Analyze level-two entries next. Should any level-two entries be level one? Can you consolidate entries? Should any level-two entries be level three?
   - Examine entries for cross references. Create synonyms and reverse the order of index terms.
   - Add *See* or *See also* references.

5. **Prune excess entries.** For example, index only the most informative discussion of a command, not every time the command is mentioned.
6. **Check for consistency of terms and symbols.** Make sure that columns that are continued have headings showing the continuation.

If you are the writer (or are working closely with the writer), many indexing experts recommend indexing each chapter as it is finished, generating master indexes at frequent intervals, and compiling a master index at the end for editing. If you are the indexer but not the writer, you may have only a week (or less) at the end of the project for indexing. You should know that indexing and editing are time intensive. Professional indexer Lori Lathrop says, ". . . ten to twelve pages an hour is a good general estimate for the amount of time it will take to create an index. Consequently, if you have a 200-page manual, you should allow at least seventeen to twenty hours to create the index. Keep in mind, however, that editing and refining the index will require approximately 25 to 30 percent more time, so you should add another four to six hours" (1997, 11). Other experts expand on this, saying that 50 percent of the total indexing time should be spent on editing, and that a 200-page manual will take a whole week or more to index.

Indexing Online Documentation

Users of online documentation also rely on indexes to access information. A survey by *PC Computing* magazine found that users of both hardware and software documentation "overwhelmingly selected the index as the most important feature for locating information" (Harris 1996, 15).

However, indexes are one area in which paper and online documents differ. Electronic documents usually are not sequential, so it would be meaningless to put page numbers after the words in an alphabetical index. Because of the hypertext nature of electronic documents, each entry in an online index must refer users to only one place, thus eliminating undefined entries. Also, like online tables of contents, online indexes are used to navigate through the document. Therefore, they are usually available from every screen (Hibsher 1995). The user simply pulls down the index, clicks on the desired term, and the relevant information appears on the screen. Figure 17.7 shows two levels of entries in an online index for Help Topics in the *Lotus cc:Mail Release 8 Administrators' Guide*. Figure 17.8 shows an optional search method.

Online documentation expert William Horton says, "In many online documentation systems, the same terms serve as both retrieval keywords and index entries. The procedures for compiling these terms are the same. The index might contain an entry *files, copying*. The user could find the same information by searching for *files and copying*. For online documents, the

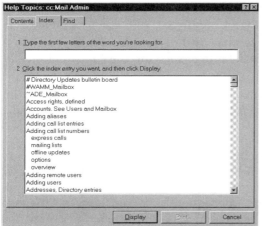 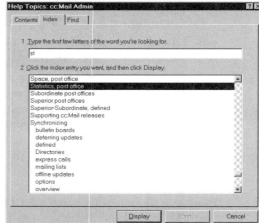

FIGURE 17.7 Online Index Showing Two Levels of Entries

FIGURE 17.8 Online Index Showing Optional Search Method

process of indexing includes both assigning search keywords and preparing a directory where the user can select from those words to jump to the associated topic. Improving the indexing and keywording is one of the best ways to make documents (paper or online) more accessible. In one test, increasing keywords from 2 to 32 per topic increased search success rates from 21 percent to 76 percent" (1994, 276). However, even though the same terms are used for both methods, an online index and keyword searching are two different ways of accessing information. Technical writer Linda Hoskins explains, "In my experience, an online index looks like a hardcopy index except that each entry is a hypertext link, and a search engine only allows for a keyword search—it is simply a prompt or field where the user enters the keyword. The advantage of an online index is that it allows users to peruse the entries even if they are not sure of the correct term" (1997).

Keywords thus are essential to information retrieval. However, if the reader is not using the same terminology as the writer, or if a concept is discussed in a paragraph where the keyword does not appear, an online index will not work. Readers may be easily frustrated by an online index when they can't find the keywords, because they cannot so easily switch to other methods of searching as they can in hardcopy. For example, in a printed manual, readers who can't find an index item might search the table of contents, look at headers and footers, or even scan the text looking for a keyword. Those access aids are not as readily available online. Editors can help readers access online indexes by taking the following actions:

1. **Choose one term and ensure that it is used consistently.**
2. **Brainstorm for synonyms, common terms, broader terms, and related terms, and include them in the index.**
3. **Anticipate the questions readers might ask and include words that will lead them to the answers.** Typical questions include:
   - How do I . . . ?
   - What is a . . . ?
   - What happens if . . . ?
   - When should I . . . ?
   - How do I fix . . . ?
   - What's wrong when . . . ?
4. **Check that each chunk of information includes only one idea.** Cross reference to related topics instead of including them in a chunk. (See Chapter 18 for more on chunking.)
5. **Schedule time for checking each entry in an online help index.**
6. **User test by asking a typical user to search for information**.

17.3 Glossaries

A *glossary* is an alphabetical list of terms with accompanying definitions. A glossary is an access aid because it helps readers find information—in this case, what terms mean. Standard practice in hardcopy writing calls for defining a new term or spelling out an acronym the first time it is used. The term is often highlighted by bolding or italicizing. Those terms and acronyms should also be listed in a glossary, both for the reader who starts reading somewhere in the text beyond where the definition was given, and for the reader who has forgotten the original definition. See the glossary at the end of this book. Glossaries are especially important for online documentation, which has no "first time" of use.

In hardcopy, a glossary can be placed in the introduction of a report if there are only a few words to be defined, all of which are essential to understanding the report. More commonly, glossaries appear in an appendix, and the reader is directed by the introduction, a footnote, or a footer to find highlighted terms in the glossary. In online documentation, the glossary must accommodate nonsequential reading. As Horton explains, "Nonsequential readers may encounter the term at any point. Do you spell it out every place, thus wasting screen space and disk storage?" (1991, 25). The typical hypertext solution is to highlight the word by color or underlining. When the user clicks on the highlighted word, a pop-up definition from the glossary appears on the screen.

> 1. key terms that will be new to readers or are used in a new way
> 2. acronyms and abbreviations
> 3. technical terms, including verbs and phrases used in special ways in the document

FIGURE 17.9 Types of Terms That Should Be in a Glossary

Figure 17.9 shows the types of terms that should be defined in a glossary to help readers access information.

The best way to define glossary terms is with a *formal definition*, which defines by classification and division. First put the word into a larger class; then, add the details that distinguish this word from other words in that class. Follow this order: *word to be defined + verb + class + distinguishing characteristics*. The verb can be omitted; then, a colon follows the word to be defined. Figure 17.10 shows the pattern of formal definitions.

You can also help the reader by expanding the formal definition in one of the following ways:

- comparing or providing synonyms
- describing
- providing examples
- classifying or dividing
- providing an analogy (comparing something unknown to something known—usually in a different class)

A glossary not only provides a valuable access tool for the reader, but it is a key element in preparing a document for international readers. Nancy Hoft, a consultant on international communication, adds two items to the list of types of terms in Figure 17.9 that should be in a glossary. For translation, the glossary should also include:

1. sources that will provide industry-standard translations of technical terms
2. words that must not be translated

Hoft quotes one consultant who says that a "glossary can save up to 20 percent of the total cost of the information translation project" (1995, 192).

What is the editor's role in preparing the glossary? Sometimes editors are responsible for the actual glossary preparation. Sometimes (especially if the editor is not familiar with the subject) the editor's job will be to go

> A *marsh* is a type of wetland that can support soft-stemmed herbaceous plants.
> *CD-ROM* (Compact Disk Read Only Memory) is a data storage and retrieval medium that can store larger amounts of information than a floppy disk.
> *Decoding:* a mechanical function of reading in which words are sounded out or recognized in the reader's head.

FIGURE 17.10 Examples of Formal Definitions

through the manual or report and mark all the words or phrases that are unknown or confusing. Then the writer will add needed terms to the glossary. Sometimes the editor is responsible for checking the accuracy of glossary entries.

17.4 Navigational Aids

Even though the term *navigational aids* can be used to refer to all access aids, more often it refers to those specific elements on a page or screen that help the readers (1) see the relationship among units in the document, (2) locate themselves within the document—in other words, keep track of where they are, and (3) move quickly to another section or information module.

The common navigational aids in hardcopy have equivalents in online documentation, and as online documentation evolves, those navigational aids and their locations begin to follow conventional placement. Common navigational aids for both hardcopy and online are shown in Table 17.1.

TABLE 17.1 Navigational Aids

Hardcopy	Online
headers and footers	topic name and path to it
page numbers	screen numbers
section or chapter number	number of related screens
maps and flowcharts	maps and flowcharts
stack of pages in the book	navigation buttons
chapter table of contents	home base or home location

Headers, Footers, and Screen Orienters

In hardcopy, document designers use *headers* (sometimes called running heads) and *footers* (running feet) to help readers locate themselves within the document and move quickly to another section by scanning.

Commercially published books most often use a header, which is a single line of type at the top of the page in a different font size and style and separated from the text by white space. On the left page of a two-page spread, a header usually includes the page number plus the number and title of the general section or part. On the right-hand page are printed the chapter number and title and the page number. Sometimes the specific topic is titled here. The three most important items that should be included in headers are page numbers, chapter numbers, and chapter titles. Figure 17.11 shows examples of headers from different books on technical communication.

Company publications frequently use *footers* (sometimes called running feet) instead of headers, though the practice varies. Some companies use both headers and footers. A footer typically includes the section and page number, for example, 4-15. In many company documents, pages are not numbered from beginning to end but within sections only for easier updating. A footer also may include the title of the document, the chapter title, the release date, the section title, and the document number. The placement and choice of information is not standardized, but a company style guide or software template will usually set the style so that all documents have a similar look.

Comparable information in online documentation would be a screen orienter like the *topic name*, usually located in the upper left of the screen. Topic names sometimes add the name of the larger topic or section of which this module is a part. In fact, some topic names include the entire path to the current screen, showing the hierarchy from the first menu selection through each subsequent selection to the current screen. Screens also include numbers—either a screen number or a number showing where the reader is in a sequence of screens, for example, "screen 3 of 4."

Left Page	Right Page
162 Part I The Process of Technical Writing	Chapter 6 Making Information Accessible 163
388 Part 2 Acquiring the Tools of Technical Communication	12 Visual Display and Presentation 389
252 Designing and Writing Online Documentation	Chapter 8: Display 253
302/Chapter 16 Online Help	Step 1: Specify the Help Environment/303

FIGURE 17.11 Examples of Headers from Different Books

Maps and Flowcharts

Maps and *flowcharts*, which show readers how the parts and wholes fit together, are effective for both hardcopy and online. "Maps showing the overall organization of the document provide a global context for users. In online documents, users can spot a topic and jump to it by selecting it on the map. The map becomes a visual menu of available topics" (Horton 1994, 215). Unlike a map, a flowchart shows by arrows the preferred sequence of topics.

Even in hardcopy, a map or flowchart can help readers see where they are in a larger sequence by visually highlighting a specific step. Figure 17.12 shows a simple flowchart in a tutorial for a complex tool for designing schematics. This flowchart is one of many; each major section opens with a similar flowchart.

The Stack of Pages and Navigation Buttons

In a book, readers use the stack of pages on the left and the right to help them determine where they are: at the beginning, in the middle, near the end, and so on. They do this subconsciously, but the physical size of the stack of pages does help in orientation.

In online documentation, navigation "buttons" fill somewhat the same role. Such buttons are usually located near the bottom of the screen; they may include *next*, *previous*, *index*, *glossary*, *find*, *home*. By clicking on a button, the user can move through the document.

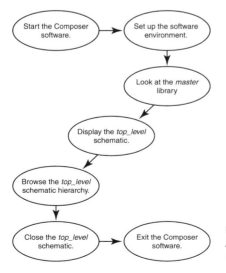

FIGURE 17.12 A Flowchart as an Access Aid (*Reprinted by permission of Cadence Design Systems, Inc.*)

The Table of Contents, the Home Location, and Bookmarks

Readers who want to move to another section in a hardcopy document usually use the *table of contents* as a base from which to launch a new search. In online documentation, the comparable *home location* or home base is often the main menu. "When all the chapters . . . start from home base, there is a common entry point for all users: those who come from the previous chapter, and those who just step in" (van der Meij 1995, 257).

Users of online documentation often want an easy way to return to information they have already read. *Bookmarks* or *bookmarked topics* provide users a way to create a list of topics of special interest, which will then appear in a pull-down menu. Clicking on the topic sends the user directly to the information.

Editing Navigational Aids

Editors can act as users in assessing how well the navigational aids work, whether in hardcopy or online. Here are some questions you can ask.

1. **Do the headers and/or footers tell the readers exactly where they are in the document?** By scanning and flipping pages can the reader find the next chapter by chapter number? By chapter title? By page number? Are the headers and footers accurate and consistent?
2. **Can readers see a picture of the whole document and the relation of the parts to each other and the whole?** If the document is complex, is there a visual means of helping the readers locate themselves? Can you suggest a flowchart or map?
3. **Does every screen in a series tell readers where they are?** Are navigation buttons visible at all times? Do the buttons provide for all the places readers might wish to go?
4. **Is it easy for readers to return to "home" or the main menu?** Can users create bookmarks?

17.5 Advance Organizers

Advance organizers help readers access information by establishing a context (or schema) into which the readers can plug new information. According to technical communications Professor Thomas R. Williams, "Advance organizers, in other words, are prepackaged schemata" (1994, 97). In one sense, both a table of contents and an online menu are advance organizers because they help readers see the whole picture. More commonly, however, the term is used to refer to the sections of prose that writers provide as an overview or preview of information. Advance organizers include abstracts, executive summaries, summary tables, and context-setting overviews in online documentation.

Abstracts are written for technical readers and are short (50 to 100 words) condensations of reports, articles, and proposals. Abstracts have two purposes: (1) to provide a capsule version—in a listing of separate abstracts or before the document—so that readers can decide whether to read the entire document and (2) to review the contents either by describing the contents (a descriptive abstract) or by giving the purpose, organization, conclusions, and recommendations (an informative abstract). For most readers, an informative abstract (which is a true summary) is more useful than a descriptive abstract. Abstracts also often list key words in the document for ease of keyword searching. Abstracts are written after the document is completed, and, though they are often written by the author, sometimes that job falls to the editor.

An *executive summary* is written for managers and is a condensation of a report or article. It precedes the report itself and provides context and the key information in the report. Because of the intended audience, executive summaries include background material, avoid technical jargon (or define it), include conclusions and recommendations, and follow the general organization of the document. Executive summaries are written in paragraph form. Like abstracts, executive summaries may need to be written by the editor.

Summary tables preview information by telling readers where to go either in an "if-then" table or in a question and answer format. These tables work especially well for manuals. Figure 17.13 shows an example of a summary table.

Context-setting overviews can provide a schema for readers of online documentation. While you cannot control where the reader will enter a cluster of topics, you can put a preview or overview at the likely beginning of a sequence. Horton says, "You can . . . define a preferred reading sequence that presents information in a coherent order that ensures context. . . . One way to

To do This	Turn to this Section
To open a document	Read Chapter 1
To find something	Use the Find command
To install a program	Look up "programs, installing" in the Help Index
To fix printer problems	Read Chapter 10

FIGURE 17.13 Example of a Summary Table

orient and prepare the reader is to give the reader a mental model of coming facts and a structure in which to organize details as they are presented" (1994, 185, 187). One starter topic would be "What is it?" Many readers are likely to start here, and the screen for this topic can contain overview information.

17.6 Citation of Sources

Citation of sources (sometimes called *documentation*) is not—strictly speaking—an access aid. However, correct and complete information about sources does allow the reader to find the cited book, journal, or Internet site in order to pursue the topic. In citing sources, writers and editors must follow the style guide recommended by the profession, the publication, or the company. Unfortunately, no one style guide is used universally.

If you must edit a document that cites sources, take the following actions:

1. **Determine the required type of citation.** If the document is an article for a journal or magazine, look at the publication for its "Guidelines for Authors" or request such guidelines. If you are working, ask to see the company style guide or a previous paper that follows the accepted citation system. If you are a student, ask your writing professor or professors in your major field.
2. **Check the citation (both in-text references and entries in the reference list) against the guidelines.** Check everything in the citation: capitalization, punctuation, spacing, use of italics, use of parentheses, dates, indentation. If possible, check the spelling of authors' names against the originals.
3. **If any information is missing, query the author or check the original source.** Citation is only useful if it is complete and accurate, allowing the reader to find the cited source. Be sure to find any missing information.
4. **For electronic and Internet sources, extrapolate guidelines for the form.** Citation of electronic sources is just now becoming conventionalized, but such citation is increasingly important. Follow the general form of the system you are using; be sure to include dates and electronic addresses.

Common Citation Systems

MLA

The Modern Language Association (MLA) style is generally used in the humanities: English, foreign languages, philosophy, and so on. It is sometimes called the *author-page* system. In-text references are placed in parentheses and include the author's last name and the page number, like this (Horton 75). The list of "Works Cited" appears at the end of the document, alphabetized by the author's last name. Three sample entries appear below:

Beason, Gary. "Redefining Written Products with WWW Documentation: A Study of the Publication Process at a Computer Company." *Technical Communication* 43.4 (1996): 339–348.
Parker, Roger C. *Looking Good in Print*. Research Triangle Park, NC: Ventana, 1997.
Wu, Corinna. "Banking on Blood Conversion." *Science News* 151.2 (1997): 24–25.

Notice that the second and third lines of the entry are indented five spaces. For detailed information, see the *MLA Handbook for Writers of Research Papers* (4th ed., New York: MLA, 1995).

The Chicago Manual of Style and APA

The Chicago Manual of Style system is widely used in the social sciences and by companies. It is sometimes called the *author-date* style, and it is similar (though not identical) to the American Psychological Association (APA) style of citation. In-text references are placed in parentheses and include the author's last name and the year of publication, like this (Horton 1994). The list of "References" appears at the end of the document, alphabetized by the author's last name. Three sample entries appear below:

Beason, Gary. 1996. Redefining written products with WWW documentation: A study of the publication process at a computer company. *Technical Communication* 43.4 (November): 339–348.
Parker, Roger C. 1997. *Looking good in print*. Research Triangle Park, NC: Ventana.
Wu, Corinna. 1997. Banking on blood conversion. *Science News* 151.2 (January 11): 24–25.

Notice that second and third lines of the entry are indented only two spaces, and only the first word of the article title is capitalized. This book uses *The*

Chicago Manual of Style citation system. For detailed information, see *The Chicago Manual of Style* (14th ed., Chicago: University of Chicago Press, 1993). For APA style, see the *Publication Manual of the American Psychological Association* (4th ed., Washington, D.C.: American Psycholological Association, 1994).

CBE

The Council of Biology Editors (CBE) citation system is commonly used in the natural sciences. This system has several variants; the one explained here is often used in journals to save space. In-text references are referred to simply by number, like this (2), and the "References" section is either alphabetized and then numbered, or citations are numbered by order of appearance. Three examples are given below.

1. Beason, G. Redefining written products with WWW documentation. Tech Comm 1997; 43.4: 339–348.
2. Parker, R. Looking good in print. Research Triangle Park, NC: Ventana; 1997.
3. Wu, C. Banking on blood conversion. Sci News 1997; 151.2: 24–25.

Notice that entries are lined up by numbers, and that the system is streamlined, using initials for authors' first names, omitting subtitles, and abbreviating journal titles and publishers' names. Forms will vary from one journal to another, so you should consult the specific journal or style guide. For details on this system, see *Scientific Style and Format: The CBE Manual for Authors, Editors, and Publishers* (6th ed., Bethesda, MD: Council of Biology Editors, 1994).

17.7 Developmental Editing of Access Aids: Major Questions

All access aids will probably not be in place at the developmental edit. However, you should ask the following questions to help the writer plan:

☐ 1. Does the table of contents show that items of equal importance are given equal status in the hierarchy? Is the number of main points fewer than nine for short-term memory? Does each entry contain key content words and organizational indicators?

☐ 2. In an online menu, are the levels layered to present only a few items at a time?

☐ 3. Does the document plan call for an index if the document has more than 10 pages? Is time alloted in the schedule for writing and editing the index?

☐ 4. Does the document plan call for a glossary? Should it? Do subject-matter experts need to be consulted about definitions? Will this document be used internationally? If so, how does that influence glossary makeup?

☐ 5. Is there a style guide that provides guidelines or templates for navigational aids like headers, footers, topic names, navigation buttons, and so on? If not, can these decisions be made now and incorporated into a style sheet?

☐ 6. Does this document require an abstract, executive summary, summary table, or context-setting introduction?

☐ 7. Does this document require citation of sources? What guidelines for citation style should be followed?

17.8 Copyediting Access Aids: A Checklist

As an editor, you are likely to be responsible for editing access aids at the second-level review. In addition to the questions listed in this chapter, check the following:

☐ 1. Do the items in the table of contents agree with the headings and subheadings in the text? Are the page numbers accurate? Is it easy to see the hierarchy?

☐ 2. Can main menu items be clicked to lead to the next level? In editing, user test the online menus.

☐ 3. Go through a printout of the hardcopy index:
  ○ Can any entries be consolidated?
  ○ Should entries be moved from one level to another?
  ○ Should synonyms be added? Cross references?
  ○ Can every item be found in at least two ways?

☐ 4. Can users find what they need in the online index? In editing, user test the index.

☐ 5. Are all needed terms and synonyms defined in the glossary? Are glossary terms defined with formal definitions? Are the definitions accurate?

☐ 6. Are all headers and footers in the document accurate and consistent? Are on-screen navigational aids clearly and consistently placed?

☐ 7. Are abstracts and summaries accurate? Do they follow the organization of the document? Do they supply key information without being too long?

☐ 8. In citing sources, is the chosen system followed exactly? Is any information missing? If possible, check each citation against the original to guarantee accuracy.

17.9 Proofreading Access Aids: A Checklist

☐ 1. For every change marked in copyediting, has the document been corrected? Check against the original.

☐ 2. In the hardcopy index, if subentries are continued to another page or another column, has the major entry been renamed at the top of the column?

EXERCISE 1

Examining Printed Indexes

Examine the indexes in two books in your major field. Test entries by looking up three items. Can you find what you want? Can items be found in more than one way? Are undefined entries limited to no more than three? Do See *and* See also *entries send you to additional information? Report to the class.*

EXERCISE 2

Analyzing Index Entries

Analyze each of the sections of indexes that follow. Apply the editing questions in section 17.2 and in the checklist in 17.8 and suggest changes. Selected answers appear in the Appendix.

Index Sample 1

Keyboard
  composing characters, E-51
  language, 3-9, H-6, H-14,
    H-15, H-26, H-28, H-29,
    H-32
  local commands, B-1
  locking/unlocking, 3-2
  numeric keypad, C-21, E-51,
    H-10
  status, E-53

Index Sample 2

multi-year contracts, 1-9—1-10
multiple exhibits, 3-12
multiple merges, 8-14
multiple quotes, 6-3, 22-9
multiple splits, 8-14
multiple SPUs, 3-16

EXERCISE 3

Editing Glossary Entries

Analyze the following glossary entries from student papers. What works well? What needs editing? Be prepared to discuss.

1. An Electronic Pickup, which differentiates an electric guitar from an acoustic guitar, depends on electromagnetic induction of electric current, which reproduces the sound of string vibrations. Electromagnetic induction occurs when the pickup's magnetic field is disturbed by the sound of string vibrations. The pickup assembly consists of a metal magnet surrounded by two coils of metal wire. The natural magnetic field creates no current, but according to electromagnetic theory, when a magnet's field is disturbed, electric current is generated if wires surround the magnet in a circular coil. (Braz 1985)

2. A link-and-lever sprinkler head is a heat-sensitive device that automatically discharges when the temperature in a vicinity reaches a specific point, usually between 135 to 150 degrees Fahrenheit. The purpose of an automatic sprinkler is to provide a warning of the existence of fire, to contain the fire, and to help control smoke, thereby allowing better visibility for exiting occupants and advancing fire fighters.

An automatic sprinkler system is composed of a piping system suspended from the ceiling and a series of sprinkler heads attached at intervals to the piping. (Ho 1995)

3. Outsourcing is similar to contracting or vending out tasks. In order to increase productivity, companies have traditionally "out-tasked" by hiring special contractors to do particular jobs or by forming complementary relationships with outside firms. However, outsourcing is different than these traditional contractual relationships; outsourcing can eliminate entire divisions within a company, rather than simply perform a portion of a particular division's tasks. Outsourcing offers its own management solutions to the company's divisional requirements; the resulting business relationship is more like a marriage than a contract. (Van Vleck 1995)

EXERCISE 4

Researching Citation Systems

Determine what citation system is used in your major field; then, find that style guide and scan it. Write a brief explanation of (1) how in-text citations are done and (2) how the reference section is organized. Give one example of each.

EXERCISE 5

Investigating Citation Software

Investigate software that automatically writes citation entries in the correct format. Report to the class what you have found, how it works, and what its availability is.

EXERCISE 6

Analyzing Workplace Abstracts

Find two examples of abstracts in your field or in the workplace. Analyze their contents and effectiveness and report your findings to the class.

EXERCISE 7

Analyzing Workplace Executive Summaries

Find two examples of executive summaries. Analyze their contents and effectiveness and report your findings to the class.

EXERCISE 8

Analyzing Navigational Aids

Examine three books or reports in your field to see how they use headers and footers. Write down what was included. Then evaluate how well they worked. How easy was it to skim? Was anything missing?

References

Braz, Bill. 1985. Technical definitions. Unpublished paper. San Jose State University.
Cadence Design Systems. 1994. *Design entry: Composer™ tutorial.* 4.3, March, 900-20023-0401. Figure 17.12 reprinted by permission.
Harris, Lynn A. and Jan Talbot. 1996. Indexes: Planning for new technologies. *Intercom* 43.3 (March).
Hibsher, David. 1995. A comparison of paper and online access aids. Unpublished report. San Jose State University.
Ho, Waiyi. 1995. Technical Description. Unpublished paper. San Jose State University.
Hoft, Nancy L. 1995. *International technical communication.* New York: Wiley.
Horton, William. 1991. Is hypertext the best way to document your product? An assay for designers. *Technical Communication* 38.1 (February).
———. 1994. *Designing and writing online documentation: Hypermedia for self-supporting products.* 2nd edition. New York: Wiley.
Hoskins, Linda. 1997. Letter to author. San Jose, CA.
Lathrop, Lori. 1997. Intro to indexing. *Intercom* 44.2 (February).
Redish, Janice C., Robbin M. Battison, and Edward S. Gold. 1985. Making information accessible to readers. In Lee Odell and Dixie Goswami, Eds. *Writing in Nonacademic Settings.* New York: Guilford.
van der Meij, Hans and John M. Carroll. 1995. Principles and heuristics for designing minimalist instruction. *Technical Communication* 42.2 (May).
Van Vleck, Sylvia. 1995. Internship. Unpublished paper. San Jose State University.
Williams, Thomas R. 1994. Schema theory. In Charles H. Sides, Ed. *Technical Communication Frontiers: Essays in Theory.* St. Paul, MN: Assn. of Teachers of Technical Writing.

18 Editing Online Documentation

Contents of this chapter:

- Electronic Approaches to Developing and Presenting Information
- Readers and Users
- Elements of Online Documentation
- Developmental Editing of Online Documentation: Major Questions
- Copyediting Online Documentation: A Checklist
- Proofreading Online Documentation: A Checklist

As we move into the twenty-first century, we live in the "information age." However, the way information is handled is changing rapidly, and perhaps nowhere is the change in information development and presentation so obvious as in the shift from printed documents to online documentation. Online documentation can appear on the Internet and the World Wide Web; on corporate intranets (private computer networks); on CD-ROMs; and on help systems, tutorials, and online reference systems. Just as writers are learning new ways to develop and present information, so must editors learn new approaches to editing that information.

Frank Guss writes documentation for software to aid the design and verification of electrical circuits. Guss says, "Editors need some introduction to issues that are especially relevant for online documentation—that is, designing of screens that can be navigated easily, effective hypertexting, the appropriate style for writing online 'chunks,' and possibly an overview of SGML [Standard General Markup Language]." This chapter will introduce you to some of these issues and provide guidelines for editing online documentation.

18.1 Electronic Approaches to Developing and Presenting Information

The computer has influenced the development and presentation of information for both the writer and the reader. For the writer, the computer allows nonsequential presentation of text, graphics, and sound; it promotes interconnections around the world; and it provides for crafting specific sections for different kinds of readers.

For the reader, according to technical communicators Grice and Ridgway, the computer allows an "information-sampling environment" where information is not just a "body of material to be absorbed and internalized." Instead, items of information provide the answers to a series of questions the reader can ask (1993, 435). "Another way of viewing the change is people's switching from learning to retrieval. The old approach is that first you learn it, then you apply it, while now we try to apply without learning or remembering. If you have a calculator, you don't need to memorize multiplication tables or understand the theory of multiplication" (1993, 432). The Grice-Ridgway statement may be too broad for what we have learned since 1993 about reader-computer interaction, but what we do know is that readers of online documentation have different expectations, and they use online information in different ways.

Defining Key Terms

Along with new approaches to information handling has come a host of new terminology—and not all experts agree on the definitions. Here are some current ways of defining key terms.

Electronic information, according to documentation consultants Hackos and Stevens, "includes any form of communication that depends on a computer for its distribution and maintenance" (1997, 2). Electronic information includes interactive demonstrations, context-sensitive online help systems, CD-ROMs, a wizard or Smart Guide to walk users through tasks, Web pages, and virtual reality programs.

Online documentation is "the use of the computer as a communication medium" (Horton 1994, 2) instead of as a number cruncher or word processor. According to online documentation expert William Horton, online documentation systems have two parts: electronically stored information and a means of accessing that information. What defines online documentation, Horton says, is both the method of storage and the means of access (1994, 2).

Documentation is the writing or creating of information, not the method of correctly and formally noting a source of information. In this book, the term for noting a source is *citation*. (See Chapter 17.)

Hypertext in online documentation means that "topics" of text and graphics (also called articles, modules, or nodes) are stored electronically and accessed by links between the topics. In *hypermedia,* the modules can contain sound, animation, and video as well as text and graphics.

Softcopy is paper-based information that has been placed on the computer. While hardcopy (paper) and softcopy (computer) documents are sometimes contrasted, the real difference is between hardcopy documents and online documentation. They are designed differently, read differently, and must be edited differently.

Editing Information Online and Editing Online Information

There is a big difference between editing information online and editing online information, a difference that illustrates what you learned in Chapter 13 about how the position of a word in a phrase can change the meaning of the whole phrase.

When you *edit information online,* you are editing any document that happens to be displayed on the computer screen: a word processing file that will be printed out as a report or proposal, a spreadsheet that will become part of a published analysis, a set of instructions that will be incorporated into a manual. You edit on the computer screen by moving sections or sentences, revising words or punctuation, correcting errors, or inserting changes. You might use actual editing marks in your software program, write comments to the author, or indicate changed passages by change bars in the margin.

When you *edit online information,* you are editing documentation that is designed to be used online: help files, hypermedia, Web pages, computer-based training, and the like. You might make the same kinds of editing changes; the difference is in the document itself.

In either kind of editing, professionals differ in their success and comfort level with on-screen editing. A study reported to the Society for Scholarly Publishing found big advantages of cost and time in on-screen editing: The switch to editing information online (as opposed to on paper) saved 22 percent of editing and production costs. In addition, the study found that editors reached non-online speeds after editing six online manuscripts (Cleary 1993, 102).

On the other hand, many of the editors I interviewed in researching this book prefer to copyedit on paper printouts, whether the document is intended for hardcopy or for online use. Jenifer Renzel, who writes documentation for software to help engineers make chips, said, "It's extremely difficult to copyedit online. I do all my copyediting on printouts. I do organizational editing on screen." Eric Radzinski, who writes for a mainframe computer software company, said, "I usually edit online, but I always go through the hardcopy version of a document before it goes for review." And Chrystal Francois, who freelances in the medical and scientific fields, said, "I always edit on hardcopy so I can have a written record of changes made."

At this time, most writers and editors consider online editing tools to be cumbersome and time consuming, and they prefer copyediting on hardcopy. Senior editor Anjali Puri says that online editing takes twice as long as hardcopy editing and that the italics for insertions are hard to see on screen. After working with online editing for two years, editors at her software company have resumed copyediting on hardcopy—each editor using a distinctive color of pen (1997).

However, these editorial preferences may change as screen resolution improves and as the editing tools themselves become more versatile.

Understanding the Similarities and Differences between Print and Online Documentation

The first 16 chapters of this book concentrate on principles and guidelines to help you edit mostly hardcopy documents. However, as more and more information moves online or becomes available in either paper or online, you need to understand the similarities and differences between the two media.

Similarities

Whether producing printed or online documentation, editors and writers need to follow these four principles, which are summarized here from earlier chapters:

1. **Understand the purpose of the document.**
2. **Identify the readers or users and understand their needs.**
3. **Provide access aids that allow readers to find information easily and quickly.** (See Chapter 17 for details on access aids.)
4. **Ensure that the writing is clear, concise, accurate, and consistent.** As Hackos and Stevens explain, "Despite the differences in the medium of presentation (on the computer screen rather than on paper), most of the rules that apply to paper documentation also apply to online documentation" (1997, 286).

Differences

Online documentation does, however, have different requirements, which will be previewed here and explained in detail in section 18.3. The four most obvious differences are as follows:

1. **Information in online documentation should be presented in short discrete topics.** Each topic represents a unit of information that "fully answers the user's question, is read entirely, can be accessed individually, and [which] the user thinks of as a unit" (Horton 1994, 100). Topics should be readable in any order, not dependent on transitions (like *next, further, in addition*) or the concept of a beginning, middle, or end.

2. **The organization of online documentation is not obvious.** Like paper documents, online documentation requires organization, which not only determines where the topics of information go but also when they are displayed on the screen. Nevertheless, the reader may never see the whole organization. Types of organization include sequence, hierarchy, grid, and web. In online documentation, extra entry points can be added for specific types of readers, and information can be layered with details added as users request them (Horton 1994).

3. **The way information is displayed differs dramatically.**
   Hackos and Stevens list six major differences.
   ○ The display area of a screen or window is smaller than a printed page. A screen will usually hold about one-third of the information on a page, or about what could be written on a three-by-five inch card.
   ○ The shape of a screen is wider (landscape orientation) than it is tall (portrait orientation). Windows can vary in shape.
   ○ The distance from the reader to the screen is greater than the distance from the reader to the page.

- The viewing angle of a screen is out in front and usually about eye level instead of down and slightly above the horizontal as it is when reading paper on a desk or in one's lap.
- The screen resolution is inferior to that of paper, making small fonts and some serif fonts hard to read.
- Monitors emit light rather than reflecting light, and on many monitors the light flickers (1997, 210).

4. **Graphics presentation is more versatile, but because of screen resolution, fine details can easily be lost.** Graphics should be confined to a single screen without the need for scrolling. However, graphics can be designed so that users can click on a part or a callout and see a closeup view—even a view from several angles.

18.2 Readers and Users

Because online documentation is interactive, *readers* also become *users*. Before you can edit, you need to know as much about those reader-users as you can.

Who Are the Reader-Users?

To determine who the reader-users are, ask the same questions you would ask about any reader (see Chapter 14):

- Where do they fall on the generalist to specialist continuum?
- Where do they work and how does their work influence what they want to know?
- What is their experience level and understanding of jargon?
- Why are they reading?
- What questions do they want answered?

In addition, Hackos and Stevens suggest that you ask at what "stage of use" are they? The model for stages of use "describes a sequence of behaviors that people proceed through as they learn to use a new tool or perform a new task. . . . [As] users learn to perform tasks and achieve their performance goals, they move through stages from the beginner who has never performed a task or used a new tool before, to those who are comfortable performing basic tasks, and then to those who become so expert in task performance that they are able to act in new, innovative ways to solve problems" (1997, 31).

The stages are novice, advanced beginner, competent performer, proficient performer, and expert performer.

Novices have no previous experience, are often worried about their ability to succeed, want simply to get a task done, don't know how to recover from mistakes, and prefer only one option for performance.

Advanced beginners have successfully performed a task and are willing to try new tasks (often without reading the instructions first). Like novices, they need help solving problems. They also need fast and easy access to information and instructions. Most users, say Hackos and Stevens, "remain advanced beginners" (1997, 36).

Competent performers have enough experience so that they begin to develop a conceptual model of how a system works. They want detailed information but prefer hardcopy for extended reading. They will diagnose and troubleshoot common problems, and they want access to the answers to Frequently Asked Questions (FAQs).

Proficient performers want to understand how the entire system works. They need access to information from subject-matter experts; they want shortcuts, not oversimplified instructions; they like to participate in online forums with others at their level.

Expert performers—"no more than 1 to 5 percent of users" (Hackos 1997, 45)—have comprehensive knowledge of a process or product and are, therefore, sources of knowledge for others. They like to exchange information with other experts and are always looking for new applications and processes.

Once you know the stage of use of reader-users, you can determine what their needs are and assess whether the writer has succeeded in meeting those needs.

Will the Information Be Printed Out or Read on Screen?

Despite the advances in screen resolution, many reader-users want hardcopy printouts. They might want the hardcopy to remind them of steps they must take while working on screen without having to continually bring up the details in a window. They might want the hardcopy for close reading of an extended passage. In fact, as explained in section 18.1, as an editor, you might want hardcopy for copyediting.

However, paper documents are expensive. As freelance technical writer Regina Roman explains, "Publication costs for manuals are, from the viewpoint of a production manager, an easy line item to manipulate. Not only does it cost less to print CD-ROMs, it also reduces the overall cost of shipping, the size of the accessory package in the shipping box, inventory storage, and so on" (1997).

One way designers of online documentation are solving the screen-paper problem is by single-source authoring. *Single-source authoring* is a method of using source information for multiple outputs. This method facilitates using the same information topics in, for example, a manual and in online help. To make this double duty possible, documents are put on a data base, separated by topic, categorized, and assigned compile directives for print and online (Hart 1996).

Single-source authoring can reduce development time by as much as 90 percent (WexTech 1997), but it presents challenges to both writers and editors. "Writing a single-source document for multiple users poses problems that are quite different than writing each version of the document separately. You must be sensitive to how any given unit of text will be comprehended and processed by readers in several different contexts. A paragraph that appears in a help panel must be written so that it can stand on its own. Yet when it appears in the narrative of the print version, it must flow smoothly and not stand out" (Ensign 1996, 39). Single-source authoring requires writing that is in discrete topics, and as an editor, you need to be able to recognize and even write effective topics. See section 18.3 for details on assessing topics.

How Do Reader-Users Use Online Documentation?

Online experts generally agree that online documentation is used for four major purposes: to provide help, to inform, to instruct, and to motivate or persuade. Within each purpose are many variants, some of which are explained below.

1. Providing help. The first extensive use for online documentation—and probably still the most widely used—is for help files. *Help files* are the "online documentation that users request and read in the middle of an online task to aid them in performing that task" (Horton 1994, 347). Help can include a reference summary (like a hardcopy quick-reference card), reference topics (through context-sensitive access), pop-up help (for definitions and instructions), and various kinds of diagnostic help: wizards, coaches, tutors, and so on.

2. Informing. Reference information that is very extensive, frequently updated, or contains urgent information benefits from being placed online. Other information for online presentation includes background material, definitions of terms, and conceptual material. This information can be presented on a company-wide intranet, on a CD-ROM, or on the Internet.

3. Instructing. Online documentation can be used successfully in various types of computer-based training (CBT). *Computer-based training* is "an interactive learning experience between a learner and a computer in which the computer provides the majority of the stimulus, the learner must respond, and the computer analyzes the response and provides feedback to the learner" (Gery 1987, 6). Types of CBT include:

- computer tutorials—teaching users how to perform common tasks
- guided tours—demonstrations, sometimes including tutorials
- embedded training—solving problems with help and instruction
- drill and practice—helping users master specific knowledge and skills
- simulation—presenting the product itself plus the tutorial or engaging the user in virtual reality
- educational games
- intelligent computer-aided instruction (ICAI)—tailoring instruction and feedback to the student's responses (Horton 1994)

4. Motivating or persuading. With the rapid proliferation of Web pages, the use of online documentation for motivating readers is obvious. Many Web pages are designed for marketing purposes: to sell one's personal skills; to sell a product or service; to sell a company or organization's public image. In addition, educational games and online reference material can be used to motivate students.

18.3 Elements of Online Documentation

Before you can edit hardcopy documents, you must understand the basic elements that constitute a document. Chapters 5 through 17 have explained the basics of these elements and given you many opportunities to practice your skills in hardcopy editing.

This section explains the elements that constitute online documentation. While some of the requirements are the same as for hardcopy, online documentation also has specific requirements. Key elements in online documentation include diction, syntax, and punctuation; topics; links; organization; navigational and access aids; and design elements.

Diction, Syntax, and Punctuation

"To write online documents, we must apply with a vengeance the principles of good clear writing while attending to the differences between paper and online documentation." In this statement, Horton (1994, 261) emphasizes the importance of knowing the guidelines for accurate word choice, effective sentence construction, and correct punctuation. Part Two of this book explains those general guidelines in detail.

In addition, you need to apply the following specific guidelines as you edit online documentation.

Diction

The guideline for jargon is to remove it unless you know that all readers will understand it. If jargon must be used, make sure it is defined. In addition, online documentation needs to avoid "computer jargon" or the language of the machine, which may be foreign to the reader. Words like "parameters," "syntax," and "metaphor" have very specific meanings for the computer but might confuse the novice user who knows these words from other fields. Horton says, "For instance, don't use 'invalid' in a document for nurses, 'terminate' in one for personnel managers, or 'default' in one for bankers" (1994, 263).

Prefixes, like "in," "un," or "an," and short but important words like "and," "one," "or," can be easily overlooked or misread on the screen. When editing, check to ensure that such words are clear, and that they are emphasized if necessary by applying a style change such as color, bold, or reverse background.

Abbreviations, like jargon, must be common enough to be understood by the reader, or they should be removed. Because you can't count on sequential reading, you or the writer will have to provide a pop-up glossary for abbreviations or jargon that must be defined.

Syntax

Two of the general rules for good sentence construction need to be emphasized in online documentation: Write short, direct sentences, and use primarily the active voice.

Writing short, direct sentences means—whenever possible—you ensure that each sentence is written in subject-verb-object order, has fewer than 15 words, and places the action in the verb. Embedded clauses, especially relative clauses using "which" or "that," cause problems for readers online. (See Chapter 13 for more on relative clauses.)

Using the active voice makes clear who or what performs the action: the user-reader (you), the system, a design tool, or a function. Active voice also promotes shorter sentences. For example:

NOT	The disk could be damaged by bending it.
BUT	Do not bend the disk.
OR	Keep the disk flat.

Punctuation

In editing, you need to ensure that the conventions of punctuation are followed. Because of the poor resolution of the screen, you also need to keep the punctuation simple. Horton says, "Avoid colons and semicolons especially. They are hard to recognize on the screen, and many users do not know the difference between them anyway" (1994, 266). In addition, turn a contraction like "don't" into the strong negative **"do not,"** and eliminate periods in common abbreviations like U.S. If your company has a style guide for online documentation, follow it.

Topics

A key element in the construction of online documentation is the self-contained building block of information called a topic. A *topic*—also called a block, a node, an article, or a module—is a unit of text, graphics, or even animation that is on one subject, expresses one thought, or answers one question. Each topic should be complete in itself. In editing, you need to know the guidelines for three descriptors of a topic: its size, its contents, and its structure.

Size

How large a topic can be depends on (1) the reader's willingness to read extended text, (2) the capacity of a screen or window (the reader shouldn't have to scroll repeatedly in order to read the topic), and (3) the reader's need to see connections among the information bits. Horton suggests that "users will read no more than three windows or screens of information to answer a question" (1994, 113).

Contents

Every topic must have a title that contains key words, and those key words should be at the beginning of the title. Every topic should also answer a question. Table 18.1 shows typical questions and the type of information that might answer that question.

Structure

The structure of the topic should be the same for each type of information so that reader-users can fit new information into an existing schema or framework. This guideline is similar to the guideline for hardcopy documents that says to structure reports in the same way, proposals in the same way, and instructions in the same way. The difference is that the topics in online

TABLE 18.1 Questions and Topics with Answers

Reader's Question	Topic Type
How do I . . . ?	list of steps for performing a task
	pictures of each step
	animation showing procedure
What is . . . ?	definition in a glossary
	index item
	pop-up in Help
	picture of examples
What is wrong?	explanation of problem
	diagram of problem
	procedure for fixing problem
	flowchart for trouble shooting

documentation are smaller. The structure of a topic is also influenced by how readers will probably read that topic.

Procedures, for example, will probably be read on screen, so each step should be numbered and confined to a single action. Steps should be in order of performance, should be short (one to two lines), and should number fewer than five.

Conceptual information may be printed out by the reader for closer reading, so it can contain scrollable text. Lists, graphics, and subheadings should be included for easier reading. Terms can be defined within the text but should be backed up by pop-up definitions.

Reference information is probably scanned, so it can be presented in scrollable text but should be in the same basic order each time. Lists and tables should be used to condense information.

Instructional (computer-based-training) information should provide learning objectives, promote interaction, and provide feedback to users. Instructional material will probably be read on screen. For details on structuring different types of topics, see Hackos and Stevens, Chapter 14.

Links

Links (also called informational pathways or jumps) connect topics of information. Reader-users interact with the documentation by choosing the links they want. Links show relationships and serve to unite topics of information. The item the user points to and activates is sometimes known as a *button.* "Buttons are specific locations in the hypertext system that permit the user to jump along a link to another node, usually by clicking a mouse or pressing a key" (Horn 1989, 9). A button can be a word in the text, a label on a graphic, or an icon of a button.

However, links need to be presented in a consistent and limited way. Nichols and Berry—one a developmental editor and the other an online information designer—say, "Perhaps the best approach is to keep all navigational (structural) information (including hypertext links) separate from actual content, potentially in separate windows. A list of 'Related Topics' is easier to understand and use than a collection of hypertext links scattered throughout the text in a window" (1996, 252).

Links occur in hardcopy documents as well, though we usually do not call them links but *pointers*. Such access aids as page references in an index or table of contents, in-text references to figures or tables, the hierarchy of headings, and phrases like "as explained in Part Three" or "as shown above" all serve as pointers.

Links in online documentation can either move the user to another location or display the new information in a secondary or pop-up window. You can apply the following guidelines in editing links:

1. **Limit the number of links to no more than five per topic.**
2. **Get the users to the information they want in no more than three jumps.**
3. **Provide buttons or controls that allow users to move forward or backward easily, and to quit when they wish.**
4. **Tell users how to recognize a link, select it, and backtrack.**

Organization

Online documentation must be organized, even though the reader-users may choose their own paths through that organization. Furthermore, the organization needs to be structured in the way the user will apply that organization. Just as hardcopy documents follow conventional organizational patterns (see Chapter 14), so online documentation follows four conventional organizations: sequence, grid, hierarchy, and partial web.

Sequence

In sequential organization, topics of information follow one another in order: by alphabet, by numerical sequence, by time. Sequences can be simple and in a straight line, but online documentation is more likely to have alternatives, backtracking, or shortcuts, as shown in Figure 18.1.

Grid

A grid is organized like a table or a spreadsheet, with rows and columns of information. The user selects one item from a column and one from a row, and

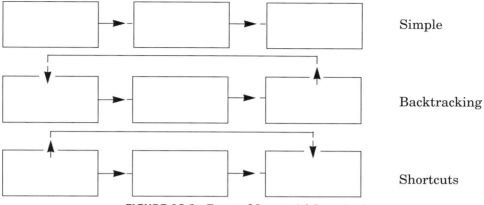

FIGURE 18.1 Types of Sequential Organization

the information at the intersection is displayed. Figure 18.2 shows a simple grid.

Hierarchy

In a hierarchy, information is divided into topics, and the topics are placed at levels depending on their importance, degree of generality, or detail. Figure 18.3 shows a simple hierarchy.

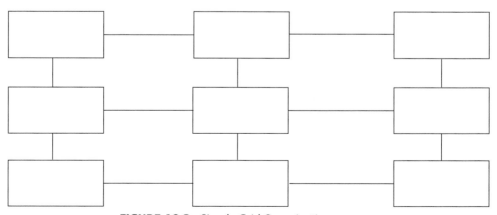

FIGURE 18.2 Simple Grid Organization

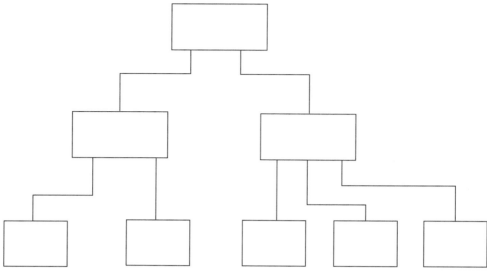

FIGURE 18.3 Simple Hierarchy

Partial Web

In a true web, every topic is linked to every other topic, and the burden on the system (and the user) is enormous. Most online documents, therefore, use a partial web—with links to only a few other topics. Figure 18.4 shows a partial web.

Horton suggests the following guidelines, which can be made into questions for evaluating organization:

1. **Does the type of organization match the reader's purpose?**
   - For learning: a sequence with possible alternatives and jumping ahead
   - For browsing: a cross-referenced hierarchy
   - For fact-finding: a grid, a hierarchy with cross references, a web—depending on the type of information
2. **Does the organization meet the needs of different types of readers?**
   - Design a short sequence for most users.
   - Add extra entry points for novices and experts.
3. **Is information layered?** Does it start with a small amount of simple information and, as users request, add more details? (1994, 194)

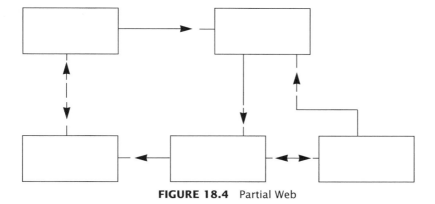

FIGURE 18.4 Partial Web

Navigational and Access Aids

This topic is covered in detail in Chapter 17.

Design Elements of Screens and Windows

Just as pages must be designed for maximum readability, so must screens and windows be designed for consistency, readability, and ease of use. Nichols and Berry (1996) suggest guidelines for online screens that can be adapted by editors.

1. **Is the amount of text per screen limited to about 30 percent, with the rest of the screen white space?** Is information in lists when possible?
2. **Is text reasonably short?** Experts do not agree on line length. Nichols and Berry recommend 12 to 14 characters in line length and six to seven lines per paragraph (1996, 25). Hackos and Stevens suggest 26 to 39 characters per line (1997, 211). Horton recommends 40 to 60 characters per line (1994, 250).
3. **Is the same sequence and placement of windows used every time?**
4. **Is related information arranged vertically, using indentation to show relationship?**

5. **Do changes need to be made in font, justification, space between paragraphs, and placement of graphics?**

Hackos and Stevens expand on the design principles of guideline five by suggesting sans serif typefaces like Arial and Helvetica, fonts of 11 to 12 points (or even larger for middle-aged users), using white space between headings and paragraphs, and keeping windows to one-fourth or one-third of the screen (1997).

As computers proliferate in business, education, and the home, online documentation grows in importance. Most of the guidelines you have learned for hardcopy editing also apply to editing online documentation. In addition, you need to learn those guidelines that apply specifically to editing topics, links, and online organization. You also need to keep up with the changing field, adding more skills to your editing repertoire by reading about online and hardcopy documentation, attending seminars, and talking to those who work with electronic information in its many forms.

18.4 Developmental Editing of Online Documentation: Major Questions

Horton suggests editing topics in random order (1994, 105). At the developmental edit, ask these questions about topics:

☐ 1. Is the topic at each level concise in scope? Does it need to be broken down?

☐ 2. Does each topic have a clear title that identifies the subject?

☐ 3. Does each topic begin with a sentence that explains the subject and its relevance?

☐ 4. Is each topic complete in itself? Is all the information relevant to the topic?

☐ 5. Is the information in each topic accurate?

☐ 6. Are there concrete examples to explain abstract subjects? Are there scenarios (typical problems in work situations) to help reader-users remember facts?

Ask these questions about links:

☐ 1. Are the links between topics logical?

☐ 2. Will users expect a link here?

- ☐ 3. Do users need the information at the end of this link?
- ☐ 4. Will users understand the information at the end of the link?

Ask these questions about organization:

- ☐ 1. Is there a map or outline of the organization showing main topics and lower-level topics?
- ☐ 2. Does the main path provide most reader-users what they will need?
- ☐ 3. Are there optional entry points and branches for novices and experts?
- ☐ 4. Does the organization allow skipping or repeating to refresh short-term memory?
- ☐ 5. Is there a natural order to the organization of topics:
    - ○ top to bottom or left to right for descriptions
    - ○ chronological for procedures
    - ○ cause to effect for arguments
    - ○ general to specific for explanations
- ☐ 6. Does the organization need revision?

Ask these questions about graphics and design:

- ☐ 1. Should text in any topic be replaced with graphics?
- ☐ 2. Do the topics have a consistent design?
- ☐ 3. Are links provided in consistent ways?

Also see the questions in Chapter 17.7, which cover the developmental edit of access aids.

18.5 Copyediting Online Documentation: A Checklist

Ask the following questions at the beta or copyedit level:

- ☐ 1. Is the language consistent, using the same word for the same object or function?
- ☐ 2. Do names of topics as referenced match the names that appear in the views?
- ☐ 3. Are most sentences in the active voice?
    - ○ Do instructions begin with verbs?

- Are descriptions in subject-verb-object or subject-verb-complement order?
- ☐ 4. Are introductions to lists clear and informative?
- ☐ 5. Are steps in instructions clearly and accurately numbered?
- ☐ 6. Are references explicit, using nouns instead of pronouns?
- ☐ 7. If all punctuation is removed, will each passage still make sense?
- ☐ 8. Is the punctuation correct?
- ☐ 9. Is the title of each topic in the correct font size and color? Is the text of each topic in the correct font size and color?
- ☐ 10. Are links properly identified by color, location, or underlining?

18.6 Proofreading Online Documentation: A Checklist

Ask the following questions at the proofreading or final editing level. If possible, read through each of the topics.

- ☐ 1. Have all marked changes from copyediting been made to the final documentation?
- ☐ 2. Have any errors been introduced?
- ☐ 3. Do all the links work? Do they display the correct information?

EXERCISE 1

Researching Online Documentation

Interview someone in your community who works in publications. Ask the following questions and others that are appropriate to the local company or organization:

- How much of your documentation is on paper? How much is online?
- How do you decide which method to use?
- How is online documentation published? On the Web? Company intranet? CD-ROM? Other?
- What do you plan for the future?

EXERCISE 2

Learning about Single-Source Authoring

In current publications, research the state of single-source authoring. What are its advantages? Disadvantages? Report to the class.

EXERCISE 3

Analyzing the Design of Web Sites

Compare two company or organization Web sites. Look at their differences in design of screens, content of topics, types of links, organization. What works best? Why?

EXERCISE 4

Researching New Developments in Online Documentation

In current publications, research new developments in online documentation. How do these new developments influence what editors do and how they do it?

EXERCISE 5

Converting Hardcopy to Online Documentation

Assume that your task is to use this book as the source for an online document. Write a paper in which you identify (1) the main topics of information, (2) the major links, and (3) the principle of organization. Graphically show the relationships. What would cause problems in converting to online? Why?

EXERCISE 6

Converting Design Features

Using Chapter 18 of this book as your source information, suggest ways in which design features used in this chapter could be converted to online design features. Consider the following features:

- type size
- type fonts
- boldface type
- italics
- capitalization
- underlining
- white space
- boxes, rules, screens

EXERCISE 7

Analyzing for User Needs

Analyze a software program you use regularly. Who is the intended user? How can you tell? How much experience does the program assume? Does it provide entry points for users with other experience?

EXERCISE 8

Analyzing Links in Hardcopy

Examine a text or printed manual and find 5 to 10 examples of cross references, indefinite references (like "in the above paragraph"), numbered headings, and the like. Analyze how well each one could be converted in an online document.

References

Cleary, J. 1993. *The editorial eye* 16.8 (August). Reported in *Technical Communication* 41.1 (February).

Ensign, Chester and Michele Cameron. 1996. Using SGML for multi-purpose documents. *Intercom* 43.3 (March).

Gery, Gloria J. 1987. *Making CBT happen: Prescriptions for successful implementation of computer-based training in your organization.* Boston: Weingarten Publications.

Grice, Roger A. and Lenore S. Ridgway. 1993. Usability and hypermedia: Toward a set of usability criteria and measures. *Technical Communication* 40.3 (August).

Hackos, JoAnn T. and Dawn M. Stevens. 1997. *Standards for online communication: Publishing information for the internet/world wide web/help systems/corporate intranets.* New York: Wiley.

Hart, Jessica, Nola Hague, and Diana Peh. 1996. Single-sourcing tools and techniques. *Proceedings of 43rd Annual Conference.* Washington: Society for Technical Communication.

Horn, Robert. 1989. *Mapping hypertext: Analysis, linkage, and display of knowledge for the next generation of online text and graphics.* Lexington, MA: Lexington Institute.

Horton, William. 1994. *Designing and writing online documentation: Hypermedia for self-supporting products.* New York: Wiley.

Nichols, Michelle Corbin and Robert R. Berry. 1996. Design principles for on-line information systems: Conclusions from research, application, and experience. *Technical Communication* 43.3 (August).

Poole, Dorothy L. and Susanne Viera. 1993. Producing online documentation. *ITCC Proceedings.* San Diego: Univelt, 186-188.

Puri, Anjali. 1997. Conversation with the author. San Jose, CA.

Roman, Regina. 1997. Letter to the author. San Jose, CA.

Schmitz, Tom. 1992. Smart paper fax lets computer users link up via letter. *San Jose Mercury News.*

WexTech Systems. 1997. *Single-source publishing using Documentation Studio™.* New York: WexTech.

19 International and Intercultural Issues for Editors

Contents of this chapter:

- Recognizing the Importance of English
- Defining Key Terms
- Clarifying Editors' Roles and Responsibilities
- Understanding the Influence of Culture
- Working Toward Global Communication: Guidelines for Editors
- Cultural Editing: Major Categories
- Developmental Editing of International Documents: Major Questions
- Copyediting International Documents: A Checklist
- Proofreading International Documents: A Checklist

Communication is about the interchange of information, and one of the great challenges of the twenty-first century will be effective communication across national and cultural boundaries. Editors and writers in the United States will bear increasing responsibility for that effective communication because of the dominant role that English plays as a communication medium. The premise of this chapter is that from the U.S. perspective, effective communication within and outside the United States begins with good written English.

19.1 Recognizing the Importance of English

Why is good written English so important? Consider the following facts:

- English is the official language in 44 countries. No other language in the world is so widely used officially.
- The number of native speakers of English is about 350 million, but as many as 700 to 750 million people use English as a national, second, or foreign language, or as a language for commerce, industry, science, or other purposes (Conner 1996, 17). Speaking and writing are, of course, two different things, but the number of English speakers worldwide indicates the concurrent importance of written English. Web sites, for example, are predominantly in English.
- English is the central language of the global economy partly because it provides a neutral language of government and commerce, and partly because American English technical terms have been so rapidly absorbed into other languages. The Japanese, for example, have borrowed as many as 20,000 words.
- English has a huge vocabulary of over 600,000 words, with about 200,000 words in common use. In addition, there are "millions more" scientific and technical terms (Long 1994, 509).
- The grammar and syntax of English are flexible, making the language versatile.

We might be tempted to say then (as, in fact, some Americans do), "If English is so important, let the rest of the world learn English." Such a narrow view ignores the realities of the global marketplace, the many non-American versions of English in the world with their own standards, and the cultural and language diversity within the U.S. population. Competence in spoken and written English might be the beginning of effective communication, but it is just that—a beginning.

The Global Marketplace

Consider the global marketplace. In the fourth quarter of 1996, 62 percent of IBM's revenue came from international sources. In fiscal 1996, Merck (pharmaceuticals) did 30 percent of its business internationally, and Johnson and Johnson (diversified health care products) did just under 50 percent internationally. At Apple Computer, more than 70 different projects were localized into more than 700 different language versions in 1993—an indication of the scope of internationalization efforts at Apple.

According to linguist Jan Ulijn and technical communicator Judith Strother, businesses need to look outside their borders to expand their potential markets. Summarizing their argument, Campbell says, "To penetrate a foreign market calls for skills of a different order: it is not enough merely to translate brochures or technical manuals into another country's language, because cultural conventions, biases, assumptions built into both languages affect the way the message is understood" (1996, 304). By "both languages," the author means the source language of the original document and the target language into which the document will be translated.

International Englishes

Readers and consumers who live in Britain, Australia, Canada, New Zealand, and parts of India may use English as their first language, but it is not American English. They spell the same words differently (in Britain, for example, *colour, theatre*), use different terms for common objects (in Britain, *gammon steak, petrol, boot*), have different money systems, address conventions, idioms and metaphors—even different ways of conventionally organizing documents. It is a mistake to assume that American English is instantly comprehensible to those who speak and read international English.

Cultural and Language Diversity in the United States

Even the speaking and writing that is for use within the United States must accommodate itself to diversity of language and culture. The 1990 census clearly showed that the United States is a multicultural society: One in seven people speaks a language other than English at home; 108 different languages are spoken by students in the Los Angeles County School District;

there are presently 25 million Hispanics in the United States (Lustig and Koester 1996, 7, 8). Many cities have large enclaves of people who communicate using their own language and who maintain cultural traditions of food, dress, and ways of doing business.

With all of these groups, writers and editors must learn to communicate in the most effective and efficient ways possible. Freelance editor Jo Levy emphasizes the need for the precise use of words and her fear that translators might choose the wrong word because the source word in English is ambiguous. Yvonne Kucher, who edits for a networking company, says, "We have to remember that often people are reading our documentation with a dictionary in hand. We have to think denotation, not connotation." Mike Belef, senior writer for a medical-equipment manufacturer, says, "Intercultural communication is vital. The workplace is very diverse. We have many qualified people in the workplace with varying communication skills. Ideally, every editor and writer would come from college armed with several courses in semantics, rhetoric, and cultural communication skills. Studying communication differences in gender, culture, occupation, and social class is more important than ever."

19.2 Defining Key Terms

The field of international and intercultural communication uses the following key terms: culture, language, translation, internationalization, localization, and globalization.

Culture means ways of living, "a set of rules and patterns shared by a given community" (Conner 1996, 101). Culture includes customs, attitudes, and beliefs, and a person can belong to several cultures—of profession, of ethnicity, of geographical community. *Languages* are ways of communicating, with people of the same geographical area or the same cultural tradition having a language in common. *Translation* is the "use of verbal signs to understand the verbal signs of another language" (Lustig and Koester, 1996, 162). Translators, who know both the *source* language and the *target* language, determine "what the author is trying to communicate in the source language and then render this accurately into the target language" (Berlitz n.d.). Machine translation is also becoming increasingly important. However, Elaine Winters, an international communication consultant, cautions: "Simple translation does not equal good communication. Do not treat translation like word processing. It is idea and concept processing" (1994, 7).

To achieve more effective communication, American companies have, since 1989, moved beyond simple translation to localization, internationalization, and globalization.

Localization is the process of adapting a document to a specific cultural context or area, thus making it seem local. Through localization, all the sounds and colors, all the religious, social, and political references of the target language are adapted, and the accent of the source language is lost (Cyphers 1991). A localized document is customized.

Internationalization, according to Nancy Hoft—an international communication expert—is the "process of re-engineering an information product so it can be easily localized for export to any country in the world. An internationalized information product consists of two components: core information [that remains the same and can be reused] and international variables [localizable elements]" (1995, 19). Core information might include product descriptions, standard warning phrases, a video clip of a product installation, or a chapter on documentation conventions for user guides. This information is carefully edited and user tested; then, it can be frozen, translated, and used multiple times. International variables might include currency format, units of measurement, graphics, writing style, and product packaging. These elements will change for each country and language. Hoft also says, "Because of diversity throughout the world, you are encouraged to internationalize as much as possible. The more you internationalize, the less you have to localize, or the easier it becomes to localize technical writing" (1995, 238).

Globalization means that a product or means of communication has universal appeal and can be used anywhere without modification. Some global symbols, like the signs for wheelchair access and men's and women's restrooms, convey a single message and are successful in most cultures. The terms globalization and internationalization are sometimes used interchangeably.

19.3 Clarifying Editors' Roles and Responsibilities

What are editors' roles and responsibilities in dealing with international and intercultural issues? The first responsibility is to understand the implications of those two terms. *International* means interaction among people from different nations, and the editor has a role in preparing documents to be used in international communication. *Intercultural* means interaction among people from different cultures; thus, it is a broader term that can include international communication. However, intercultural issues can also surface within the United States. The two terms are used in this chapter to emphasize that editors must be prepared to work with both of the following groups:

1. **International and intercultural *writers*.** Editors often work with documents written in English by non-native speakers and must, therefore, understand typical second-language sentence or organizational problems and know how best to solve them.

2. **International and intercultural *readers*.** Editors also often work with documents written in English that will be read by non-native speakers or translated into other languages. To be effective in either situation, these documents must follow specific guidelines.

International and Intercultural Writers

According to the U.S. Bureau of Labor Statistics, by the year 2000, immigrants will hold 26 percent of all jobs in the United States, and they will bring with them their cultural practices and their communication styles. A study done by the U.S. Census Bureau indicates that the fastest growing segment of the U.S. population consists of immigrants from various parts of Asia—an increase of 107.8 percent between 1980 and 1990 (Thrush 1993). A high rate of Asian immigration has continued since 1990. In addition to second-language writers in the United States, you may well be required to edit documents written in English from native writers of international English or second-language writers from abroad.

Non-native speakers, when they write in English, need to know the rules (or grammar) of the language, have time to apply those rules, and be able to focus on the rules. Iloni Leki, an expert in English as a Second Language (ESL), says that ". . . when an editor receives material from a non-native speaker of English and finds sections that are awkward or difficult to understand or which contain errors in English, the editor is seeing evidence that the non-native speaking colleague did not know the rule governing the error, if in fact an articulated rule even exists, did not have time to apply that rule, or did not focus on the rule" (1990, 149).

Leki explains that one cause of problems may be the result of first-language interference. Another is that some forms of English are acquired late by all non-native speakers—for example, the form of "-s" on the end of singular present-tense verbs (*she chooses, the study includes*). A third problem may be vocabulary: A bilingual dictionary may give only a single meaning of a term that is used differently in different contexts. At the sentence level, ". . . errors in the construction of complex sentences may come, as they do with native speakers, from the writer's loss of control of the sentence" (150). Other sentence-level problems may come from first-language interference.

Leki suggests that editors can help non-native writers in the following ways:

1. Do not underestimate the intelligence of a writer because of errors in writing or differences in rhetorical style.
2. Recast problem sentences or paragraphs into standard English. Then, encourage the writer to locate and describe the differences between the two versions. This technique helps non-native writers understand how the problems occurred.

International and Intercultural Readers

Editors also play a role in preparing documents to be translated and localized or to be read in English by second-language readers. General guidelines for success in this role appear in this section; specific guidelines for internationalizing documents appear in section 19.5.

To successfully communicate with international and intercultural readers, follow these guidelines:

1. For cultural sensitivity,
   - learn about beliefs, practices, methods of reading, and methods of organizing text in other cultures. For localization, perform an international user analysis of the target country. (See section 19.4.)
   - avoid geographical, social, and environmental references.
   - avoid attempts at humor.
2. For effective organization and document design,
   - avoid using the alphabet as an organizing factor.
   - develop information in small, discrete modules.
   - with each type of document, develop a generic organization that fits the purpose of that document (for example, in organizing research reports, follow this organization: introduction, methods, results, and discussion).
   - keep paragraphs short.
   - design pages to allow for text expansion in translation. (See section 19.5 and Chapter 16.)
   - design graphics and pages to allow for different reading patterns.
3. For effective syntax,
   - set up a style guide indicating punctuation and stylistic choices.
   - keep sentences short (approximately 15 words).
   - include only one thought or action per sentence.
   - use the active voice whenever possible.
   - avoid over-modified nouns. (Keep noun and adjective strings to no more than three modifiers by changing some modifiers to prepositional phrases.)

For example,

> NOT Connect the pressurized nitrogen supply to the VDS pump seal purge quick disconnect fitting.
>
> BUT Connect the pressurized nitrogen supply to the quick disconnect fitting on the VDS pump seal purge.

4. For effective diction,
   - use only one word for one thing, concept, or action.
   - avoid acronyms, abbreviations, jargon, and contractions. If acronyms and jargon must be used, define them the first time they are used and include them in a glossary.
   - establish a glossary of all key terms.
5. For effective transfer of information,
   - make sure that the software used to create the document is also available to translators or localizers.
   - provide complete information about the software that was used for the text and graphics (including the revision number). Provide the information both as a readme file and in hardcopy.

For further information, consult the sources listed in the reference section or one of the following references: Deborah Andrews, Ed. 1996. *International dimensions of technical communications*. Arlington, VA: STC Press.

Rob Sellin and Elaine Winters. *Cross-cultural communication: Tips for those who develop material or products for translation or export*. A Web site at http://www.bena.com/ewinters/xculture/html.

19.4 Understanding the Influence of Culture

Depending on the degree of your involvement with international and intercultural communication, you may need to research one or more other cultures and learn how they differ from your own.

Robert G. Hien, a documentation specialist in Ohio, says, "Our accumulated knowledge and experiences, beliefs and values, attitudes and roles—in other words, our cultures—shape us as individuals and differentiate us as peoples. . . . Most important for communicators, our cultures manifest themselves in our information needs and our styles of communication. In other words, our cultures define our expectations as to how information should be organized, what should be included in its content, and how it should be expressed" (1991, 125).

Examples of Cultural Variables

One example of a cultural variable is the amount of common knowledge shared in a culture. When people have very similar ethnic, religious, and educational backgrounds, they also share attitudes, feelings, and values. Such cultures are called *high-context cultures*. Japan and France are two countries considered to have high-context cultures. "Writers in high-context cultures do not need to produce a great deal of background information or to spell things out precisely. In fact, it is considered insulting to the reader to include too many details as if the reader were incapable of filling in the gaps" (Thrush 1993, 275).

When regional, ethnic, religious, and educational backgrounds differ, people have little shared knowledge and few shared attitudes, feelings, and beliefs. Such cultures are called *low-context cultures,* and the United States and Germany are examples. In the United States, we teach that in good writing, generalizations must be supported by examples, statistics, and expert testimony. German documents tend to give a great deal of background information (Thrush, 1993, 276).

A second cultural variable concerns *rhetorical strategies,* or the way that documents are organized. Linguists who research contrastive rhetoric maintain that each language has rhetorical strategies unique to it. In 1966, linguist Robert Kaplan identified five types of paragraph development for five cultural groups, as shown in Figure 19.2.

Characteristics of each of the rhetorical strategies can be summarized as follows:

1. Writing in English in the United States is called "linear" because it starts with the point to be made and then supplies evidence to support that point.
2. Writing in Semitic languages is based on a series of parallel coordinate clauses. Repetition may provide emphasis, and the citing of authorities provides the primary evidence.
3. Oriental (Asian) languages use an indirect approach and come to the point only at the end. For example, Mahalingam Subbiah says that "whereas American readers expect that clarity is an important, perhaps the most important, element of technical writing, Japanese perceive beauty, surprise, and 'flow' as desirable measures of good technical writing" (1992, 15).
4. In some Latin cultures (called Romance in Kaplan's sketches), "proof" of a point is accomplished through an appeal to emotions, while persuasion calls only for the expression of feelings (Thrush 1993, 277).
5. ". . . in Russian, essays are permitted a degree of digressiveness and extraneous material that would seem excessive to a writer of English" (Conner 1996, 15).

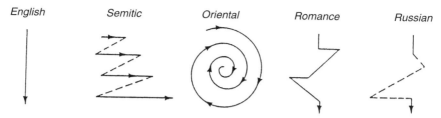

FIGURE 19.1 Cross-Cultural Differences in Paragraph Organization (*From R. B. Kaplan. Reprinted by permission of* Language Learning.)

Kaplan's theory has been criticized for including too many cultures in one broad category ("Oriental," for example) and for ignoring other variables. Kaplan modified his position in articles in 1987 and 1988; still, the "differences may reflect different writing conventions that are learned in a culture" (Conner 1996, 16). If you remember that Kaplan's sketches are oversimplified, you can use them as a starting point to assess how different cultures develop rhetorical strategies.

Designing an International-User Analysis

As an editor, you might be part of a team that needs to learn the ways in which the information needs of a reader or user in one country or culture differ from the information needs of readers in your own or another country. International communication expert Nancy Hoft suggests a method for designing an international-user analysis. Because only 10 percent of a culture's characteristics is easily visible, Hoft recommends creating a worksheet for the target country and the source country and noting the differences and similarities between the two. A brief summary of Hoft's approach follows. For details, see Chapters 2 and 4 in *International Technical Communication: How to Export Information about High Technology* (1995).

An international-user analysis should include three general areas:

1. *surface characteristics*—for example, currency, date and time formats, units of measurement, graphical conventions, and writing style
2. *unspoken rules*—for example, business etiquette and protocol
3. *unconscious rules*—for example, nonverbal communication, personal space, and rate and intensity of speech

Hoft's worksheet contains seven basic variables, and she recommends that others be added as the business or communication situation warrants. The seven variables and examples of their components appear in Table 19.1.

TABLE 19.1 Hoft's International Variables

Differences and Similarities	Examples
1. political	trade, law, political tradition and symbolism
2. economic	cost of common materials and services, cost of living, average gross income
3. social	business etiquette, family and social interaction, forms of discrimination and prejudice, popular culture
4. religious	beliefs, icons, colors, food, major documents
5. education	literacy rate, common knowledge, learning style
6. linguistic	the target language, text directionality and orientation on the page, writing style
7. technological	available computer hardware and software, telephone service

Hoft recommends filling in the details of each variable on a table and then using that information to localize the product and the documentation.

Learning about Other Nations and Cultures

Large companies and organizations with a strong international presence will have localization and marketing groups whose job it is to secure data about other countries and cultures. However, if you are working at a small or start-up company, you may be responsible for gathering data and for learning about other cultures. You may find the following sources of information helpful.

Written Information

General information about countries can be obtained from the United Nations, from embassies, and from foreign trade organizations. The U.S. Department of State publishes *Background Notes* and *Post Reports,* which are available online from the National Trade Data Bank, in depository libraries, or from the Government Printing Office. The most current *Background Notes* are available on the Department of State's Web site at http://www.state.gov. *Background Notes* cover more than 170 countries; each report contains information about a country's people, land, history, government, political conditions, economy, and foreign relations. Information is updated every three or four years. *Post Reports* are prepared primarily for U.S. government employees and their families. Information includes geography and climate; food, clothing, and religious activities; currency, banking, and weights and measures; local holidays and recommended reading.

The Government Printing Office (GPO) also publishes an area handbook series. Each book is titled *[name of country]: A Country Study.* The handbooks are compiled by the Foreign Area Studies of American University; they are available for sale by the GPO or are located in depository libraries.

Two other good sources of information are (1) the Ernst & Young International Business Series, each title beginning with *Doing Business in . . . ,* and (2) *Culturgrams,* short introductions to a country, its culture, and facts about doing business there. *Culturgrams* are available from Brigham Young University.

General books like Roger Axtell's 1993 *Do's and Taboos Around the World,* 3rd ed. (New York: Wiley) are also very helpful. In addition, Intercultural Press publishes a wide variety of books.

Oral Information

You can tap into the knowledge available at your local college or university by interviewing international students, attending meetings of international groups, or hosting an international student. You can also interview international employees of local companies.

A very good way to learn about another culture is to study its language—either formally in the classroom or informally by learning simple greetings, how to say "please" and "thank you," and so on.

Personal Experience

Personal experiences will not provide you with instant information, but they will provide the basis for a deeper understanding of the values and ways of communication in other cultures. Rajinder Gupta, director of an international executive search firm, suggests living in another culture "anywhere in the world, for a month or two or even a summer, working part time and not living in a hotel. Learn about the country 'on the ground.' Reading isn't enough" (Kleiman 1996, 2PC).

Tom Peters, author and entrepreneur, suggests that when you travel, you "Slow down. Tuning in to body language and spoken language simply takes more time in another culture. We instantly 'understand' thousands of subtle cues in our own environment, from physical trappings to the nuances of language. But in foreign settings, getting even a hint of what's going on calls for intense concentration." Peters also says to "Walk the streets. Nothing helps you soak up the culture more than a two-hour stroll down Zeil in Frankfurt, the Galleria in Milan, or Regent Street in London. Look at the toys, the appliances, the foods, the posters in the travel shops, the houses, the people.

Buy a few papers and magazines and thumb through them—you'll be surprised at how much you can 'get' even if you can't read the language" (1992, 2D).

19.5 Working Toward Global Communication: Guidelines for Editors

As an editor, one of your main tasks will be to promote effective communication with international and intercultural readers. As noted in the introduction, the premise of this chapter is that effective communication begins with good written English. However, as Marlana Coe of Coe Communications explains, "Whether you are developing information for non-native speakers of English or information that a vendor is going to translate, you must write in an international style that transcends culture. . . . Designing and developing international information is challenging, but strong, clear writing that anyone can access is the hallmark of a good communicator" (1997, 17, 19). The previous chapters in this book show you how to edit to ensure strong clear writing. This section includes specific guidelines that you can follow for editing English that will be read by second-language readers and will be translated into other languages, either by translators or by machine. Remember that "it may take the same amount of time or even more to properly translate a document, as it did to create the original" (Berlitz n.d.). Since translation is often charged by the word, a tighter original document will also be cheaper to translate.

The following guidelines are for diction, syntax and punctuation, and technical elements. Guidelines for editing graphics intended for international and intercultural readers appear in Chapter 15.2.

Diction Guidelines

When editing, **check for and encourage** the following practices during the development of a document:

1. Use of only one term for one concept, object, or action. Though U.S. college writing students are sometimes told to use synonyms for variety, workplace writers need to seek clarity over variety. For example,

USE	INSTEAD OF
select	click on, choose, pick
start	begin, initiate, invoke, activate
users	participants, learners, readers, audience
device	unit, equipment, system

2. Compilation of a glossary of terms as the document is developing. Include technical terms; needed jargon and acronyms; and verbs that can be used in more than one way, like *can, set, backup.*
3. Development of a style sheet that ensures consistent capitalization—for example, always capitalizing proper nouns.

Eliminate or discourage the following usage:

1. Slang, idioms, and culturally bound metaphors, which have little or no meaning outside the United States or which cannot be translated. Eliminate or replace terms like these:

   ground rules
   what you need to know
   shotgun approach
   on a roll
   latebreaking
   got its work cut out for it

2. Industry jargon that may be untranslatable, like *ramp up, get a dump, port it to.* . . . If the jargon is necessary, define it the first time it is used and include the definition in the glossary.
3. Abbreviations, acronyms, and contractions. American business and technical English is full of abbreviations like *FYI, ASAP,* and *FAQ* that are better written out; acronyms like *CAD, MOS, LAN,* and *SEM* that might have different meanings in different industries; and contractions like *don't, that'll,* and *shouldn't* that are difficult to translate and may be too informal.
4. Nouns and verbs that have multiple meanings in English, like *run, drive, see, time, pay, can, help, file, input.* For example,

   You can *pay* the bill tomorrow. (verb)
   The manager is waiting for his *pay.* (noun)
   Look in your *pay* envelope. (adjective)

   Suggest more specific replacements for words like these, or ensure that the word is used as one part of speech only.

Syntax and Punctuation Guidelines

When editing, **check for and encourage** the following practices:

1. Noun strings of no more than three words in a row. Convert longer strings into a series of prepositional phrases. For example,

NOT   Remove radiation shield hoist rings and radiation shield-to-outer top cover bolts.
BUT   Remove the hoist rings on the radiation shield and the bolts on the outer top cover of the radiation shield.
NOT   Shut off the VDS vacuum seal purge nitrogen supply at its regulator.
BUT   Shut off the nitrogen supply for the VDS vacuum seal purge at its regulator.

2. Short, subject-verb-object sentences averaging 15 words. Short sentences are easy to translate, easy for non-natives to read, and avoid the embedded clauses of complex sentences.
3. Active voice whenever possible.
4. Replacing vague or ambiguous pronouns with nouns.
    NOT   The engineers archived the materials. They were outdated and needed to be stored offsite.
    BUT   The engineers archived the materials. The materials were outdated and needed to be stored offsite. (Synopsys 1996)
5. In complex sentences with relative clauses, including the correct relative pronoun (*which* or *that*) in the clause and ensuring that the clause conveys the intended meaning. For example,
    NOT   The instructions indicated a CD-ROM was required.
    BUT   The instructions indicated *that* a CD-ROM was required.

Differentiate between restrictive and nonrestrictive clauses:

Restrictive: Those are the corrections that must be made today.
    (Among the corrections, make those today.)

Nonrestrictive: Those are the corrections, which must be made today.
    (Make all of the corrections today.)

Even better, reword to avoid the clause.

6. Use of punctuation that will help clarify related word groups. For example, introductory phrases should be followed by a comma.
    NOT   To eliminate overheating the fan must be turned on.
    BUT   To eliminate overheating, turn the fan on.
7. Bulleted lists instead of long series of words or phrases in a sentence. List items are easier to identify.
8. Hyphens in compound modifier phrases. The hyphens help the reader identify how the modifiers are to be understood. For example,
    real-time program
    short-term assignment
    text-editing tools

Eliminate or discourage the following practices:

1. Use of possessives for objects. Possessives are difficult to translate. Replace the possessives with prepositional phrases. For example,

      NOT   the document's glossary
                 the instrument's reading
      BUT   the glossary of the document
                 the reading on the instrument

2. Use of the conjunctions *while* and *since* because they are ambiguous. *While* can mean either *because* or *concurrently*. *Since* can mean *because* or *time past*. Replace *while* and *since* with more specific terms.

If you need further explanation of any of these terms, consult Chapters 12 and 13.

Guidelines for Graphics, Technical Terms, and Design Issues

When editing, **check for and encourage** the following practices:

1. Use of graphics (including videos) whenever possible, to present concepts, show relationships, and so on. See Chapter 15 of this book for details. Also consult William Horton's November 1993 article "The Almost Universal Language: Graphics for International Documents" in *Technical Communication 40.4*.
2. In graphics, use of arrows or numbers to show readers the direction in which you want them to process the information. Remember that not all cultures read from left to right or clockwise.
3. Clear use of numbers:
   - metric numbers for measurements, with U.S. equivalents if necessary
   - dates written out with month, day, and year (March 3, 2000) instead of number equivalents (which may appear in a different order in other countries)
   - identification of billion by 10^9 (American) or 10^{12} (British)
   - clear indication of what are decimal points and what are commas to separate thousands
   - indication of what currency system is being used; for example, for the United States $1,000 USD
   - use of the 24-hour clock for time to reduce ambiguity
   - addresses and phone numbers that are understandable outside of the United States. For telephone numbers, for example, you may need to

include the international access code and the country access code. Consult or create a style guide.
4. Wide margins and generous white space to allow for differences in paper sizes and for text expansion in a translated target language of as much as 30 percent.
5. Consistent typography and page size.
6. A plan for usability testing with potential readers.

Eliminate or discourage the following practices:

1. Alphabetic organization for reference material, lists, or other factors. Translation into languages with different alphabets requires complete reorganization. Another method of organization that would not require complete transformation in translation is preferable.
2. Use of these special symbols, which may have different meanings in other cultures. Replace the symbols with the equivalent word.

    # for pound   ' for feet
    $ for currency   ? for help
    " for inches or ditto

19.6 Cultural Editing: Major Categories

The term *cultural editing* is used by Nancy Hoft to explain how to check a document for cultural bias. Hoft says that if culturally based information is not needed in the document, remove it. If it is needed, "start compiling a list of all the cultural variations in that category that are relevant to the countries to which your company is exporting" (1995, 129). In addition to many items covered in section 19.4 of this book, Hoft's list includes these items:

- accounting practices
- historic events
- forms of address and titles
- legal information (warranties, copyrights, trademarks, warnings)
- illustrations of people
- architecture
- gender roles in the workplace
- learning styles (for example, the relationship between the instructor and the students)

19.7 Developmental Editing of International Documents: Major Questions

At the developmental edit, ask the following questions in addition to those for organization, graphics, design, and access aids. See Chapters 14, 15, 16, and 17.

- ☐ 1. Is a glossary being compiled as the document is being developed? If not, start such a glossary.
- ☐ 2. Is there a style guide or style sheet that includes relevant categories from the guidelines for editors in section 19.5? If not, start such a style sheet.
- ☐ 3. Is culture and gender neutrality maintained in language, graphics, videos, music, and the like?
- ☐ 4. Are there core information modules that can be used over without modification? Identify them.
- ☐ 5. Is the same kind of information presented in the same way?

19.8 Copyediting International Documents: A Checklist

Both readers and translators depend on accurate, consistent, and easy-to-read sentences. Check each of the following areas very carefully.

Diction

- ☐ 1. Is only one term used for one concept, object, or action? If not, revise.
- ☐ 2. Is capitalization consistent and definable? Is the rationale for capitalization included in the style guide for reference by translators?
- ☐ 3. Have all idioms, slang terms, and culturally bound metaphors been removed?
- ☐ 4. Is industry jargon eliminated or—if necessary—defined?
- ☐ 5. Are all abbreviations, acronyms, and contractions eliminated or—if needed—defined?
- ☐ 6. Does each common noun and verb have only one meaning (and one part of speech) within the document?

Syntax and Punctuation

☐ 1. Are noun and adjective strings limited to three words? If not, revise.
☐ 2. Are most sentences short and simple? Can any complex sentences be divided or simplified?
☐ 3. Do any passive constructions need to be made active?
☐ 4. Are all pronouns clear? Replace any ambiguous pronouns with nouns.
☐ 5. Is the relative pronoun included in all relative clauses? If not, add it.
☐ 6. Is the correct relative pronoun (*which* or *that*) used in nonrestrictive or restrictive clauses? Is the message clear?
☐ 7. Are introductory and interrupter phrases set off by commas?
☐ 8. Are long series of items set in bulleted or numbered lists?
☐ 9. Are hyphens used in all compound modifier phrases?
☐ 10. Are possessives replaced with prepositional phrases?

Graphics, Technical Terms, and Design Issues

☐ 1. Can graphics replace any text?
☐ 2. Do arrows or numbers show the order in which graphics should be read?
☐ 3. Is the use of numbers clear and free of cultural bias? Check all the items in this list.
  ○ measurements in metric
  ○ dates written out
  ○ clear identification of billion
  ○ decimal points and commas in large numbers
  ○ indication of the currency system being used
  ○ time with the 24-hour clock
  ○ addresses and phone numbers understandable outside the U.S.
☐ 4. Is there enough white space in margins and around figures to allow for text expansion?
☐ 5. Is information organized by a means other than the alphabet?
☐ 6. Are ambiguous symbols replaced by words?

Also see the checklist for graphics in Chapter 15 and the checklist for document design in Chapter 16.

19.9 Proofreading International Documents: A Checklist

Ask these questions at the final edit:
- ☐ 1. Have all the corrections marked in the copyediting pass been made to the document?
- ☐ 2. Are all the graphics properly oriented on the page?
- ☐ 3. Have any errors been introduced?

EXERCISE 1

Learning about Another Country

Interview a student, professor, or employee in your community who is from a country different from your own. Choose at least three of Hoft's categories (from section 19.4) and create a list of about 20 questions to guide your interview. Report to your class orally or in writing.

EXERCISE 2

Researching Information about Another Country

Choose a country about which you know little. Consult two of the sources listed in section 19.4 and write a brief report summarizing what you have learned about its culture. As part of the report, discuss what cultural elements you would consider in writing a document that would be used in that culture.

EXERCISE 3

Learning about Localization and Internationalization

Interview an internationalization or localization expert in your local community. Determine into what languages the company's documents are translated.

Find out what guidelines the company uses in preparing documents in English for translation and localization.

EXERCISE 4

Editing for International Readers

Examine an article in a technical or business journal to see if it has any idioms, slang, or metaphors that would be unclear to a person using an international form of English, a second-language reader, or a translator. Suggest alternatives if you can.

EXERCISE 5

Editing for International Readers

Analyze the following sentences to determine what phrases are culture bound. Suggest alternatives. Selected answers appear in the Appendix.

Example A

"Compared to products like *Word* or *AmiPro, XyWrite* is not a pretty face. It's also difficult to learn. But on the plus side, it is extremely fast—an important characteristic given the size of our documents—and it can be customized to beat the band" (Ensign 1993, 389).

Example B

"In 1986, six engineers left General Electric Co.'s research center in Research Triangle Park, N.C., to form a software company—the kind of venture the state had hoped for when it made a major effort in the '70s and '80s to copy Silicon Valley by attracting and growing high-tech companies.

"The company, Synopsys, has been a solid success and now has 210 employees and annual sales of more than $40 million.

"But Synopsys was no feather in North Carolina's cap; it was a thumb in the eye. A year after it was founded, Synopsys officials moved the company to Silicon Valley to improve their chances of success by being in a place where money, a pool of engineers and potential customers were much more readily available.

"Despite the current slump in the computer industry, Silicon Valley remains unique as a magnet and incubator for high-tech industrial development" (Mitchell 1991, 1A).

References

Apple Computer. 1995. *Instructional products.* Cupertino, CA.
Berlitz tips: Writing copy for better translations. n.d. New York: Berlitz Translation Services.
Campell, Chuck. 1996. Review of Communicating in business and technology: From psycholinguistic theory to international practice. *Technical Communication* 43.3 (August).
Coe, Marlana. 1997. Writing for other cultures: Ten problem areas. *Intercom* 44.1 (January).
Conner, Ulla. 1996. *Contrastive rhetoric: Cross-cultural aspects of second-language writing.* Cambridge, UK: Cambridge University Press.
Cyphers, Tony. 1991. *Writing for an international audience.* Unpublished paper. San Jose State University.
Ensign, Chet. 1993. SGML by evolution. *Technical Communication* 40.3 (August).
Hein, Robert G. 1991. Culture and communication. *Technical Communication* 38.1 (February).
Hoft, Nancy. 1995. *International technical communication: How to export information about high technology.* New York: Wiley.
IBM. 1996. *Annual report.* Armonk, N.Y.
Johnson & Johnson. 1996. *Annual report: Improving people's lives.* New Brunswick, N.J.
Kaplan, R. B. 1966. Cultural thought patterns in intercultural education. *Language Learning* 16. Figure 19.1 reprinted by permission of *Language Learning.*
Kleiman, Carol. 1996. A world of opportunity in the global marketplace. *San Jose Mercury News,* September 8.
Leki, Ilona. 1990. The technical editor and the non-native speaker of English. *Technical Communication* 37.2 (April).
Long, Michael D. 1994. English at large. *Technical Communication* 41.3 (August).
Lustig, Myron W. and Jolene Koester. 1996. *Intercultural competence: Interpersonal communication across cultures.* 2nd ed. New York: HarperCollins.
Merck and Co. 1996. *Annual report: Partners for healthy aging.* Whitehouse Station, N.J.
Mitchell, James. 1991. Silicon valley's firm grip. *San Jose Mercury News,* August 25.
Peters, Tom. 1992. Tips for traveling abroad: Be sensitive, patient, outgoing. *San Jose Mercury News,* November 30.
Subbiah, Mahalingam. 1992. Adding a new dimension to the teaching of audience awareness: Cultural awareness. *IEEE Transactions on Professional Communication* 35.
Synopsys. 1996. *Technical publication style guide.* Cupertino, CA.
Thrush, Emily. 1993. Bridging the gaps: Technical communication in an international and multicultural society. *Technical Communication Quarterly* 2.3 (Summer).
Winters, Elaine. 1994. Thoughts on achieving international technical communication success. *Intercom* 41.3 (March).

Glossary

abbreviation: a shortened form of a word; like a contraction, an abbreviation is formed by omitting some of the letters of the word.
abstract word: a word that indicates a general condition, quality, concept, or act.
abstracts: short (50 to 100 words) condensations of reports, articles, and proposals written for technical readers.
access aids: those elements in the whole document and on individual pages that help readers find the information they need.
action schemata: frameworks in long-term memory that capture general information about routine and recurring kinds of events.
active voice: the sentence construction in which the subject of the sentence *performs* the action of the verb.
acronym: an abbreviation formed from the first letter or letters of several words. Acronyms are pronounced as words.
adjectives: words that modify nouns and pronouns.
advance organizers: introductory material that provides an overview and thus helps readers access information by establishing a context (or schema) into which the readers can plug new information.
advanced beginners: readers of online documentation who have successfully performed a task and are willing to try new tasks (often without reading the instructions first).
adverbial connectives: adverbs that can be used, with a semicolon, to join independent clauses. They are also called conjunctive adverbs.
adverbs: words that modify verbs, adjectives, and other adverbs.
alignment: whether lines of type are "squared off," or fully justified, at the right and left margins.
alpha drafts: first drafts that are ready for developmental editing (an examination of purpose, audience, content, and organization).
alphabetical organization: a sequence arranged by the alphabet, often used for reference material.
ambiguous usage: the use of general or abstract words when specific or concrete terms are available.
antecedent: a word that comes before a pronoun and to which the pronoun refers.
apostrophe: a mark of punctuation to show possession, to show the omission of letters or numbers, or to prevent misreading of numbers or figures used as words.
applicability: appropriateness or "fit" of the document to the purpose for which it is written and to the intended reader or readers.
author-date style of citation: a system of presenting source information explained in *The Chicago Manual of Style* and used in the social sciences and by

companies. It is similar (though not identical) to the American Psychological Association (APA) style of citation.

author-page style of citation: a system of presenting source information explained by The Modern Language Association (MLA) and generally used in the humanities.

balance: a principle governing page design.

bar graphs: graphics that compare amounts and show proportional relationships with horizontal or vertical bars.

beta drafts: second drafts that are ready for technical review and copyediting (an edit of sentences, words, and punctuation).

block diagrams: graphics that provide an overview by chunking information into blocks or symbols and connecting the blocks to show relationships.

blueline: camera-ready copy.

boilerplate: those portions of written documents that writers can reuse by making only slight modifications.

bookmarks or bookmarked topics: topics marked on the computer to provide users a way to create a list of topics of special interest, which will then appear in a pull-down menu.

bottom-up editing: eliminating errors in spelling, punctuation, and capitalization so that readers can concentrate on organization and effectiveness of presentation.

boxes: thin lines that surround a block of text or a figure. Boxed copy is usually related to the main text but can also be read and understood by itself.

boxhead: the top line of a table formed by one horizontal line (called a "rule") above and one below the column heads.

brackets: punctuation marks used within sentences to indicate words that are marked off because they are not part of the sentence itself.

bulleted list: a list in which each item is set off by a typographical symbol, usually a circle or square.

buttons: specific locations in a hypertext system that permit the user to jump along a link to another node, usually by clicking a mouse or pressing a key.

callouts: word identifications of parts of a graphic.

case: the change that pronouns undergo based on their function in the sentence. The three cases in English are subjective, possessive, and objective.

cause-and-effect organization: hierarchical-sequential organization used in analysis, accident reports, and proposals.

CBE citation system: a system of presenting source information explained by The Council of Biology Editors (CBE) and commonly used in the natural sciences. Also called a *number* system of citation.

charts: graphics that show parts and their relationship to the whole on the land, sea, or in the sky. Aeronautical and nautical maps are usually called charts.

chronological organization: sequential organization in which the points are arranged on the basis of time, from first to last.

chunking: the grouping of information into manageable units like sections or paragraphs to reduce the load on short-term memory.

chunks: segments or logical units of information. Chunks can be individual chapters, single pages or screens, or even paragraphs or lists.

circle graph: another name for a pie chart.

citation of sources: (sometimes called *documentation*) information about sources that allows the reader to find the cited book, journal, or Internet site in order to pursue the topic.

classification: hierarchical organization in which items are grouped by their relationships.

clause: a group of related words that contains a subject and a verb. The subject may be understood. A short sentence is called a clause when it becomes part of a larger sentence.
clustering: the process of grouping items and connecting them with lines to show relationships.
collaboration: the participants work together.
collegial: all participants bear collective responsibility for the project.
comma-spliced sentences: another name for run-on or fused sentences.
comparison: hierarchical organization in which the organization is achieved by showing similarities or differences between two or more objects or courses of action or by explaining an unfamiliar subject by comparing it to a familiar one.
comparative report: a type of recommendation report.
competent performers: readers of online documentation who have enough experience that they begin to develop a conceptual model of how a system works.
completion report: a document that presents information gathered by research in the laboratory or field or at the close of a project.
complex sentence: a sentence containing one independent clause and one or more dependent clauses.
compound sentence: two or more independent clauses joined together.
comprehension: the reader's ability to understand information.
computer-based training: an interactive learning experience between a learner and a computer in which the computer provides the majority of the stimulus, the learner must respond, and the computer analyzes the response and provides feedback to the learner.
concrete word: a specific place, object, action, or person. Concrete words help readers understand by creating pictures in the reader's mind.
conjunctions: words that join two or more words, phrases, or clauses.
conjunctive adverbs: another word for adverbial connectives.
connotative meaning: meaning that includes associated meanings of a word that are indirect, subjective, and often emotionally loaded.
content plan: a description of the manual or individual module for an editor.
context sensitivity: a retrieval device in online documentation.
context-setting overviews: introductory material that provides a schema for readers of online documentation.
coordinators: connecting words. They are also called coordinating conjunctions. Coordinators join items of equal value or importance. Coordinators can join words, phrases, or clauses.
copyediting: the process of analyzing and correcting a document to ensure that it is well written, consistent in style, and error free.
critical reading: reading for evaluation.
cropping: marking a photograph to indicate what should be cut away before the photograph is used.
cross references: in indexing, references to preferred terminology, synonyms, and added information.
cross-sectional drawing: a graphic that shows what is under the surface. In a cross-section, the object or place has been sliced through the center, revealing the constituent layers or the internal makeup.
cultural editing: a method of checking a document for cultural bias.
culture: a set of rules and patterns shared by a given community.
cutaway drawing: a graphic that shows what is under the surface. In a cutaway drawing, part of the surface has been removed to show what is inside.

dangling modifier: a modifier at the beginning of a sentence that has a different implied subject from the real subject of the sentence.

deadwood: the use of two words that mean the same or nearly the same thing. Also called repetition.

denotative meaning: the direct, objective, and neutral meaning of a word.

dependent clause: a clause that cannot stand alone as a complete sentence. It is subordinate to an independent clause.

descriptive abstract: a short overview that describes the contents of a document.

developmental editing: the process of planning a document—working with the writer to determine the overall content, define the readers, and plan the general organization and method of presentation.

diction: the choice of words and the force and accuracy with which they are used.

direct object: (also called the object of the verb) the thing or person that receives the action of the verb.

division: hierarchical organization in which a topic is broken into its constituent parts.

document design: the process of determining the way the manual, report, proposal, or other document looks.

document style: the look of a document based on type size and style, page design, use of white space, headings, and so on.

documentation: in this book, the writing or creating of information, not the method of correctly and formally noting a source of information.

documentation of sources: see citation of sources.

documentation plan: a plan for documenting the software. It usually describes all of the manuals and other information modules planned.

electronic information: any form of communication that depends on a computer for its distribution and maintenance.

ellipses: three spaced dots that indicate where cuts have been made in quoted text.

em dash: a punctuation mark used to set off words. The name comes from the width of the em dash—about the same width as the capital M. If the em dash is not available, it can be shown by two hyphens. No spaces are used before or after the dash.

emphasis: a principle of page or screen design that helps readers who skim, scan, or search-read by providing visual cues about what is important and by influencing the direction the readers' eyes travel over the information.

emphasis by position: the placement of words at the beginning or end of the sentence to give them more weight.

ethical and legal exposures: potential problems that editors should correct. Ethical problems can include ambiguity or unclear wording, understatement, and plagiarism. Legal problems can include copyright, trademark, or patent issues; inadequate or missing warnings or cautions in instructions.

euphemism: a word that tries to soften unpleasant facts.

executable: a module in a help system that will be displayed when the user clicks on a link.

executive summary: an overview written for managers that condenses a report or article.

expert performers: readers of online documentation who have comprehensive knowledge of a process or product and are, therefore, sources of knowledge for others.

expletives: introductory words *there* and *it* which appear in the subject position of a sentence as placeholders. The real subject follows the verb.

exploded drawing: a graphic that shows the order of assembly of parts.

false implicature: writing that will lead a reader to infer something that was never directly stated.

feasibility report: a type of recommendation report that evaluates the advisability or possibility of some action.
field report: a type of completion report that provides information learned on the job or in the field.
figures: all graphics that are not tables.
filtering: a means of helping reader comprehension by creating layers of information within a hierarchy. Filtering visually identifies and differentiates various types of information, so that readers can find what they need.
final drafts: documents that are in final form and ready for proofreading (the last edit) and then release to the public.
flowcharts: graphics that use boxes and other symbols connected with arrows or lines to provide an overview, orient readers to particular steps, show how a step branches or loops back to a previous step, or indicate how many steps are involved.
font: (also called a typeface) a complete set of type of a distinctive design, for example, Helvetica or Times Roman.
footer: (or running foot) a line of type at the bottom of the page that helps readers locate themselves within the document and move quickly to another section by scanning.
form schemata: frameworks in long-term memory that help readers process and remember written documents.
formal style: writing that is highly structured, impersonal, and deliberate and thus distances the reader from the writer.
format: the arrangement of information on every page.
formatted résumé: a résumé designed to be read on hardcopy and using bolding, italics, and white space for ease of reading.
fragment: a group of words punctuated as a sentence but lacking a verb, or a subject, or an independent clause.
fused sentence: a run-on or comma-spliced sentence: that is, two or more independent clauses punctuated like a simple sentence.
Gantt charts: graphics that explain schedules or tasks by showing an overall project timeframe and its individual components.
gatekeepers: those individuals who control whether or not a document is published.
general-to-specific organization: hierarchical-sequential organization that provides the broad picture before the specifics.
genre: a *kind* of writing recognized as distinctive by writers and readers.
gerund: a verbal: that is, the *-ing* form of a verb functioning as a noun.
globalization: the process of ensuring that a product or means of communication has universal appeal and can be used anywhere without modification.
grammar: the system of rules governing the use of a language.
graphics: visual elements in documents.
graphs: graphics that condense large amounts of information and compare items; show relationships; and indicate trends, distributions, cycles, and changes over time.
grid: division of a page or screen into segments to determine the number of columns, margin size, heading placement, text placement, and so on.
hanging-indent head: a heading that protrudes into the left margin. Also called outdenting.
hardcopy: a version printed out on paper from a computer file.
header: (or running head) a line of type at the top of the page that helps readers locate themselves within the document and move quickly to another section by scanning.
help files: the online documentation that users request and read in the middle of an online task to aid them in performing that task.

hierarchical diagrams: graphics that use labeled boxes connected by rules to show the whole, the parts of that whole, and how the parts relate to one another.

hierarchy: the result of discriminating between important points and less important points and showing their relationship by means of grouping.

high-context cultures: cultures in which people have very similar ethnic, religious, and educational backgrounds and share attitudes, feelings, and values.

highlighting techniques: in document design, choices of boldface, all capitals, italics, or underlining to emphasize text. Also called type style and emphasis.

home location or home base: in online documentation, the starting point for reading, often the main menu.

hypermedia: modules of online documentation that can contain sound, animation, and video as well as text and graphics.

hypertext: topics of text and graphics in online documentation (also called articles, modules, or nodes) that are stored electronically and accessed by links between the topics.

icons: visual representations of ideas, things, or actions used to provide shortcuts in both online documentation and hardcopy.

illustrations: graphics that are either examples or pictorial representations. In common use, the term *illustrations* usually refers to presentations like line drawings or photographs, but not to graphs, flowcharts, or tables.

IMRAD structure: in a scientific report, organization by introduction, materials and methods, results, and discussion.

independent clause: a clause that can stand alone as a complete sentence.

indirect object: the thing or person that indirectly receives the action of the verb. An indirect object answers the questions *to what? for what?* or *to whom? for whom?*

infinitive: a verbal: that is, *to* plus a verb. Infinitives can be used as nouns, adjectives, or adverbs.

inflection: a pronoun's shift in form depending on its use and position.

informal style: writing that is casual and conversational.

informational pathways: another name for links in online documentation.

informative abstract: a short condensation of a document that gives the purpose, organization, conclusions, and recommendations.

initialism: an abbreviation formed from the first letter of each word. The letters are pronounced individually.

intercultural communication: interaction among people from different cultures.

international communication: interaction among people from different nations.

internationalization: the process of re-engineering an information product so it can be easily localized for export to any country in the world.

Internet: a global system of public and private computer networks that enable universities, governments, businesses, and consumers to share files, post notices, and converse via computers, modems, and phone lines.

intranet: a private computer network within a company or organization that promotes daily business, file sharing, and collaborative work.

jargon: technical words used in a specific profession.

justification: the technique of squared-off lines of type at the right and left margins.

jumps: another name for links in online documentation.

kerning: the adjustment of space between selected pairs of letters.

key words: words—primarily nouns—that tell what the important information is.

landscape style of presentation: lengthwise on the page.

languages: ways of communicating, with people of the same geographical area or the same cultural tradition usually having a language in common.

leader line: a line that points directly to the part indicated for labeling a graphic.

leading: the space above and below a line of type.
levels of editing: (or levels-of-edit) systems that define how thoroughly a document or series of screens is edited.
line drawings: graphics that can be surface, three-dimensional, or shaded perspective, showing an object or place as it is or will be when built.
line graphs: graphics that are plotted on a horizontal and vertical axis. Usually variables of distance, time, load, stress, or voltage are placed on the horizontal axis, and variables of money, temperature, current, or strain on the vertical axis.
linking or condition verbs: *is, was,* and other forms of *be; appear, become, get, look, remain, seem, feel, taste, smell.* Linking verbs join the subject to a subject complement.
links: (also called informational pathways or jumps) connectors of topics of information in online documentation. Reader-users interact with the documentation by choosing the links they want.
listing: placement of items in vertical arrangement. Listing takes advantage of the defining quality of white space and also (if the list items are parallel) of the patterning of similar syntactic structures.
localization: the process of adapting a document to a specific cultural context or area, thus making it seem local.
long modifier string: a phrase of more than three modifiers.
long-term memory: the part of memory that stores information and experiences in schemata (frameworks or structures of related knowledge).
low-context cultures: cultures in which regional, ethnic, religious, and educational backgrounds differ, and people have little shared knowledge and few shared attitudes, feelings, and beliefs.
lower-case type: type with no capital letters.
manual: a document that usually gives both information and instructions or procedures.
maps: graphics that show parts and their relationship to the whole, usually on the land.
medium: the type or form of presentation of information, either online documentation (a computer screen, a Web page, a video, a film, animation) or hardcopy (a book, a manual, a report, a series of slides, a journal article).
milestone chart: a graphic that explains schedules of tasks by showing an overall project timeframe and its individual components.
misplaced modifier: a modifier that appears by its position in a sentence to modify the wrong thing.
modifiers: words, phrases, or clauses that limit or describe another word. Some modifiers act as adjectives; they modify nouns or pronouns. Other modifiers act as adverbs; they modify verbs, adjectives, or other adverbs.
module: in online documentation, another name for a topic or node.
modular design: page design in which each step appears on a single screen or page or on two facing pages.
most-to-least-important organization: hierarchical-sequential organization in which the items of primary importance are discussed first.
navigational aids: the specific elements on a page or screen that help readers see the relationship among units in the document, locate themselves within the document, and move quickly to another section or information module.
node: in online documentation, another name for a topic or module.
nominalization: the excess use of nouns made from verbs or adjectives.
nonrestrictive clause: a relative clause that should be surrounded by commas because it functions as a secondary identifier.
noun clause: a dependent clause used as a noun (the subject or object) in a sentence.

novices: readers of online documentation who have no previous experience, are often worried about their ability to succeed, want simply to get a task done, don't know how to recover from mistakes, and prefer only one option for performance.

number system of citation: a system of presenting source information in the sciences, for example, the CBE system.

object of the verb: another name for the direct object.

online documentation: the use of the computer as a communication medium instead of as a number cruncher or word processor.

organizational diagram: a graphic that provides an overview of relationships by simultaneously showing the whole, the parts of that whole, and how the parts relate to one another.

organizational indicators: words in headings and titles that tell the reader the method of organization and the purpose of the section, for example, "instructions for" or "descriptions of."

orphans: lone words at the bottom of a page or column.

outdenting: a heading that protrudes into the left margin, also called a hanging-indent head.

page design and page layout: the look of each page based on spacing, specifications of type, location of graphics, and so on.

paragraph spacing: the white space between paragraphs.

parallelism: the use of similar syntactic structures to help readers understand organization of similar items.

partial web: in online documentation, one way that topics can be linked. Most online documents use a partial web, with links to only a few other topics.

passive voice: the sentence construction in which the subject of the sentence *receives* the action of the verb.

past participle: a verbal: that is, a verb form ending in *-ed*. When combined with a subject and auxiliary verb, it is a verb. Otherwise, it is an adjective.

phrase: a group of related words that does not contain a subject and verb.

pictograph: a variation of a bar graph that uses a single simple drawing to represent one unit and then repeats that unit to make up a bar.

pie chart: (also called a circle graph) a graphic that uses a segmented circle to show the relationship of a whole and its parts, especially the percentage composition of a whole.

point: a measurement of type, with 72 points to the inch.

pompous language: language that calls attention to itself and the cleverness of the writer.

portfolio: a collection of editing and writing samples organized and annotated so they can be read and assimilated quickly by an interviewer.

portrait style: placement vertically on the page.

predicate: the verb plus related words that complete the meaning of a sentence.

preposition: a word that joins with a noun or pronoun to form a prepositional phrase.

present participle: a verbal: that is, a verb form ending in *-ing*. When combined with a subject and auxiliary verb, it is a verb. Otherwise it is an adjective or noun.

primary identifier: a relative clause necessary to establish the specific identity of the noun it follows. A primary identifier is called *restrictive*, and no commas surround it.

pro-and-con organization: hierarchical organization that presents information by grouping advantages and disadvantages, presenting one group at a time, and follows the presentation by conclusions and recommendations.

problem-solution organization: hierarchical-sequential organization in which the presentation of the problem precedes an explanation of the solution.

production edit: the last editing pass before a document is released. Includes editing of graphics and design.

proficient performers: readers of online documentation who are experienced and want to understand how the entire system works.
pronoun: a word that can take the place of a noun.
pronoun-antecedent agreement: a pronoun must agree in both number and gender with its antecedent (the noun that comes before it and which it replaces).
pronoun reference: the relationship between a pronoun and its antecedent.
proofreading: the final sentence-level edit a document or screen receives before it is produced.
proposal: a written offer to solve a problem or provide a service.
pull quotes: key statements pulled from the text and set in larger type or boxed.
purpose: the reason for writing the document.
question-answer organization: hierarchical-sequential organization in which questions and answers provide the structure.
queuing: a layering technique that enhances comprehension by showing the reader the relationship of items of information.
ragged-right text: lines of type that are unjustified on the right margin.
readability: the extent to which a document's meaning can be easily and quickly comprehended for an intended purpose by an intended reader.
receptive reading: reading for thorough comprehension.
recommendation report: a report that either examines alternative products or courses of action or the feasibility of action.
redundancy: wordiness caused by unnecessary repetition.
relative clause: a clause functioning as an adjective that stands after the word it modifies.
relative pronoun: a pronoun that begins a relative clause.
restrictive clause: a relative clause that is a primary identifier. No commas surround it.
rhetorical editing: substantive editing that usually includes two parts: review of technical content and review of emphasis, logic, and organization.
rules: in page design, horizontal or vertical lines that can emphasize or link two related columns or isolate the text below or next to them, thus guiding the reader's eye.
run-on: two or more sentences punctuated as one sentence. Other names for run-ons are comma-spliced or fused sentences.
running foot: (or footer) a line of type at the bottom of the page that helps readers locate themselves within the document and move quickly to another section by scanning.
running head: (or header) a single line of type at the top of the page that helps readers locate themselves within the document and move quickly to another section by scanning.
sans serif fonts: type without serifs (the short cross-strokes at the top and bottom of a letter).
scannable résumé: a résumé designed to be scanned into a data base. It emphasizes key terms.
scanning: reading quickly to find specific items of information.
scenarios: hypothesized situations in which a human agent performs a particular action.
schemata (singular *schema*): the structures or frameworks in which information and experiences are stored in long-term memory.
schematic diagram: a graphic that shows specialists, technicians, and operators details in manuals of assembly, test, and maintenance by showing the *logical* connections and current and signal flow.
screen snapshots, screen shots, screen captures, or screen dumps: names for literal representations of what appears on the user's computer screen or in a window on that screen.

scope: a document's intended coverage.
screens: the portions of a page set off by pale dots over which type is printed.
search reading: scanning with attention to the meaning of specific items.
secondary identifier: a nonrestrictive clause that merely adds extra information to the noun that it describes.
***see also* reference:** in indexing, a reference that sends the reader to additional or related information in another entry.
***see* reference:** in indexing, a reference that sends the reader to a synonym or the standard (or preferred) company term.
segmented bar graph: a graphic in which the segments of a whole are stacked vertically in a single bar starting with the largest segment at the bottom.
semiformal style: writing in which the writer uses the pronouns *I* and *you*, uses active voice and short sentences whenever possible, but structures the document tightly and follows the conventions of Standard Written English.
sentence: an independent unit of words that contains a subject and a verb and closes with a mark of punctuation.
sentence-level editing: (also called copyediting or proofreading) the process of examining and correcting a document at the individual sentence and word level.
sequence: a principle of page or screen design that helps readers who skim, scan, or search-read by providing visual cues about the relationship of items.
sequential organization: a method of organization in which topics of information follow one another in order: by alphabet, by numerical sequence, or by time.
serif fonts: type with serifs, the short cross-strokes or extensions on the top and bottom of a letter.
sexist language: language that assigns roles or characteristics to people on the basis of gender.
short-term memory: working memory that can hold seven or fewer items at one time.
sidebar: boxed explanatory material on a page.
single-source authoring: a method of using source information for multiple outputs.
skimming: reading for the general drift of the passage.
softcopy: paper-based information that has been placed on the computer.
spatial organization: organization in which the points are arranged as they appear physically, either in relationship to one another or the surroundings.
squinting modifier: a modifier that could modify either the preceding word or the following word; it's unclear which.
stub: in a table, the first column on the left, which contains the list of items.
style: the total effect achieved in writing by word choice, sentence length, types of sentences, sentence patterns, rhythm, sensory images, and tone.
subject: what or who is talked about in a sentence.
subject complement: a word following a linking or condition verb that renames or describes the subject.
subject-matter expert: any person who has specialized knowledge that he or she wants or needs to share with others through the written word.
subject-verb agreement: the required match between a subject and verb in number and person.
subordinator: a conjunction that can begin a dependent clause.
substantive editing: (or rhetorical) editing that usually includes two parts: review of technical content and review of emphasis, logic, and organization.
summary table: a preview of information that tells readers where to go either in an "if-then" table or in a question-answer format.
syntax: the structure or pattern of the words in a sentence, clause, or phrase.

table: a graphic that contains ordered columns of numbers and/or words.
technical reviewers: subject-matter experts who review a document's content.
text expansion: the increase in physical space required in translation from a source language to a target language. The average expansion across all languages is 30 percent.
thesis: the main point of the document.
thumbnail proofs: pages reduced in size and on the same sheet of paper.
top-down editing: starting to edit with content, organization, and presentation before moving to details at the sentence and word level.
topic: (also called a block, a node, an article, or a module) in online documentation, a unit of text, graphics, or animation that is on one subject, expresses one thought, or answers one question.
topic name: in online documentation, a screen orienter that tells the reader the contents of a particular screen.
translation: the use of verbal signs to understand the verbal signs of another language.
typeface: another name for font.
type style and emphasis: in page design, the highlighting techniques of boldface, all capitals, italics, and underlining.
undefined page entries: in indexing, page numbers listed after an index term without any further identification.
unity: the design principle that says that graphics, sections of text, headings, and subheadings must look like they belong together.
unjustified type: lines of type that are uneven on the right margin (also called ragged right).
upper-case type: capital letters.
usability: the success of the document in enabling the readers to carry out a task.
usage: the customary manner in which words are used. Usage can change over the course of time or from one locale to another in which the same language is spoken.
vague usage: words that are unclear to the reader.
verb: a word or words that show action, occurrence, state of being, or condition.
verbal: a word derived from a verb but not functioning as a verb, though a verbal can be part of a verb phrase.
verbal devices: techniques like headings and subheadings, parallelism, and access words that help readers understand organization.
verbalizing the action: finding the action in a sentence and then lodging that action in the verb.
visual devices: design features such as chunking with white space, showing hierarchy with type size and style, listing, and using icons to help readers understand organization.
visuals: the projected pictures, graphs, and so on that accompany a speech, or the picture elements (as opposed to the sound elements) in a film or video.
vogue language: trendy language that begins within one industry and then moves out into general circulation.
voice: the verb form that shows whether the subject is acting or acted upon.
Web: (World Wide Web) a subset of the Internet that promotes interaction between the user and the content and that shares a format called hypertext markup language (HTML).
widows: lone words at the top of a page or column.
wiring diagram: a graphic that shows specialists, technicians, and operators details in manuals of assembly, test, and maintenance. Wiring diagrams show the actual point-to-point connections.
word spacing: the amount of space between words.

Appendix

CHAPTER 6

Exercise 1 Proofreading by Comparing a Proof Copy to the Original Master

Proof Copy To Be Corrected

Cujus genas ac faciem, omneaque undique totius venarandi capatis superficiem, pannus subtilissimus operienda obtegit, qui ita omnibus membris subpositis districtissima sollictudinis arte cohaesit, quasi casariei, pelli, temporibus, ac barbae, conglutinatus sit. Qui ex nula parte, alicujis arte, altius aliquantulm a cute vel carne elaveri, divelli, vel sub-rigi, potuit.

Exercise 2 Proofreading for Accuracy

1. The contract called for reports four times a year: March 31, June 3, September 30, and December 31.
3. Send your response directly to Senator Ardis McManus at the state capital building in Hartford, CN.
5. Because Winnipeg is adjacent to the Wisconsin border, the machine parts can be transferred to barges at Duluth.
7. This license is issued solely as a license to drive a motor vehicle in this state; it does not establish eligibility for employment, voter registration, or public benefits.
9. The price of a safety seat should be mainly a secondary consideration.
11. The deficits were absolutely unexceptable.

Exercise 6

Topic: THE CORPORATE LIFE
Source: Power, Ruth M. "Who Needs a Technical Editor?" IEEE Transactions on Professional Communication PC-24 NO. 3. (September 1981): 139–140.

Ensuring quality control in written communication is the job of a technical editor. With a better understanding of the responsibilities of a technical editor, an engineering manager might want to hire either a full-time or part-time technical editor. The 1981 article "Who Needs a Technical Editor?" provides information to engineering managers about the responsibilities, qualifications for technical editors.

Power begins her article with a scenario in an engineering office. Old and young engineers believe that they do not need technical editors. The engineers believe that their writing and editing skills are adequate. However, when the manager receives poorly written reports from engineers, he or she must spend hours to correct simple grammatical and typographical errors. The manager's conclusion: Engineers can't write. However, Power says that engineers are not the only ones who have problems writing correctly and clearly. Government workers and lawyers also use too much jargon in their writing. Power says that a technical editor can solve many problems with unclear and incorrect writing.

According to Power, a technical editor should be able to perform a variety of job functions such as organizing data, checking references, writing copy, organizing layouts, copy editing, proofreading, and making sure that the whole project runs smoothly. Power concludes that the technical editor's job is one of quality control for written communication.

After providing her definition of a technical editor, Power breaks the rest of her article into three topics: editor's functions, qualifications, and hiring options (for employers).

A technical editor's function should start at the early stages of development on a written document. Although he or she is not actually writing the document, the editor can help authors start writing the document by giving suggestions on style and organization. The editor can provide continuing guidance as the project develops. Power says that asking an editor to come in at the last minute and "fix-up" a paper would be frustrating to the editor and the writer. To give engineering managers a better idea of what technical editors do all day, Power lists 21 jobs (see Table 1) that editors are performing in industry today. The author also gives several scenarios where technical editors can be helpful. The topics of the scenarios include technical editing of a journal article, assistance in preparation of a technical article, individual coaching, assistance in preparing a speech, and assistance with translations.

When listing the qualifications of technical editors, Power uses the Society for Technical Communication as a source of information. She lists characteristics of technical editors that include gender, age, education, work experience, average salary, and technical background.

In her final topic, hiring options, Power lists several places where employers can start their search for a technical editor. Power cautions the engineering mangers that technical editors may not be the solution to all their written communication problems. If the manager, engineers, and technical editor cannot work as a team, the editor will be very little help. The editor may even cause more controversy because he or she might have an opinion totally different than the manager and engineers. Power concludes by answering the title of her article—"Who Needs a Technical Editor?" The answer to the question is "anyone who is responsible for communicating technical information and who needs help in getting projects organized, following up on details, coordinating groups of writers, untangling convoluted sentences, suggesting just the right word, settling questions of grammar, or finding that one last typo."

Although this article was intended for engineering managers, I found this document very informative and useful for technical writers and editors. Power says that in the last 10 years industry has recognized the need for better quality control in written documents. She presents many strong arguments to managers to hire technical editors. The author gives a very clear definition of a technical editor. However, there was other information I found valuable to me as a writer and editor.

If the engineering manager decides he or she needs a technical editor, Power lists several places (telephone directories, colleges, or the Society for Technical Communication) where the manager can begin searching for an editor. I found this information very valuable. When I begin looking for a job, I will have some idea of where managers are looking for potential employees.

When I begin looking for a job, I want to be sure my qualifications are what the manager is expecting. Power lists qualifications and probable backgrounds of technical editors. I would like to know what my employer expects from me on the job. Also, it would be nice to know what type of background my competition (other people who are applying for the job) might have.

After I find a job, a salary must be determined. Power lists several different salaries with corresponding amounts of experience. This information is nice to have also. When I find a job, I would like to know what my peers are earning, so I have some idea of how much the job might pay.

The author stresses one point about technical editors that I like. She reminds managers that editors will not solve all the problems of projects. Editors may cause even more controversy among themselves, engineers, and managers during a project. She says that managers, engineers, and editors must work together as a team for a project to be successful.

The most valuable part of Power's article was the list managers might use to determine if they need a technical editor. After reading this list, I have a better idea of what managers are going to expect from me in the "real world." This is very valuable information. What a student learns in the classroom is sometimes very different from what management wants in the real world.

TABLE 1
Do You Need a Technical Editor?
Use This Checklist to Write a Job Description

Prepare or supervise preparation of reports, manuals, letters, proposals, memorandums, forms, procedures, scripts, articles, charts, and tables.
Work with engineers to achieve effective written and oral communications.
Design and present in-house writing courses.
Coach individuals in effective writing techniques.
Edit and proofread prepared copy at the request of an author.
Provide authors with outlining and editing assistance.
Write or help authors write abstracts, introductions, summaries, and articles for publication.
Review, edit, and rewrite as necessary, in cooperation with the authors, all technical copy.
Prepare articles for publication according to the editorial requirements of specific journals.
Prepare and maintain a style manual specifically designed for the needs of the organization.
Maintain a reference library to help referee differences of opinion on grammar, spelling, and punctuation.
Help authors prepare speeches, select visual aids, and rehearse presentations.
Review and edit all technical documents for general readability, consistency of style, logical flow, and correctness of spelling, punctuation, and grammar.
Advise on graphics, printing methods, and binding practices.
Supervise filing of technical documents.
Write procedures for a word processing area.
Supervise and manage a word processing area.
Maintain overall quality of materials processed through a word processing area, editing and proofreading as required.
Assist in preparation of materials for a source documentation library.
Prepare a publications guide to ensure uniformity of layout, typing, and printing.
Provide timely and effective services as required to save engineering time.

© 1981 IEEE

APPENDIX **451**

CHAPTER 8

Exercise 1 Editing for Capitalization and Spelling

1. After chapter 10 was reviewed for technical accuracy, the Technical Writing Manager asked senior editor P. E. Barazza to copyedit the document.
3. The Northeast suffered its worst storm of the season, and the government offices canceled their office hours for two days.
5. The gauge was in error, but an accurate reading was not necessary to the analog readout.
7. The principal agency involved is the Department of Agriculture; however, the Budget Bureau is also peripherally concerned.
9. The analog-to-digital converter was a good use of the current technology's capability.

Exercise 2 Revising Numbers for Clarity and Consistency

1. In the Northeast, water temperatures near the surface during the spring and summer range from 50 degrees F to 70 degrees F. Bottom temperatures at 100 feet range from 8.9 degrees C to 12.2 degrees C. *(Convert all measurements to either Fahrenheit or Celsius, or both.)*
3. Visibility in Great Lakes waters ranges from 100 feet in Lake Superior to less than 1 foot in Lake Erie.
5. The singular hose regulator may freeze in the free-flow position after 20 to 30 minutes exposure in polar waters.
7. The phylum *Coelenterata* includes about 10,000 species. Only 70 species have been involved in injuries to humans.
9. The operation with the atmospheric diving system began at 8:00 A.M. and lasted for 5 hours and 59 minutes below the ice with minimal discomfort for the operator. By 2:00 p.m. (small caps), the operator was back on the surface. With conventional diving methods, the decompression obligation would have been more than 8 days.

CHAPTER 9 A REVIEW OF BASIC GRAMMAR

Answers to Diagnostic Quiz on Grammar

1. Yes, this is a sentence because it contains a subject and a verb, is a complete thought, and ends with a period.
2. companies

3. those eight companies, which require writers to use sophisticated software publishing tools
4. are. Yes, there is also the verb *require* in a dependent clause.
5. a linking or state-of-being verb
6. are leaders in the industry
7. predicate noun or predicate nominative
8. to use sophisticated software publishing tools: infinitive phrase; in the industry: prepositional phrase
9. those companies are leaders in the industry: independent clause; which require writers to use sophisticated software publishing tools: dependent or subordinate clause
10. to use: infinitive

Score Yourself

8–10 correct. You know the basics; move on to the next chapter.

6–7 correct. A short review would help you.

5 or fewer. Study Chapter 9 before moving on.

Exercise 1   Recognizing Subjects and Verbs

1. Hydraulic mining / was invented by Edward Mattison at a gold mine in Nevada County, California, in 1853.
3. The strong water jet / washed gold-bearing gravel into his sluices.
5. Sixty million gallons of water / were used daily at the Malakoff mine alone.
7. The canal and reservoir / cost $750,000.
9. Floods / occurred downstream and destroyed both crops and fields.

Exercise 2   Recognizing Subjects and Verbs

1. The belt sander / is the best all-around sander for a woodworker.
3. However, belt sanders / can be heavy and awkward on edges or small pieces.
5. The stand / accepts 3-inch and 4-inch belt sanders.
7. The stand / can be used with the belt running vertically.
9. The stand's worktable / is ribbed for easy collection of sawdust between the ribs.

Exercise 3   Recognizing Subjects, Verbs, Direct Objects, and Subject Complements

1. In 1851, Henry Burden built the largest water wheel in America. [D.O.]
3. It produced 278 horsepower [D.O.]; in theory, it could produce 1000 horsepower [D.O.].

APPENDIX **453**

5. The iron <u>spokes</u> <u><u>were</u></u> radially laced; [S.C.] they <u><u>were</u></u> [S.C.] not crossed.
7. A segmental <u>gear</u> at the rim <u><u>supplied</u></u> the drive. [D.O.]
9. Burden's <u>machinery</u> <u><u>was</u></u> a wonder [S.C.] of the age.

CHAPTER 10

Exercise 1   Sentence Combining: Writing Compound Sentences

Your choice of the most effective sentence may differ. The important thing is that you can explain *why* your choice is effective.

1. * . . . United States, but most city dwellers . . .
   . . . United States; most city dwellers . . .
   . . . United States; however, most city dwellers . . .
   But or *however* set up the contrast between the two clauses.
3. . . . the wire, and they must . . .
   * . . . the wire; they must . . .
   . . . the wire; in addition, they must . . .
   The two clauses provide a simple explanation.
5. * . . . one method, but Joseph Glidden . . .
   . . . one method; Joseph Glidden . . .
   . . . one method; nevertheless, Joseph Glidden . . .
   But or *nevertheless* establish the contrast. In a continuous paragraph, you would want to vary the means of joining the clauses.
7. . . . the wire, and a second . . .
   * . . . the wire; a second wire then . . .
   OR . . . the wire; then, a second wire . . .
   The word *then* acts as an adverbial connective. It can be placed right after the semicolon or later in the sentence.
9. . . . fence posts, for these woods . . .
   . . . fence posts; these woods . . .
   . . . fence posts; indeed, these woods . . .
   The choice here depends on the emphasis desired on the second clause.

Exercise 2   Recognizing and Punctuating Compound Sentences

1. In software technology, the <u>word</u> *object-oriented* <u>can be used</u> to refer to computer interfaces; it <u>can</u> also <u>be used</u> to refer to computer programming.   *Semicolon joins two independent clauses.*

APPENDIX

3. For example, a picture or <u>icon</u> <u>can represent</u> a report∧and the <u>user</u> <u>can print</u> the report by dragging the icon and dropping it on an icon of a printer. *Comma plus "and" joins two independent clauses.*

5. In computer programming, *object-oriented*, however, refers to a particular way of organizing the internal elements of a program. *Simple sentence. Commas surround the interrupter word "however."*

7. The <u>stock</u> <u>represents</u> the computer's data, and the <u>robot</u> <u>represents</u> the operative force manipulating the data. *Comma plus "and" joins two independent clauses.*

9. The robot's <u>tasks</u> <u>are</u> complex, so it's difficult to revise all the instructions to accommodate a new situation. *Comma plus "so" joins two independent clauses.*

Exercise 4 Correcting Fragments and Run-Ons

1. Recording/playing <u>time</u> <u>is</u> the major difference between the two formats; a maximum of 8 hours for VHS compared with 5 for Beta. *(dash or comma)*

3. Please (you) <u>send</u> me a technical report about the EXXXTRA ski; a report dealing with the design of your ski, the materials used in building the ski, and how your skis compare to your competitors' skis.

5. However, the <u>IRS</u> <u>has caused</u> problems with the law-abiding citizen by auditing a percentage of tax returns every year; this <u>creates</u> problems of finding records and taking time off to correct any problems found; in general, the <u>IRS</u> <u>invades</u> an individual's privacy.

7. All <u>recorders</u> <u>are equipped</u> with a carrying handle. ~~Which~~ It <u>lies</u> flat against the unit for listening or storage purposes and <u>swings</u> up for easy transport.

Exercise 6 Sentence Combining: Writing Complex Sentences with Subordinators Showing Cause or Logic

1. Chemically, Silly®Putty is a liquid, *even though* it resembles a solid.

3. Silly®Putty will shatter *if* it is struck with a hammer.

5. *Although* The early Silly®Putty was sticky, in 1960 it was reformulated to be nonsticky.

Exercise 8 Sentence Combining: Writing Complex Sentences with Subordinators Showing Time or Place

1. Silly®Putty was discovered by accident in 1945, *when* James Wright dropped boric acid into a test tube containing silicone oil.

3. *After* Wright examined the resultant compound, he discovered its bouncing characteristics.

5. *Once* The catalog contained a page of adult toys including Silly®Putty, Silly®Putty's popularity grew.

7. Peter Hodgson borrowed $147 to market the first Silly®Putty; ^while Lack of funds forced him to use plastic eggs to hold the product.

Exercise 11 Recognizing Relative Clauses and Relative Pronouns

1. Scientists searching for the event (that) killed the dinosaurs think (that they have found it.) — noun clause
3. The researchers (who) worked on the study included geologists and paleontologists.
5. Using sophisticated new tools, the scientists examined a crater (that) was formed in Mexico's Yucatan Peninsula.
7. A team from NASA, to (whom the information was given,) used satellite images to map the crater. — noun clause
9. Geologist Walter Alvarez, (whose) theory seems to be supported by the crater, speculates (that the dust and smoke from the impact blocked the sun and led to climate changes.) — noun clause

Exercise 13 Recognizing and Punctuating Restrictive and Nonrestrictive Clauses

1. Word processing is available with computer systems ~~which~~ that range in size from personal microcomputers to huge commercial installations. — restrictive
3. Word processing provides a mechanism to prepare material that would otherwise be typewritten. — restrictive
5. Information entered on the keyboard is not put on the paper but is collected and put into memory, the word processing component that holds that particular data. — restrictive
7. Once a text has been corrected, the user can strike a function key, which directs the word processor to print that text. — nonrestrictive
9. To show the characters that are to be removed, a cursor is supplied on the screen. — restrictive

CHAPTER 11

Exercise 1 Abbreviations, Acronyms, and Initialisms

1. Alternative fuels for energy are making a "surprisingly significant" contribution to US energy needs, according to R.S. Martin.
3. However, S. Freedman of the Gas Research Institute, speaking at an energy conference at MIT, pointed out the relatively high cost and lack of proven large-scale capacity of alternative fuels.

5. In 1981, it cost the first Luz solar plant 24 cents to produce 1 kilowatt hour (Kwh) of electricity. By 2000, experts anticipate costs will be down to 6 cents per Kwh.
7. The mirrors focus the sun's rays on oil-filled glass tubes, heating the oil to 735 deg and turning water into steam.
9. There are more than 17,000 wind turbines in the USA with a total capacity of 1,500 mw, as much as a large nuclear plant.

Exercise 2 Apostrophes

1. Cartographers' greatest challenge is to map the round Earth on a flat surface.
3. The four mapmakers' projections were produced over a period of 400 years of cartography.
5. In Robinson's projection, Canada's, Greenland's, and the former Soviet Union's shapes are fairly true to relative size.
7. All regions of the Earth are depicted in correct relative size in Mallweide's 1805 map, but the higher latitudes show compression, elongation, or warping.
9. Gerardus Mercator's map exaggerates the size of landmasses in the high latitudes. For example, Alaska and Brazil look about the same size, while they're vastly different: Brazil is six times larger.
11. The projection's based on an ellipse, with lines of longitude curving toward the poles. However, the latitude lines are straight, evenly divided by the longitude lines, as on the globe's surface.

Exercise 3 Editing for Colons

1. As the nation's central bank, the Federal Reserve System's current goal is the following: to balance expanding economic growth against increasing pressures of inflation. *To introduce an explanation of the first word group.*
3. Congress established the Federal Reserve System after the Panic of 1907 to create a pool of money: a reserve from which banks could borrow in a crisis. *To introduce an explanation of the first word group.*
5. Soon the directors of the Fed learned that the reserves could be used to control banking activity and to influence the business cycle. *Do not use colon to separate the verb from its object.*
7. The sellers of those securities begin the following cycle: they deposit their proceeds in banks, increasing bank reserves. The banks can then lend more money, which stimulates demand and raises prices. *To introduce an explanation of the first word group. A complete sentence should be capitalized.*
9. Alan Greenspan became the chairman of the Fed in 1987: all seven members of the Board of Governors serve 14-year terms. *Start a new sentence for a new idea.*

APPENDIX **457**

Exercise 4 Introductory and Interrupter Words and Phrases

1. Arbitration, a method of settling labor disputes, is a substitute for costly and time-consuming litigation.
3. In fact, until Executive Order 10988, union recognition and collective bargaining were not found in the federal service.
5. On rare occasions, courts have returned cases to an arbitrator for reconsideration.
7. To bring a dispute to an arbitrator, the union must "make a case"; in other words, show probable cause that the contract has been violated.
9. The documentation must include names, dates, times, events, and witnesses. In addition, the union must show specific violation of the agreement and the specific relief sought.

Exercise 5 Commas and Semicolons in Series

1. The Internet can connect a user to electronic mail, news, or libraries of information.
3. A user needs a personal computer, (PC), and a modem or connection to a local area network, (LAN), and a communications program.
5. The account identifies the user to the computer system in two parts: the username, userid, or loginname, and the password.
7. The terms *loginname*, *userid*, and *username* are all common names for the user's identity.
9. With communication established, you can log in by typing your loginname/ and pressing the ENTER key.

Exercise 6 Editing for Dashes, Commas, Parentheses, and Brackets

1. The candidate company for the merger is the Ocean Containers Group—consisting of Ocean Containers, Inc. and Oceanco. Dash for summary.
3. Ocean Containers owns 200,000 cargo containers—it specializes in refrigerated containers—and 25 container ships. Use em dash to emphasize information.
5. These facilities—in England, Chicago, and Singapore—enable the company to market its products easily to more than 70 countries. Em dashes emphasize information.
7. In addition, Ocean Containers has operations not related to the transportation industry: hotels, a train service, a large condominium complex, and undeveloped real estate. Use colon to introduce a list.

Exercise 7 Editing for Hyphens

1. As U.S. companies do increasing business with international clients, writers must consider second-language readers of documentation. Adjective plus noun as a modifier.

3. Management is concerned that an overly optimistic forecast is contained in the program planning guide. *No hyphen with adverb ending in -ly. Use hyphen with adjective plus a present participle as a modifier.*
5. Thirty-five contributors are listed in the program notes. *Use hyphen in numbers when written as words.*
7. No Salt®, on the other hand, contains potassium chloride crystals and potassium bitartrate. *No hyphen with names of chemicals.*
9. Use of the blood-pressure gauge was explained in a five-page brochure. *Use hyphen with adjective plus noun as a modifier. Use hyphen with adjective plus noun as modifier.*

Exercise 11 Quotation Marks

1. The following is an excerpt from a larger report titled "Laser Cutting of Sheet Metals."
3. "Because the output wavelength of the Nd:YAG laser is 1,064 nm," Hudson continued, "it can produce narrower widths than the CO_2 laser, which operates at 10,600 nm."
5. According to Hudson, "The cutting efficiency of a laser can be improved by applying a jet of oxygen through a nozzle located coaxial with the beam."
7. "When used with a reactive gas such as oxygen, both CO_2 and Nd:YAG lasers can produce burr-free cuts with a minimum heat-affected zone."
9. "The Computer Numerical Control (CNC) unit controls the table's movement, the type of cut, and the depth, width, and angle of the cut," Hudson explained.

CHAPTER 12

Exercise 2 Checking for Accurate Diction

The following answers are examples; your changes may differ. Be sure you can explain why the words are inaccurate.

1. Work assignments not included in the listed shift hours will be established on an individual basis [or individually] with a supervisor.
3. Our freedom decreases as the price of gas increases. (The sentence should also indicate what kind of freedom—freedom to travel, for example.)
5. Monitor the record levels and tone adjustments to make sure they stay within acceptable bounds.
7. Fewer than 100 students were eating lunch in the dining room at the time of the explosion.
9. She assumed the mantle of leadership at XYZ Corporation, and her subsequent actions proved she was a good choice.

Exercise 6 Removing Euphemisms and Abstract Language

The following sentences illustrate possible solutions.

1. We must investigate the procedure for acquiring spare parts to improve our operations.
3. The two heads of state drafted a statement that the problems of nuclear arms and space-based defense systems will be resolved as a single issue.
5. The senator said he would not be present.

Exercise 8   Choosing Appropriate Nonsexist Revisions

The following is one possible revision.

1. Alcohol intoxication, whether due to an acute overdose or to prolonged abuse, is treated as follows:
    a. If the person is sleeping quietly, the face color is normal, breathing is normal, and pulse is regular, no immediate first aid is necessary.
    b. If the person shows such signs of shock as cold and clammy skin, rapid and thready pulse, and abnormal breathing, or does not respond at all, obtain medical aid immediately.
    c. Maintain an open airway, give artificial respiration if indicated, and maintain body heat.
    d. If the victim is unconscious, place him or her in the coma position (Fig. 58) so that secretions may drool from the mouth. This position will usually allow for good respiration.
    e. Remember that an intoxicated person may be violent and obstreperous and will need careful handling to prevent harm to self and others.

    The alcoholic should be encouraged to seek help from Alcoholics Anonymous or from a drug abuse treatment center.

CHAPTER 13

Exercise 2   Correcting Subject-Verb Agreement and Other Number Problems

1. The cost of the ovens was rated as follows: under $300, 5 points; $300–$400, 3 points; over $400, 1 point.
3. The link between the hardware (the physical components of a computer system) and the software (programs that are composed of high-level language instructions designed to solve a problem or perform a certain task) is the operating system. (Sentence is grammatically correct.)

APPENDIX

5. On all VCRs there exist certain standard features, which vary little from model to model.
7. The findings of this report indicate the advantages and disadvantages of the personal computer.
9. In the previous section of this report, each of the four instruments in question was compared relative to each other in a number of different areas of interest or criteria.
11. The possibility of TSC machining its own tools in the future exists; hence, a need for N/C tools and CAM is established.

Exercise 3 Solving Pronoun-Antecedent Agreement Problems

1. The <u>products</u> will be compared (in each category), and (each) of the units) will be rated according to ~~their~~ *its* performance and/or availability.
3. There <u>is</u> a <u>need</u> to study different brands (of microwave ovens) because, (with so many brands) available (to the consumer), (one) <u>has</u> to be careful about what ~~they~~ *he or she* <u>purchase</u>*s*.
5. The swing (shift), which begins (at 3:00 P.M.,) <u>has</u> no opportunity to buy food (during their breaks) because the food truck's last <u>stop is</u> (at 5:00.)   *"shift" is considered here as plural, so "they are can be used*
7. The (oscilloscopes) were given one point (for each feature) that ~~it is equipped~~ *they are* with.
9. The <u>audience</u> (for our documentation) ~~are~~ *is* adult (learners) who <u>have</u> a problem-centered approach (to learning.)

Exercise 6 Moving Modifiers for Clarity

1. Entering through the front door of the house, you will see the living room to your immediate left.
3. The long carpeted hallway eventually comes to an end at a rather large white door.
5. When I asked how he manages such highly technical concepts without being an engineer himself, he explained, "It's easy to learn the jargon in any field."
7. When my grandmother was on the wagon train, he hitched up his livestock.
9. On the other side of the room is a small table.
11. A California native, Nancy has lived in the San Francisco Bay Area.

Exercise 8 Rewriting to Eliminate Redundancies

1. It has a built-in tractor paper feed.
3. We hope that with these improvements a situation like the boiler room fire will never happen again.
5. This model measures 17x4x13 inches and weighs 12.8 pounds. It is made of smooth metal silver in color. *(The metal may or may not be silver.)*

APPENDIX **461**

7. Certain elements of automobile performance are critical for a particular car. ⟨The elements should be named.⟩
9. I have reviewed the initiatives and am glad to report that they will influence our plans.

Exercise 10 Evaluating Expletives

1. The proposal's introduction has three key paragraphs.
3. Multimedia components can be evaluated in many different ways.
5. There are three reasons for changing to the new word processing package: ease of use, compatibility with the graphics software, and the ability to mount on many platforms. [This expletive use effectively moves the three reasons to the end of the sentence where they can introduce the detailed information.]

Exercise 12 Verbalizing the Action in a Sentence

1. The preceding line graphs illustrate the fluctuations in power consumption.
3. Correct the draft with standard copyeditors' marks.
5. Figure 5-4 clarifies the maintenance procedure.
7. The leads on the wafer do not contact the pins of the second layer. ["Make contact" might be acceptable jargon in some professions.]
9. Bldg. D houses the adult units.
11. In the translation, I was forced to choose just one of those words and imply the other by context.

CHAPTER 14

Exercise 3 Recognizing the Existing Organization: A Definition

The organization is hierarchical by division. The last sentence in paragraph one establishes the three parts, so the reader expects one paragraph to deal with each part. However, sentence one in paragraph two is a more general statement that should either begin paragraph one or be omitted. Paragraph two should begin with sentence two. "The next step in using a database is data accessing" should begin paragraph three. "Data reporting . . ." should begin paragraph four. The organization could be emphasized with subheadings over each of the three subpoints.

Exercise 6 Rewriting for Parallelism

1. For ordinary writing, follow these three recommendations:
    1. Know the audience and the subject.

    2. Respect the length requirement.
    3. Prefer short words, keeping the number of long words to a minimum.
 3. Ski bottoms are cleaned, dried, and smoothed before treatment.
 5. Other reasons for choosing these criteria are that they are easy to understand, applicable to all portable generators, and practical.
 7. The thermal printer has the following problems:
    1. Printing quality is poor.
    2. Thermal paper is expensive.
    3. Printed copy will fade over a short period of time.

CHAPTER 15

Exercise 1 Choosing Effective Graphics

1. organizational diagram: shows the job assignments of workers
   table: shows resources with specific numbers
3. table: summarizes criteria applied to each one
   line, cutaway, or exploded drawings: shows individual features
5. line or bar graph: demonstrates the changes over time
   table: provides specific numbers

CHAPTER 17

Exercise 2 Analyzing Index Entries

Index Sample 1

Undefined page entries under *language* and *numeric keypad* need to be further differentiated at the third level with entry names.

Exercise 3 Editing Glossary Entries

1. The term *electronic pickup* needs to be defined by a formal definition. After the formal definition, information in the existing paragraph can be used to extend the definition.

CHAPTER 19

Exercise 5 Editing for International Readers

Example A

The phrase "not a pretty face" should be replaced by "not attractive" or "not popular." [The writer's intention is not clear.] "On the plus side" would be better as "Its advantage is . . ." "To beat the band" is a meaningless idiom; better words would be "easily" or "extensively."

Index

abbreviations. *see also* acronyms; initialisms
  common
    not requiring period, 140–141
    requiring period, 139–140
  definition, 139
  lists, 46
  in online documentation, 399
abstracts, 16, 381
abstract words, 177, 178
access aids
  advance organizers. *see* advance organizers
  chapter overviews, 268
  copyediting, 385–386
  developmental editing, 384–385
  documentation of sources. *see* citation of sources
  flowcharts, 268
  glossary. *see* glossary
  headers and footers, 268. *see also* headers and footers
  indexes. *see* indexes
  maps, 268
  menus, online documentation, 268. *see also* menus; online documentation
  navigational aids. *see* navigational aids
  proofreading, 386
  table of contents. *see* tables of contents
  tabs, 268
accuracy, content, 5
  compromise, 14, 15
  editor's role in ensuring, 15–16
  levels expected by clients, 15
acronyms. *see also* abbreviations; initialisms
  definition, 141
  lists, 46
action schemata, 244
active voice
  common usage, 227–228
  definition, 226
  vs. passive voice, 226
administrators, writing for, 247
advance organizers
  abstracts, 381
  context-setting overviews, 381–382
  definition, 380–381
  executive summaries, 381
  summary tables, 381
adverbial connective, 114

alignment (text), 338
all capitals, 90
all caps, 90
alpha draft, 55, 252
ambiguous language, 179. *see also* abstract words; concrete words
American Medical Writers Association, 47
analytical ability, 38–39
antecedent, 203
apostrophe
  for clarity, 144
  in contractions, 144
  misreading, used to prevent, 144
  omission, 144
  personal pronouns, 143–144
  plural words ending in s, 143
  plural words not ending in s, 143
  possession, 143–144
applicability, 5, 39–40
Associated Press Stylebook and Libel Manual, 44
attention lines, 183
audience, *see also* readers
  designers, writing for, 247
  developers, writing for, 247
  generalists, graphics preferred by, 315
  installers, writing for, 247
  intercultural. *see* intercultural editing
  international. *see* international editing
  maintenance workers, writing for, 247
  managers, graphics preferred by, 315
  production workers, writing for, 247
author-editor relationship, 34
awareness, language, 79–80

bar graphs, 301–303, 310, 312
beta draft, 55
block diagrams, 314
boilerplate, 46
boldface type, 272, 341
bookmarks, online documentation, 380
"bottom-up" approach to editing, 8
boxes, 342
brackets, 153
budget constraints, 12, 14–15
  cost estimating. *see* cost estimating

in determining expenditures on readability, 16
bulleted lists, 97, 236, 346–347

CAD. *see* computer-aided design systems, 332
capitalization
  all capitals emphasis, 90
  corporate style guides, following, 88
  general rules, 88
  list items, 89
  names, 88
  specific rules, 89–90
  titles, 88
  trade names, 89
case, of pronouns, 203
CBE Style Manual: A Guide for Authors, Editors, and Publishers in the Biological Sciences, 44
CD-ROMs, 391
  business uses, 318
  cost effectiveness, 396
change bars, 31, 81
chapter overviews, 268
checklists, 347, 349–350
Chicago Manual of Style, 32, 44, 83, 88
chronological organization, 254
chunking
  modular design, 350
  in online documentation, 252, 271
  as retention aid, 349
  short-term memory capacity, 246, 271, 345, 346
  as a visual device, 271
circle graphs, *see* pie charts
citation of sources
  common systems
    APA, 383–384
    CBE, 384
    Chicago Manual of Style, 383–384
    MLA, 383
  editing, 382
clause
  definition, 102, 113, 122
  dependent
    adjective, used as, 128–129
    definition, 123
    moving to make stronger sentences, 234
    noun, used as, 129
    primary identifier, 130
    punctuation used, 129–130

clause *(cont.)*
  dependent *(cont.)*
    secondary identifier, 130
    subordinate conjunction use with, 121, 122
  independent
    definition, 113, 123
    punctuation with, 114
  nonrestrictive, 130
  relative, 129
  restrictive, 130
coherence, 8
college continuing education courses, 47
colon
  compound sentences, use with, 115
  as connector, 154
  with *following,* 146
  introducing word groups, 147
  list introductions, use with, 146
  typographical distinctions or divisions, used to show, 147
  uses, 146–147
color, in documents, 342, 343, 348
comma
  dates, used in, 151
  dependent clauses, used with, 129–130
  inserted words, used to set off, 153
  interrupter words and phrases, used with, 150
  introductory phrases, used with, 149–150
  introductory words, used with, 149
  parenthetical information, used with, 154
  place names, used in, 151
  series, 151–152
communication
  globalization, 415, 416
  skills, 39
  successful, tools necessary, 2
complex sentence, 121–123
  punctuation of, 123
compound sentence
  colon use, 115
  punctuation rule, 114
compound words, hyphens with, 157–158
computer-aided design systems, 332
computer-based training, 397–398
computers
  indexing software. *see* indexes
  industry terminology, 190
  skills, 48
concrete words, 177, 178
conjunction, 106
  coordinate, 113
  subordinate, 123
conjunctive adverb, 114
connotative meaning, 180
content accuracy
  compromise, 14, 15
  editor's role in ensuring, 15–16
  levels expected by clients, 15
contractions, apostrophe use with, 144

coordinating conjunction, 113
coordinator, 113
copyediting
  access aids, 385–386
  definition, 6
  document design, 351
  hardcopy vs. screen, 80–81
  marks. *see* editing marks
  online documentation, 407–408
  as part of writer-editor interaction, 17
  prior to computer age, 54–55
  process, 55–56, 81
  screen vs. hardcopy, 80–81
  sentence level, 8
  skills required, 80
copyright, 42
corporate editing procedures, 23
corporate style guides, 96
cost constraints, 14–15
cost estimating, 25
country names, periods with, 140
courtesy titles, use of periods with, 140
critical reading, 248
cropping, photographs, 298
cultural sensitivity, 418. *see also* intercultural editing; international editing
culture. *see also* intercultural editing; international editing
  definition, 415
  high-context, 420
  low-context, 420
currency, numbers with, 93

dangling modifier, 210
dashes
  break in thought, 154
  as connector, 153–154
  em, 154
  emphatic idea, 154
  parenthetical information, used with, 154
  setting off material, 155
  summary, used to show, 154
  usage, 153
dates
  comma use with, 151
  use of periods with, 140
deadline
  editor's role, 34
  importance of meeting, for writers, 30
  levels-of-edit, impact on, 17
  pressure
    affect on time spent editing, 2, 12
decimals, vs. fractions, 94
denotative meaning, 180
dependent clause
  adjective, used as, 128–129
  comma with, 129–130
  definition, 123
  moving to make stronger sentence, 234
  nonrestrictive, 130
  noun, used as, 129

primary identifier, 130
punctuation used, 129–130
relative, 129
restrictive, 130
secondary identifier, 130
subordinate conjunction use with, 121, 122
design, documents. *see* document design
designers, writing for, 247
desktop publishing, 55
developers, writing for, 247
developmental editing
  choosing packaging/presentation genre, 248–252
  definition, 6
  document design, 350
  importance, 244
  online documentation, 251–252, 406–407. *see also* online documentation
  organizational plans, 245
  as part of writer-editor interaction, 17
  roles of writer and editor, 243–244
  writer, relationship between, 252
diagnostic quiz, grammar, 100–101
diagrams, 303–305, 314–315
diction. *see* word choice
dictionary
  general, 44
  importance of, 90
  specialized, 44–45
direct object, 102
document design, 38. *see also* document style
  alignment, 338
  boxes, 342
  color, 342, 343, 348
  conventional designs, 333
  copyediting, 351
  developmental editing, 350
  evaluating
    access aids, 343–344
    packaging, 333
    page design, 335–336
    page level construction, 333
    readability, 337–343
    six questions to ask, 333
  factors influencing, 331–333
  format, 331
  grid divisions, 337–338
  highlighting, 341
  IMRAD structure, 333
  intercultural issues, 418
  international issues, 418
  leading, 337
  letter spacing, 337
  line length, 337–338
  organization, 330
  page design, 331
  page layout, 331
  proofreading, 351
  purpose of document, 331
  reader methods (of reading), 332
  reader needs, 331–332

INDEX

reference books, 45
rules, 342
screens, 342
spacing, 341–342
templates, 330
type size, 338–341
types of documents, 333
type style, 338, 339, 340
word spacing, 337
document editing, as part of larger picture, 8
document style. *see also* document design
  italicization, 96
  list style. *see* lists
  per corporate style guide, 96
  quotation marks for setting off words, 97
drawings, 299–301

editing
  asking questions, 5
  boilerplate, using, 46
  "bottom-up" approach. *see* bottom-up approach
  citation of sources, 382
  computer skills. *see* computers
  corporate procedures, 23
  developmental. *see* developmental editing
  document, 8
  drafts
    alpha, 55
    beta, 55
    final, 55
  ethical and legal exposures, 42
  fact-checking, 243
  figures. *see* figures
  final. *see* proofreading
  flexibility, 38
  graphics. *see* graphics
  indexes. *see* indexes
  industry awareness, enhancing, 47
  initial vs. final, 55
  intercultural issues. *see* intercultural editing
  international issues. *see* international editing
  levels. *see* levels-of-edit
  lists, 348
  management skills, 42–43
  marks. *see* editing marks
  multi-layered approach, 6. *see also* top-down approach
  navigational aids, 380
  number of passes through a document, 6, 55
  objectivity, 38
  order of editor's analysis, 6
  organizational. *see* organizational editing
  page per hour rate, 25, 42–43
  planning checklist, 23–24
  reader's advocacy, 5
  self-editing. *see* self-editing
  sentence-level. *see* sentence-level editing
  skills, 38–39
  softcopy, 80–81
  tact, 261
  team approach, 17
  "top-down" approach. *see* top-down approach
  types of
    Jet Propulsion Laboratory, 13
editing checklist, 23–24
  customizing, 25
editing marks, 56–60
  where made when editing, 81
  where made when proofreading, 81
editor
  -author relationship, 34–35
  clarifying role, 39–43
  collaborative environment, maintaining, 43
  collegial environment, maintaining, 43
  content review as part of job, 15–16
  deadline, importance of meeting, 34
  "gatekeeper," 31
  groups who carry out the role of, 31
  interaction with writer, 17, 28
  lack of in companies, 7
  management skills, 48
  manager's role, 6
  offering direction to writers, 33
  as part of team, 15, 17
  positive feedback, giving to writers, 33
  professional, 32
  professional growth, 47
  as reader representative, 33
  role, 2–3
  skills needed, 3
  team approach, 17
  to writer ratio, 4, 32
  -writer relationship, 34
ellipses, 163, 165
em dash, 154
emphasis in text
  boldface type, 272, 341
  boxes, 342
  bulleted lists, 97, 236, 346–347
English-as-a-Second Language graphics in documents for readers, 316
  reader characteristics, 94
English language
  diversity, U.S., 414–415
  importance, 413–414
  international Englishes, 414
errors
  grammar, 16
  percentage of per document, 15
  substantive vs. typographical, 14
  typographical, 14
ESL. *see* English-as-a-Second-Language
ethical and legal exposures, 42
euphemisms, 182
exclamation mark, 111
executive summaries, 249, 381
expletives
  appropriate usage, 224
  inappropriate usage, 223–224
  weak sentences, 230

fact-checking, 243
figures
  editing, 16
  for international readers, 348. *see also* graphics
filtering information, 346
final editing, *see* proofreading
first draft, 252
flexibility, editor's need for, 38
flowcharts, 268, 305
fonts, 338. *see also* type style
  in online documentation, 406
  sans serif, 340–341
  serif, 340–341
footers, 268, 377–378
formal style, 53
format, 331
form schemata, 244
fractions, 94
fragments, 111
fused sentence, 111

"gatekeeper," 31
generalists, 247
generalists, graphics preferred by, 315
gerund, 106
global communication
  design issues, 427–428
  diction, 424–425
  graphics, 427–428
  punctuation, 425–427
  syntax, 425–427
  technical terms, 427–428
globalization of communication, 415, 416
global marketplace, 414
glossary
  definition, 375
  highlighting words for, 375
  importance, 46
  international communication, use in, 376
  types of terms, 376
  where to place, 375
Government Printing Office (GPO) Style Manual, 92, 143
grammar
  diagnostic quiz, 100–101
  importance, 100
  minor errors, 16
  reference books, 45
  sentence parts
    adverbial connective, 114
    clause, definition, 102
    clauses. *see* clauses
    condition verb, 102
    conjunctions, 106
    coordinating conjunctions, 113
    direct object, 102
    indirect object, 102
    linking verb, 102
    modifier, 105

grammar (cont.)
    sentence parts (cont.)
        object
            position in sentence, 198–199
            predicate, 102
            preposition, 105–106
            recognizing, 103
                in complex sentences, 108
            subject, 101, 108
                position in sentence, 198–199
                understood, 122
            subject complement, 102
            subordinating conjunctions, 123
            verb, 101–102
                condition, 102
                linking, 102
                position in sentence, 198–199
            verbals, 106
        traditional, 100
grammar checkers, 4
graphics, choosing, 295–296, 315–319
    animated sequences, 293
    content, 317
    copyediting, 321–323
    cost, impact of on choice, 318
    degree of visualization, 319
    developmental editing, 320–321
    drawings, 293
    editing
        bar graphs, 310, 312
        bar graphs, segmented, 301–303
        block diagrams, 314
        charts, 303–304
        diagrams, 303–304
        document purpose, 296
        drawings, 299–301
        examples, 312–313
        flowcharts, 305
        Gantt charts, 306
        line graphs, 310, 312
        maps, 303–304
        milestone charts, 306
        photographs, 298–299
            cropping, 298
        pictographs, 310, 312
        pie charts, 301–303
        schematic diagrams, 314
        tables, 308–310
        task-breakdown charts, 305
        wiring diagrams, 314
    effectiveness, 320
    factors influencing, 295
    figures, 295
    flowcharts, 293
    graphs, 293
    icons, 293
    illustrations, 294–295
    importance, 293–294
    intended readership, 321
        cultural background, 316
        native language, 316
        for international audiences, 316–317
    line drawings, 299
        cross-sectional, 299

cutaway, 299
exploded, 299
matching reader type to graphic type, 315–316
medium, 317–318
online documents, 395
photographs, 293, 298–299
placement, 320–321
    online documentation, 406
proofreading, 323–324
reading ease, 321
stereotypes, 317
tables, 295
time, impact of on choice, 318
visuals, 294
graphic design
    reference books, 45
graphs, 310–312

hardcopy
    presentation forms, 318
    vs. online documentation, 317–318, 393–395
headers and footers, 268, 377–378
headings, 265–266
    characteristics of effective, 343–344
    organizational indicators, 265–266
    point size, 339
    syntax, 266
    type size, 272
    visual prominence, 346
headlines, 334. see also headings
hierarchical organization, 254
hyphen
    compound words, used with, 157, 157–158
        adjectives, 157–158
        nouns, 157–158
    as em-dash, 154
    syllable breaks, 156
    usage, 156, 158–159
hypertext, 392

icons, 271
illustrations, see graphics
immigration, communication issues related to, 417. see also intercultural editing; international editing
IMRAD structure, 333
independent clause, 113
    adverbial connective, 114
    comma use with, 114
    definition, 113, 123
    punctuation with, 114
indexes, 268
    consistency of terms, 373
    cross-references, 370
    double entries, 370
    editing, 372
    hierarchical entries, 369–370
    importance, 45
    importance of being an expert, 45–46
    keywords, 374
    online documentation, 373–375

process, 372–373
purpose, 369
software, 45
time to complete, 373
undefined page entries, 370
indirect object, 102
indirect quotations, 165
infinitive, 106
informal style, 53
initialisms, 141. see also abbreviations; acronyms
initials, periods with, 139
installers, writing for, 247
instructions
    organization of, 254
    poorly written, 2
    user-testing, 16
intercultural editing. see also international editing
    copyediting, 429–430
    definition, 416
    design issues, 418
    developmental editing, 429
    high-context cultures, 420
    influence of culture, 419–424
    low-context cultures, 420
    major categories, 428
    methods for learning about other nations
        oral information, 423
        personal experience, 423–424
        written information, 422–423
    proofreading, 431
    readers, 417, 418–419
    rhetorical strategies, 420–421
    surface characteristics, 421
    unconscious rules, 421
    unspoken rules, 421
    user analysis, 421–422
    vs. international editing, 416
    writers, 417–418
international editing. see also intercultural editing
    copyediting, 429–430
    dates, 95
    definition, 416
    design issues, 418
    developmental, 429
    figures, 348
    glossaries, 376
    international Englishes, 414
    major categories, 428
    metric system, 95
    number use, 94–95
    proofreading, 431
    readers, 417, 418–419
    tables, 348
    telephone numbers, 94–95
    vs. intercultural editing, 416
    writers, 417–418
international Englishes, 414
internationalization of communication, 415, 416
Internet, 391
interpersonal skills, 39
interviewing skills, 47

INDEX

introductory phrases, 149–150
italics, 96, 341
   document style, 96
   new terms, used to call attention to, 164

jargon, 190–192
   in international documents, 419
   in online documentation, 399
Jet Propulsion Laboratory
   levels of edit, 13–14
   original purpose, 13
job titles, 183
Journal of the American Medical Assn. Manual for Authors and Editors, 44
JPL. *see* Jet Propulsion Laboratory
justification of text, 338, 406
justified type, 338

kerning, 341

"ladder of abstraction," 177, 178
language
   awareness, 79–80
   definition, 415
leading, 337, 341
legal exposures, 42
letter spacing, 337
levels-of-edit
   agreement, reaching before start of project, 33
   approaches, 16–17
   determining before start of project, 30
   guidelines, 12
   Jet Propulsion Laboratory standards, 13–14
   task specification and agreement, 21
   technical review, importance, 15
line drawings, 299. *see also* graphics
   cross-sectional, 299
   cutaway, 299
   exploded, 299
line length, 337–338
linking verb, 102
links, online documentation, 401–402
lists
   bulleted, 97, 346–347
   capitalization of first word in list items, 89
   checklists, 347
   colon use with, 146
   editing, 348
   numbered, 97, 346
   as organizational device, 272
   style choices, 97
localization of communication, 415, 416
logic, 243
long-term memory
   event sequencing, 244
   information storing, 344–345
   schemata, 244

maintenance workers, writing for, 247
management skills for editors, 42–43

manager, as editor, 31
managers, definition of, 247
managers, graphics preferred by, 315
manuals, 12, 248
   "active" learning, relationship between, 245
   definition, 249
   typical form, 249–250
   user-testing, 16
maps, 303–304
   as access aids, 268
marketing professionals, writing for, 247
marks, editing, 56–60
measurement
   numbers with, 93
   use of periods with, 140–141
memory
   long-term
      event sequencing, 244
      information storing, 344–345
      schemata, 244
   recalling, 348
   remembering, 348
   retaining, 348
   retrieving, 348
   short-term
      affect on readers' absorption of subject, 220, 233–234
      "chunk capacity," 246, 271
      "chunking," 345, 346
      limitations, 344
      limits, 246
      organization, affect on, 246
   techniques for helping reader, 349–350
menus
   levels of entries, 367
   online documentation, 268
metric system, 95
misplaced modifier, 210–211
modifier, 105
   long string, 219
modifier, 209–211. *see also* prepositional phrase; verbal phrases
   dangling, 210
   misplaced, 210–211
   moving to strengthen sentence, 234–235
   squinting, 210, 211
modular design, 350
money, number use, 93
money constraints, 12
   cost estimating, 25
   in determining expenditures on readability, 16

navigational aids
   editing, 380
   flowcharts, 379
   footers, 377, 378
   headers, 377, 377–378
   maps, 379
   navigation buttons, 378
   online, 252
   screen orienters, 377, 378

   stack of pages, 378
negative sentences, 232
New York Public Library Writer's Guide to Style and Usage, 44
nominalization, 229
nonrestrictive clauses, 130
nonsexist language. *see* sexist language
numbered lists, 236
numbers
   approximations, 94
   beginning a sentence, 94
   counting items, 92, 93
   currency, 93
   decimals, vs. fractions, 94
   fractions, 94
   guidelines, 92–95
   international audience, 94–95
   international telephone numbers, 94–95
   international usage, 94–95
   large, 93
   measurement, 93, 140–141
   metric system, 95
   money, 93
   numeral usage, 92–93
   ordering items, 92, 93
   in sentences containing 2 or more numbers, 92
   spelling out, 93–94
   telephone, international, 95
   time, 93
   words, 93–94
numerals, 92–93. *see also* numbers

objectivity, 38
online documentation, 12
   bookmarks, 380
   chunking, 252, 271
   computer-based training, 397–398
   copyediting, 407–408
   definition, 392
   design elements, screens and windows, 405–406
   developmental editing, 406–407
   diction, 398, 399
   editing, 393
   electronic information
      definition, 392
      factors affecting choice, instead of hardcopy, 317
   glossaries, 375
   graphics, 252, 395
   hierarchical organization, 403
   hypertext
      definition, 392
   icons, 273, 293
   indexes, 373–375
      keywords, 374
   key words, searching by, 252
   links, 401–402
   menus, 252, 268
   navigational aids, 252
   navigation buttons, 378
   onscreen editing, 392–393

objectivity (*cont.*)
  organization, 394
    grid, 402–403
    hierarchy, 403
    partial web, 404
    sequence, 402
  proofreading, 408
  punctuation, 398, 400
  readability, 252
  reader-user, relationship between, 397–398
  screen format, 252, 264
  screen orienters, 378
  single-source authoring, 397
  softcopy
    definition, 392
  syntax, 398, 399
  tables of contents, 367
  text amount, 252
  topics
    contents, 400
    size, 400
    structure, 400–401
    type style, 406
  typical organization, 251–252
  vs. print documentation, 393–395
  "white space," 272
online editing
  change bar usage, 31, 81
  effect on copyeditor and proofreader roles, 55
  vs. hardcopy editing, 80–81
online reference systems, 391
operators, graphics preferred by, 316
operators, writing for, 247
organization, 5
  "big picture," 38
  of complex projects, 6
  as part of bottom-up approach, 8
  vs. disorganization in writing, 4
organizational editing. *see also* developmental editing
  access aids, 268. *see also* access aids
    headings, 265–266
    parallelism, 266–268
    visual devices, 264
      chunking, 271
      hierarchy, showing, 271–272
      icons, 271, 273
      listing, 271
  clarifying
    page or screen design, 264
    verbal devices, 264–265
  editor's participation, 256–257
  by hierarchy, 254–255
    by division or classification, 255
    pro and con, 255
  justifying changes, 260–261
  methods, 253–254
    combinations, 255–256
    outlines, 258–259, 260–261
  by sequence, 254
    alphabetical, 254
    chronological, 254
    spatial, 254
outlines
  evaluating, 258–259
  reorganizing, 259

packaging documents, 334
page design
  balance, 335–336
  emphasis, 335, 336
  layout, 336
  sequence, 335, 336
  unity, 335
page layout
  factors determining, 336
  hierarchical organization, 349
  repeated patterns, 349
  trial design sketches, 337
paragraph spacing, 341–342
parallelism, 266–268
parallel structure, 79
parentheses, 153
parenthetical information, 154
participle, 106
passive voice, 223
  common usage, 227–228
  definition, 227
  expletive use, 235
  vs. active voice, 226
patenting, 42
peer review, 31
period
  country names, 140
  courtesy titles, 140
  use at end of sentence, 111
persuasive writing, 181, 248, 249. *see also* proposals
photograph cropping, 298
phrase, grammatical, 102
phrase, introductory, 149
pictographs, 310, 312
pie charts, 301–303
place names, 151
plural words, use of apostrophe, 143
point size, 339
pompous language, 188
portfolio, editing samples, 48–49
position in sentences, 233–236
positive feedback, 33
positive sentences, 232
postal abbreviations, 141
predicate, 102
prefixes, emphasizing, in online documentation, 399
preposition, 105–106
prepositional phrase, 105–106
  as modifier, 209
prewriting, 2
production workers, writing for, 247
professional societies, 47
pronoun
  agreement with noun replacing, 202–203
  -antecedent agreement, 203–204
  cases, 203
  definition, 202
  phrase/clause substitution, 205–206
  relative, 129
  who-whom-whose, 203, 207–209
pronoun reference, 205
proofing. *see* proofreading
proofreading
  access aids, 386
  comparing final to draft, 65
  definition, 6
  document design, 351
  errors, 62–63
  grouping items to be checked, 65
  importance, 62
  marks. *see* editing marks
  online documentation, 408
  as part of writer-editor interaction, 17
  prior to computer age, 55
  process, 56, 64–66
  sentence-level, 8
  as training for editors, 64
  vs. reading, 62
proposals, 12, 248
  definition, 249
  typical form, 249
Publication Manual of the American Psychological Association, 44
punctuation
  apostrophe. *see* apostrophe
  comma. *see* comma
  of complex sentences, 123
  of compound sentences, 114, 115
  dependent clauses, used with, 129–130
  exclamation mark, 111
  in online documentation, 400
  period. *see* period
  question mark, 111
  semicolon. *see* semicolon

question mark, 111
questions, asked at end of unit as memory-enhancing device, 349–350
queuing information, 346
quotation marks
  exact words, setting off, 163–164
  new terms, used to call attention to, 164
  published works, 164
  punctuation marks with, 163–164
  for setting off words, 97
  single, 164–165
  usage, 163–165

"ragged right" type, 338
readability, 5
  ensuring for reader, as editor, 33
  evaluating, 40
  money and time considerations, 16
readers. *see also* audience
  appropriateness for when writing, 5, 8
  cultural background graphic choices affected by, 316–317
  expectations of, 198–199
  graphics, relationship between, 315–316
  language, graphic choices affected by, 316–317
  novice vs. experienced, 245
  short-term memory of. *see* short-term memory
  tasks, analyzing, 247
  type, 40
    continuum, 246
    generalists, 247

INDEX 469

managers, 247
operators, 247
specialists, 247
technicians, 247
-users. *see* reader-users
reader-users
  analyzing, 395–396
  hardcopy printout for, 396
  online documentation, relationship between, 397–398
reading
  critical, 248
  receptive, 248
  scanning, 332
  search, 248, 332
  strategies, 248
  style, 248
reading style
  changing to proofreading, 63–64
  reading vs. proofreading, 62
receptive reading, 248
redundancy
  general plus specific pairs, 216–217
  general vs. specific words, 215
  grandiose pairs, 216
  implied plus stated pairs, 217–218
  repetition, 216
reference books, grammar, 45
reference materials, 44–45
relative clauses, 129
reports, 12, 248
  completion type, definition, 250
    typical form, 251
  ensuring technical accuracy, 15–16
  recommendation type
    definition, 250
    typical form, 250–251
  research, definition, 250
    typical form, 251
researchers, writing for, 247
restrictive clauses, 130
resume
  keeping current, 48
revising, 2
rhetoric
  reference books, 45
rules, 342
running heads, 377. *see also* headers
run-on sentence, 111

salutations, 183
san serif type, 340, 406
scanning, when reading, 248, 332
schemata
  action, 244
  definition, 244
  form, 244–245
  links, via icons, 273
schematic diagrams, 314
screen orienters, 377–378
screens (gray, in text), 342
screens, online, 405–406
search reading, 248, 332
self-editing, 5, 23, 25
  distancing self from copy, 80
semicolon
  in series, 152
  use with independent clauses, 114

semiformal style, 54
sentence-level editing, 31. *see also* copyediting
  definition, 54
  vs. document editing, 55
sentence parts
  adverbial connective, 114
  clause, 102. *see also* clause
  condition verb, 102
  conjunctions, 106
  coordinating conjunctions, 113
  direct object, 102
  indirect object, 102
  linking verb, 102
  modifier, 105
  object, position in sentence, 198–199
  predicate, 102
  preposition, 105–106
  recognizing, 103
    in complex sentences, 108
  subject, 101, 108
    position in sentence, 198–199
    understood, 122
  subject complement, 102
  subordinating conjunctions, 123
  verb, 101–102
    condition, 102
    linking, 102
    position in sentence, 198–199
  verbals, 106
sentence
  comma-spliced, 111
  complete, 112
  complex
    definition, 123
    punctuation with, 123
    with relative pronouns, 128–130
    when used, 121
  compound
    colon use with, 115
    definition, 113
    examples, 112
    punctuation rule, 114
    when used, 121
  compound-complex
    when used, 121
  definition, 101
  fragment
    common types, 112
    use in advertising, 111
  fused, 111
  length, 215
  passive
    word position as indicator, 235
  pluralizing, 184
  positive vs. negative constructions, 232
  revising, for emphasis, 233–236
  run-on, 111, 112
  schema, 198
  structure. *see* syntax
  vs. phrase, 102
  word order, 220–221
  word position, 233
sequential organization, 254
series comma, 151–152
serif type, 340
sexist language, 183–184

short-term memory
  affect on readers' absorption of subject, 220, 233–234
  "chunk capacity," 246, 271
  "chunking," 345, 346
  limits, 246, 344
  organization, affect on, 246
SI units. *see* Systeme International units
skimming, when reading, 248, 332
SME. *see* subject-matter expert
Society for Technical Communication, 47
softcopy editing, 80–81
software
  grammar. *see* grammar checkers
  screen captures, 295
  screen dumps, 295
  screen shots, 295
  screen snapshots, 295
  spelling checkers. *see* spelling checkers
spatial organization, 254
specialists, definition of, 247
specialists, graphics preferred by, 316
specialized language. *see* jargon
spell checkers. *see* spelling checkers
spelling
  consistency, 90
  context, 90
  guidelines for checking, 90–91
spelling checkers, 3
  limitations, 4
  use by professional writers, 4, 62
squinting modifier, 210, 211
stereotypes, avoiding in graphics, 317
style
  definition, 53, 197
  document. *see* document style
  formal, 53
  guide. *see* style guide
  informal, 53
  semiformal, 54
  sheet. *see* style sheet
  types, 53–54
style, writing. *see* writing style
style checkers, 3
  limitations, 62
  use by professional writers, 62
style guide
  abbreviation and acronym lists, 46
  author's responsibility to use, 30
  corporate, 32, 43–44, 82, 84
    document style choices, 96
    starting small, 85
    using as first rule, 88
  creating, 32
  developing own, 83–84
style sheet, 32
  preferences noted, 82
  process of creating, 82–83
subject, 101, 108
  position in sentence, 198–199, 220–221
  understood, 122
subject, hiding action, 230
subject complement, 102

subject-matter expert
  clarifying technical terms, 174
  definition, 29
  as editor, 31
  role of, 6, 15
  time considerations, 7
subject-verb agreement, 200–201
subordinate conjunction
  definition, 123
  use with dependent clauses, 121, 122
substantive editing, as part of writer-editor interaction, 17
substantive errors, vs. typographical, 14
summary tables, 381
syntax
  definition, 197
  of headings, 266
  modifier problems, 209–211
  in online documentation, 399
  pronoun-antecedent agreement, 203–204
  pronoun-noun agreement, 202–203
  pronoun phrase/clause substitution, 205–206
  subject-verb agreement, 200–201
  wordiness. *see* wordiness
Systeme International units, 141

tables, 308–310
  for international readers, 348
tables of contents, 268, 380
  compiling, 366
  online documents, 367
technical accuracy
  methods for ensuring, 15–16
technical review
  definition, 6
  importance, 15
technicians, definition of, 247
technicians, graphics preferred by, 316
templates, design, 330
terminology accuracy, methods to ensure, 174
text alignment, 338
thumbnail proofs, 334
time
  numbers with, 93
  use of periods with, 140
time constraints, 12, 14–15. *see also* time management
time estimating, 25
time management, 48
titles, courtesy, use of periods with, 140
"top-down" approach, 5
  ideal, 8
  vs. bottom-up approach, 8
topics, online, 400
trademark, 42
traditional grammar, 100
translation, 13
  definition, 415

trendy language. *see* jargon
tutorials, 391
type
  size. *see* type size
  style. *see* type style
type size
  headings, 339
  point size, 339
  showing hierarchy, 272
  subheadings, 339
type style
  all capitals, 341
  boldface, 272, 341
  font choice, 339
  italics, 341
  lower-case, 340
  in online documentation, 406
  sans serif font, 340–341
  serif font, 340–341
  showing hierarchy, 272
  underlining, 341
  upper-case, 340
typographical errors, 14
  vs. substantive, 14

unclear writing, 2
United States Government Printing Office Style Manual, 32, 44, 83, 88
UNIX, 48
usability, 5, 350
  definition, 41
  importance, 173
  laboratories, 41
  questions for assessing, 41
  user-testing, 33
usage, 173, 190, 191
user-testing, 33, 41
  of instructions, 16
  of manuals, 16

vague language, 179. *see also* abstract words
verbal phrases, 108
  definition, 106
  as modifiers, 209
verb, 101–102
  condition, 102
  linking, 102
  position in sentence, 198–199, 220–221
  "verbalizing," 229
  weak, 229–231
video production, 48
visuals, *see* graphics
vogue language. *see* jargon
voice
  active. *see* active voice
  definition, 226
  passive. *see* passive voice
  reader expectations, 227

Web pages, 318
white space, 341

white space, separating information items, 334
who-whom-whose dilemma, 203, 207–209
windows, online, 405–406
word choice
  abstract vs. concrete, 178–179
  connotative meaning, 180–181
  definition, 173
  denotative meaning, 180–181
  importance, 173
  pompous language, 188
  sexist language. *see* sexist language
word choice, online, 398, 399
word emphasis, 234
wordiness
  editing for, 214
  modifier strings, long, 219
  positive vs. negative constructions, 232–233
  weak verbs, 229–231
word processing
  effect on copyeditor and proofreader roles, 55
word spacing, 341
word usage
  definition, 173
  importance, 173
  jargon, 190–192
World Wide Web, 391
writer
  addressing editor's suggestions, 30–31
  to editor ratio, 4, 32
  -editor relationship, 34
  as freelance professional writer, 29
  importance of solid editing skills, 8
  interaction with editor, 17, 28
  meeting deadlines, 30
  positive feedback, receiving from editor, 33
  respect for editor, 30
  responsibilities, 28–31
  role, 2–3
  skills needed, 3
  style guide, importance of using, 30
  as subject-matter expert, 29
  taking editor's suggestions, 29
writing, 2
  analyzing reader tasks, 247
  applicability. *see* applicability
  organized vs. disorganized, 4
  persuasive. *see* persuasive writing
  plain vs. gradiose, 216
  process
    prewriting. *see* prewriting
    revising. *see* revising
    writing. *see* writing
  purpose, clarity of, 173
  readability. *see* readability
  readers, relationship between, 173–174
  style, 197–198
  unclear, 2
  usability. *see* usability